Plant Systematics and Evolution

Entwicklungsgeschichte und Systematik der Pflanzen

W0019328

Supplementum 1

Flowering Plants

Evolution and Classification of Higher Categories

Symposium, Hamburg, September 8-12, 1976

Edited by K. Kubitzki

Springer-Verlag Wien GmbH

Prof. Dr. Klaus Kubitzki

Institut für Allgemeine Botanik und Botanischer Garten,
Universität Hamburg, Federal Republic of Germany

© 1977 by Springer-Verlag Wien

With 118 Figures

Library of Congress Cataloging in Publication Data.
Main entry under title:
Flowering plants. (Plant systematics and evolution — Entwicklungsgeschichte und Systematik
der Pflanzen: Supplementum; 1). 1. Botany-Classification-Congresses. 2. Plants-Evolution-Con-
gresses. 3. Phanerogams-Congresses. I. Kubitzki, Klaus, 1933—. II. Series: Plant systematics and
evolution: Supplementum; 1. QK95.F58. 582'.001'2. 77-25200.

ISBN 978-3-211-81434-5   ISBN 978-3-7091-7076-2 (eBook)
DOI 10.1007/978-3-7091-7076-2

Preface

The original suggestion to organize a symposium about the classification and evolution of the Flowering Plants was made at the
International Botanical Congress at Leningrad in 1975, and the idea
was so well accepted by several colleagues that plans for such a
symposium quickly took shape. An organizing committee consisting
of Professor H. MERXMÜLLER, München, Professor V. H. HEYWOOD,
Reading, and Professor K. KUBITZKI, Hamburg, was set up. The
conference took place on 7–12 September 1976 in the Institut für
Allgemeine Botanik of the University of Hamburg under the auspices
of the International Association for Plant Taxonomy and was attended by 80 participants from 14 countries.

There have been several meetings in recent years which have
dealt with the origin and evolution of the Flowering Plants so that
it might be questioned whether yet another symposium dealing with
more or less the same subject were really ·justified. As the reader
will see from the contents of the book, this symposium differed
from similar ones held recently in two respects: 1. Emphasis was
given to methodological aspects of the classification of higher taxa,
and 2. much classificatory and evolutionary evidence relating to
the higher taxa of Flowering Plants was presented. It would have
been foolish in the extreme to expect from this meeting a general
agreement on one universally acceptable classificatory system; on
the contrary, it transpired very clearly that there is still a great
deal of dissent about the higher taxa of the Flowering Plants and,
for example, even the apparently well-established separation between
Monocotyledons and Dicotyledons is challenged.

If this meeting was successful it was because it had the important
function of permitting communication and promoting understanding
not only among taxonomists but also between different kinds of
practitioners, and there is no doubt that the impact of the newer
sub-disciplines such as phytochemistry, anthecology and ultrastructure
on classification and evolution is still increasing.

The conference was made possible through the financial aid obtained from various organizations and institutions to which we are
most grateful: The Deutsche Forschungsgemeinschaft; the Freie und
Hansestadt Hamburg; the Hamburgische Universitätsstiftung; and the

Universitätsgesellschaft Hamburg. It should be pointed out here that the two last-mentioned organizations depend wholly on private funds and have taken over the role of private Maecenas whose importance is only fully appreciated in these days of butgetary restrictions.

The President of the University of Hamburg offered a reception for the participants on 7 September, and I am grateful for this and for allowing the meeting to be held at the University. Also the combined efforts of staff members and research students of this institute are gratefully acknowledged.

Finally, I want to express my very grateful thanks to all contributors for so extensively displaying their knowledge of the subject, and to Professor V. H. HEYWOOD for revising the English of several contributions.

Hamburg, October 25, 1977   K. KUBITZKI

Contents

General Principles and Methods

Evolutionary Aspects and Taxonomic Evidence

Evolution and Classification of Major Taxa

Summary Lecture

Plant Syst. Evol., Suppl. 1, 1—12 (1977)

Department of Botany, Plant Science Laboratories,
University of Reading, England

Principles and Concepts in the Classification of Higher Taxa

By

V. H. Heywood, Reading

Abstract: Despite the very great interest shown by botanists during the last 100 years in the "phylogenetic classification" of the Flowering Plants, much more attention seems to have been paid to the phylogenetic components of such schemes (despite the lack of adequate evidence) than to their taxonomic/classificatory components (for which much evidence is available). There is in fact an inbuilt conflict involved in constructing a scheme that attempts to express both the dynamic/historical phylogenetic component and a static horizontal present day classification, which makes most of the conventional phylogenetic tree-like diagrams conceptually suspect.

In the Flowering Plants, only the classes and families are widely accepted whereas there are numerous plausible ordinal classifications. The kind of questions we have to ask in the case of higher categories and taxa such as the subclass and order are: are the criteria to be primarily phylogenetic, or taxonomic in the sense of aiming at better circumscriptions of the taxa, greater predictive value, stability, etc.? Only when we have answered these and similar questions can we decide whether we should be aiming at cladogenetic schemes or anagenetic grades or an explicit combination of both.

Introduction

In introducing a symposium on the evolution and classification of higher taxa, I am conscious of the pitfalls that lie ahead if we proceed without clarifying certain basic assumptions and establishing principles to be followed. At the same time, I realise both the futility of attempting this and my own inadequacy. Moreover, I am aware that I am out of sympathy with much of the work on systems of classification that has been done in the past and that is discussed later in this symposium, largely on the grounds that it is un-scientific, misconceived and conceptually unsound. If, however, my remarks have the effect of stimulating a discussion that will lead to clear and logical thinking and action, then I shall have fulfilled my role.

1

Classification

I think it is generally agreed that one of the major roles of taxonomy is to produce a classification (or system of classification) of organisms that best reflects the totality of their similarities and differences. At the level of the genus and above the units of classification are basically a filing and retrieval device, or language of communication for use on a single time-plane as HUGHES (1976) put it recently. The basic method used is that of comparative morphology and is concerned with both the content of the taxa recognised and their delimitation from others of the same rank. This can be, and in the majority of cases is, undertaken without using any phylogenetic information, although evolutionary concepts may inform the taxonomist in his task. It is, as CRONQUIST (1975) says, possible to perceive what we call natural groups without thinking in terms of evolution, but such perception is facilitated by an evolutionary frame of reference. This is not the same, however, as using factual phylogenetic data, which is seldom possible. The evolutionary frame of reference is, as regards extant plant groups, restricted to a loose monophyletic requirement which, as we will see later, is determined from comparative morphology and based on inference, so that it seldom, if ever, affects the actual procedures of classification.

Before considering in detail the role of phylogeny in classification, I think we should look at the current classificatory situation in the Flowering Plants (Angiosperms), at the level of the family, order and above.

The number of Angiosperm families recognised varies from about 300–450, according to which system is followed. In purely practical terms, the family is as basic a unit in taxonomy as is the species. In other words, in the majority of cases a consensus is reached as to the specific identity of a given plant, while there may be considerable debate as to its generic position due to the vagaries of generic/subgeneric classification, but the moment any generic classification is accepted, there is usually an automatic attribution to a family. Indeed in many, if not most, cases one can place a plant in a family without knowing to which genus or species it belongs—this is certainly true of the 200 or so families recognised by BENTHAM & HOOKER and which formed part of our basic training. The question that we must pose is how well we know the families? As regards delimitation from other families there is still considerable debate, especially in the case of many of the smaller families recognised in recent years and there is still debate as to the familial or subfamilial position of the major components of many of the larger families—the *Rosaceae, Leguminosae, Saxifragaceae,* etc. To the non-taxonomist, this may seem very strange—after many decades of apparent intensive study, there is still debate as to which families should be recognised, especially when we remember the relatively small number of families involved!

We should perhaps enquire into the reasons for this. Basically they are that we have seldom made adequate comparative studies (a point I shall return to later), which in turn depends on a detailed knowledge of the content of families (again a point to be considered later), and also because there is no agreed principle for the delimitation of families other than comparability of status in relation to allied families. We have to remember that most families were recognised and described long before the majority of the genera they contain today were described. They have, therefore, been built up by a process of accretion. Originally they were recognised, at least in the case of a natural family such as the *Compositae, Labiatae, Umbelliferae, Palmae*, etc., by a process of mental/visual correlation of plants with similar characteristics. In a very non-scientific way, they were recognised by a process of synthesis, i.e. from below. Since then, thousands of genera have been added into families when they appeared to be accommodated without too much difficulty but the process of revision of families from below, on a synthetic basis, has not been repeated, largely because it is too big a task to undertake and we are not even sure how to proceed. A consequence of this is that the family today is a somewhat vague abstraction. It is, in fact, virtually impossible for us to envisage a family, let alone see or handle one.

This does not, however, prevent us from talking about families being advanced or primitive, or related, etc. when what we really mean is our appreciation of certain component genera with which we are familiar or about which some special piece of information has been available. Even more basic is the fact that in very few cases do we know the actual content of a family because, as noted above, so many additions and alterations have been without any overall revision having been effected. There is not even available an up-to-date account of families giving their generic content, or even a series of detailed family descriptions. The average family description is at such a level of abstraction as to be useless except perhaps for diagnostic purposes (and even then of dubious value). As a basis for information-processing or evolutionary interpretation it is of little value. When a phylogenist tells about the positioning of a family or its status or origin, he is really talking in terms of his personal appreciation of those members of the family with which he is familiar. I need not elaborate on this point except to emphasize the need for: (1) the accumulation at family level of a more scientifically organised data-base; (2) at the same time, the need for detailed reviews (if not revisions) of the Angiosperm families. How this is to be achieved is something to which we should address ourselves.

Readers will doubtless be aware of a paper by Dr. JACOBS of Leiden (1969) entitled "Large Families—not alone" in which he lamented the neglect of the families from a revisionary point of view. Only by co-operative effort can an adequate data base be achieved, possibly by using a computer-based system recording the data or, if not, some centralised international centre for the accumulation, listing, recording and storing of data. Without an adequate data base, no major familial revision will be possible. In the case of the *Compositae*, we attempted last year (1975) at Reading, as part of the work of a symposium, to revise the tribal classification of the family. The method we adopted was for a specialist (or group of specialists) to accept the responsibility of preparing for each tribe a review to generic level, including an assessment of the content and sub-division of the tribe and its status, relative to other tribes. Even the task of finding out which genera existed and to which tribe they belonged proved to be quite a major task, let alone the question of the validity of the genera and the tribe it-self. There are still too many genera whose tribal position is uncertain and quite major realignments of the tribes themselves were suggested. It would be idle to pretend that we have solved most of the problems or that a consistent tribal or generic ranking has been achieved, but at least we are now in a position to assess what the content of the family is in terms of genera in the first overall assessment since BENTHAM's masterly mono-graph over a century ago (1873). The results will be published as part of the proceedings of the symposium (HEYWOOD et al. 1977). It is proposed to hold a similar review of the family *Leguminosae* next year organised by the Royal Botanic Gardens, Kew, and the Missouri Botanical Garden in association with the University of Reading.

As regards the rank or taxon o r d e r, it has to be said that in practical terms, it is of little value. There are at least ten current ordinal classifica-tions available today and the order has no precise meaning unless one adds sensu CRONQUIST, sensu ENGLER, sensu SOÓ, sensu HUTCHINSON, sensu TAKHTAJAN or sensu THORNE, etc. I know of no principles or norms or even conventions that apply to the circumscription and ranking of orders (except, of course, their relative positional value in the hierarchy) and I cannot find any coherent explanation in the litera-ture. At least at the family level one can seek certain comfort in talking about the family representing some conspicuous grouping, recognisable in some cases even by the layman, or about it representing some evolu-tionary peak or adaptive radiation. The order, on the other hand, seems to be purely arbitrary as a taxonomic tank and apart from being supra-familial, there is little to say about it. Is it even needed? It is certainly not a useful information retrieval device unless qualified as noted above. The diversity of content of the orders in the different classificatory systems makes a nonsense of talking about their evolutionary rela-tionships. How can we seriously discuss the phylogeny of units which in view of the divergence of opinion as to their circumscription may not, in the majority of cases, have an evolutionary coherence!

Certainly it may be useful to have some supra-familial groupings for ease of reference, but I wonder whether an informal system might not be the answer, or even an avowedly artificial one. The prevalent insistence on the use of clades (the result of evolutionary divergence) in classification is understandable but there is a case to be made for the occasional use of artificial (and presumably polyphyletic) grades, a practice previously adopted by ENGLER and (with hindsight) by BENTHAM & HOOKER and others.

Before leaving the classificatory aspects of higher taxa, perhaps a word ought to be said about the two classes of the Angiosperms—the Dicotyledons and the Monocotyledons. In a recent review EL-GAZZAR & HAMZA (1975) have considered a number of characters in addition to the conventional ones used to separate the two classes. These new characters include pollen and stomatal morphology, distribution of calcium oxalate crystals, lignin chemistry, duration of the mitotic cycle, etc. and they find that the traditional distinction between the Monocotyledons and the Dicotyledons is supported but that the *Alismataceae, Butomaceae,* and *Limnocharitaceae* are better associated with the *Nymphaeaceae* and *Nelumbonaceae* in the Dicotyledons while, on the other hand, the *Dioscoreaceae* fit better with the *Aristolochiaceae.*

Others, including MEEUSE, are of the opinion that the Monocotyledon-Dicotyledon groups will have to be dismembered. So even at the highest levels in the Angiosperms there is room for considerable debate about content and circumscription.

Evolutionary or Phylogenetic Classification

It is probably misleading to talk of evolutionary or phylogenetic classifications since one is forced to combine in such schemes both a vertical dimension, representing the historical-evolutionary dynamic elements and a horizontal dimension, representing present-day relationships, i.e. the actual classification to be used. Two-dimensional, horizontal classifications, based largely on comparative morphology of living Angiosperms, are perfectly valid and are an expression of the traditional box-within-box methodology normally employed in biological taxonomy.

On the other hand, trees are basically a false representation of the pathways and pattern of evolutionary processes. There is seldom any discernible basis for the branching pattern, not even for the major branches which in the Angiosperms are in dispute. For the full representation of evolutionary relationships multi-dimensional diagrams would be needed as remarked elsewhere (HEYWOOD 1966).

When a tree (or shrub) diagram is purported to represent both the evolutionary and the classificatory components, the result is untenable.

Unfortunately, the majority of recent phylogenetic/evolutionary schemes of classification have been presented in the form of a branching tree. While I agree with STEBBINS (1975) that the evolutionary tree is better represented as a much-branched shrub, I disagree with him that this form is characteristic of the CRONQUIST-TAKHTAJAN system. In fact most of such tree or shrub systems can be either placed on their sides or have the connecting lines removed and they become expressions of present-day relationships. They are little more than a graphic extension of phenetics. In this connection the attitude of CONSTANCE (1964) is a correct one, i.e. that such phylogenetic/evolutionary schemes or diagrams are aids to teaching or serve as an intellectual stimulus. Harm is caused, however, when they are regarded as more than this and as statement of factual or even putative evolutionary (i.e. cladistic, patristic, chronistic) relationships.

Systems such as those of DAHLGREN (1975) which are graphic representations of phylogenetic trees in transection are more satisfactory, although it must be noted that the tree itself is imaginary and the extent to which the imagined branches determine the spacing of the present-day groups clearly detracts from their usefulness. DAHLGREN writes, "To present orders of families of Angiosperms as a two-dimensional model is no innovation. Where this has been done, the relative position of the groups has been determined by the degree of mutual similarity. One disadvantage is that the reader, and sometimes even the constructor of the system, has been inclined to look upon the system of now living plants as an evolutionary tree, where some groups are regarded as descendants of others in the diagram. This applies in particular to HUTCHINSON 1969. Evidence for this type of evolutionary tree is usually sparse or lacking. The present Magnoliales, in particular, is often regarded as an ancestral group, other groups being frequently indicated as shooting out of it like lateral buds. The introduction here of a third dimension, time, is intended to prevent any such misinterpretation. It must be said that practically nothing is known about the course of evolution in the Angiosperms, so that the tree must be presented in such a generalised form that no evolutionary details are shown. Even the two-dimensional representation of the Angiosperms involves a tremendous number of problems."

It seems to me quite clear that the vertical dimension which makes the diagram a "tree" is derived from an interpretation of present-day relationships between the groups and not vice versa. The question has to be asked, therefore, if the time dimension adds anything of value or simply misleads by giving apparent information. DAHLGREN does not give a satisfactory answer to this point. His justification—that of avoiding misinterpretations about the direct descent of contemporaneous groups, one from another—seems gratuitous in that he has not made a case for the use of a tree at all. He does, however, discuss the problems of constructing a "reasonably functional two-dimensional diagram" for the orders and families of the Angiosperms. It is worth quoting his arguments in full since such two-dimensional "bubble" diagrams seem likely to acquire popularity as methods of presenting Angiosperm relationships and perhaps supplant the more customary tree diagrams.

DAHLGREN does not in fact believe that it is possible to construct reasonably functional two-dimensional diagrams but continues, "To place groups in exactly those positions that reflect their affinities becomes increasingly difficult when consideration has to be taken to the number of species in each group. For example, large 'bubbles' may prevent other, related, groups from meeting in the model, and small groups cannot be extended so as to approach sufficiently close to other groups showing great similarity. In any system, it seems, some families or orders apparently appropriately placed at the same time show several perhaps phylogenetically important similarities to one or more remotely placed group which in turn appears to occupy an appropriate position." He concludes by saying that it is imperative that botanists should persevere with the attempt to produce systems of this or similar types to survey the many groups of Angiosperms. The need, he says, is pedagogic rather than scientific. I cannot help observing, however, that his self-confessed difficulties of juxtaposing "bubbles" as wished, due to the physical constraints of the system, the failure to explain exactly why any given set of juxtapositions is employed, plus the acute difficulties mentioned above of determining the circumscription and content of orders, makes a nonsense of the tree or time-dimension implied and I would suggest that for pedagogic reasons, never mind scientific ones, it be ignored.

If the problem of presenting relationships in a two-dimensional diagram can be overcome, then what is theoretically possible by way of an evolutionary-phylogenetic classification is a series of horizontal classifications in time, like a stratigraphy, culminating in the present-day time-scale classification which would receive its explanation from the past but not necessarily its form. It has to be stressed that even if the full chronological fossil history of the Angiosperms were known, the classificatory problems would not be automatically solved as these depend on questions of ranking, of taxonomic judgement, criteria such as relative size of gaps between taxa, taste, tradition, etc. Classifications are made by man, although made possible by evolution.

Another point that is fundamental here, although often overlooked, is that the rank of any taxonomic group followed through evolutionary time will change. This has been clearly stated recently by HUGHES (1976) as follows, "If, however, a genus or higher taxon-concept is used to include 'species' from a range of geologic time-planes, its significance in terms of limits against other taxa (or even in terms of cluster centres) will be different on each time-plane studied depending on the general state of biological evolution at each of these times. This is not a measure of the number of fundamental taxa in each higher taxon-concept which may either increase or decrease with time, but a statement of the position of the family (or order) in the context of all those other families (or orders) existing at the time selected."

This leads to another aspect, what HUGHES calls "backwards systematics" (with perhaps a vague inference about systematists themselves!) i.e. the persistent use of living taxa of all ranks to accommodate Palaeogene and Mesozoic fossils. In fact nearly all fossil Angiosperms have been treated in this way, although it virtually precludes any meaningful discussion of evolution within the families or groups concerned. HUGHES asks, how can the details of descent of a living plant taxon from some Cretaceous fossils be determined when the latter have already been identified with the former? The lack of an adequate fossil record, and the inadequate way in which fossils have been classified in terms of extant taxa rather than in relation to members of their own time-zone, has led to comparative morphology of extant plants being used as a substitute for genuine fossil historical information by various methods of extrapolation backwards.

It is, nowadays, standard practice to use information from the morphology of living Angiosperms to attempt the reconstruction of the "primitive" flower or even the ancestral Angiosperm in all its characteristics. The difficulty about all these ingenious (or, in some cases, ingenuous) theories is that there is no way they can be tested, short of finding actual fossils for confirmation, and, what is more, they have fed on each other to such an extent that the body of a p p a r e n t evidence (much of it derived in a circular manner) is so impressive that it is difficult at times to remember that there is no factual basis for it beyond the present-day starting point and general trends in the fossil record. If one adds to this certain other pieces of dogma such as the improbability (= virtual impossibility) of the eight-celled female gametophyte and associated double fertilisation having evolved more than once (although this hypothesis can be challenged and has no factual basis—the eight-celled gametophyte has not been seen in pre-Quaternary material!), we are then more or less forced to accept another dogma, the monophyletic origin of the Angiosperms, which in turn makes it easier to concentrate our thinking on a single archetype or archetypal group. This is what one might term canalisation of conceptual thinking.

The methods which may be employed for attempting to determine ancestral or primitive characters by extrapolation from the features possessed by contemporaneous organisms are several. While they have been familiar to zoologists for several decades, few Angiosperm taxonomists have applied them explicitly. There is a marked difference between the botanical and zoological literature in this respect. This is very evident if one consults the journal "Systematic Zoology", for example, which has no botanical equivalent.

The main methods are correlation of characters used in conjunction with what HENNIG (1966) calls "reciprocal illumination" or what WALKER (1974, 1976) calls "spiral reasoning". As WALKER (1976) says, in view of the wholly inadequate fossil record of Angiosperms, a combination of character correlation and spiral reasoning frequently represents the only available way of determining the primitive character-state of characters peculiar to Angiosperms.

Character correlation may be what WALKER calls random at one particular taxonomic level or time-scale. This is the basis of the approach of SPORNE in a series of papers (see p. 33). Character correlation may also be directional, which consists of individually observing the direction of character trends within a number of different lower taxa included in the taxon for which it is hoped to determine which character-states are primitive. This involves working out character phylogenies or semophyleses. If there is a high correlation of many characters, some of which are thought to be primitive on independent grounds, then, so the argument goes, there is a strong probability that each character represents an ancestral or primitive state.

But characters are not taxa and while a primitive character is usually ancestral, a primitive taxon is not. As WALKER says, a primitive taxon is simply one that retains a large number of primitive characters relative to some other taxon. Unfortunately, in the absence of fossils, it is not easy to deduce the direction of character trends and it is impossible to prove. STEBBINS (1975) has some very pertinent comments to make on this question in relation to the developmental basis of the individual characters concerned.

The definitive way of determining primitive characters (but not primitive taxa) is the study of character-states sequentially in a time-sequenced fossil record. This can seldom be done except, for example, in the case of pollen (cf. WALKER 1976). Even when, as in pollen grains, there is a reasonably complete fossil series, it tells us nothing about the other characters of the plant produced the pollen. There is certainly no guarantee that pollen similarity means taxonomic identity. Unfortunately, there are virtually no cases where the "primitive" characters found in fossil pollen are associated with other independent "primitive" characters because of the lack of linked fossil material.

Walker (1976), by applying the methods of character correlation, character trends and spiral reasoning, arrives at the conclusion that the ancestral pollen possessed the following characteristics: anasulcate, with a long sulcus; heteropolar; bilateral; boat-shaped-elongate; atectate and primitively columella-less; more or less psilate; solitary; and large. In the face of this, the most primitive pollen found in living Angiosperms occurs in the *Magnoliaceae, Degeneriaceae* and *Annonaceae.*

When we look at the fossil record, however, we find that the earliest fossil pollen accepted as Angiosperm is that of the genus *Clavatipollenites* but this is too specialised in several features (such as the possession of columellae) to have been ancestral and is probably most similar to that of the *Chloranthaceae* amongst extant Angiosperms.

Walker concludes that while a tremendous diversity has been found in the pollen of living "primitive" Angiosperms, it has to be remembered that most of it is at the level of monosulcate-derived pollen. And while tricolpate or tricolpate-derived pollen is essentially restricted to the Angiosperms (dicotyledons) and can therefore be more readily identified in the fossil record as being angiosperm rather than gymnosperm, it is too advanced to inform us about the early evolution of the Angiosperms. At the critical period in geological history, pollen of Angiosperms and Angiosperm ancestors was at the monosulcate level of evolution and consequently difficult, if not impossible, to distinguish as being angiosperm rather than gymnosperm in the absence of other evidence.

Palaeobotanists interested in the origin of Flowering Plants must look, Walker says, for fossil Gymnosperms that produced large, more or less psilate, boat-shaped-elongate pollen grains with a long anasulcus and an atectate, primitively columella-less exine. "For it was probably pollen of this type that some extinct groups of seed plants crossed the line from gymnosperm to angiosperm."

Whether this line was crossed by one or several lines—whether the Angiosperms were a monophyletic clade or a pleiophyletic grade—remains to be seen. To me it seems improbable that the whole syndrome of angiospermy evolved neatly once in a single evolutionary line.

In summary, the fossil record even as it is available today, has much to tell us if studied more thoroughly. So far it does little to confirm current views as to the actual origins of the Angiosperms although it supports the relative primitive nature of the woody Ranalean families in respect of some (but not all) features.

As Stebbins (1975) comments, in the absence of significant fossils, so many unknown factors exist that the method of extrapolation is inadequate as a means of deducing what ancestral forms were like on the basis of any living Angiosperms. He draws attention to the fact that if one considers the fossil record of the conifers, the common ancestors of the main modern families, *Pinaceae, Taxodiaceae, Cupressaceae, Taxaceae,* and *Podocarpaceae,* are so different that they could not possible be reconstructed except on the basis of fossils: this seems to be generally true of other groups for which there is an adequate fossil record and there is no reason to believe that the situation in the Angiosperms will prove to be different. A recent report of a fossil dicotyledonous Angiosperm flower from the Upper Cretaceous (TIFFNEY 1977) tends to confirm this: its characteristics are distinctly angiospermous yet their exact association does not appear to occur in any modern group and attempts to place it in an extant angiosperm taxon have failed.

Conclusion

While methods do exist for the extrapolation of phylogenetic trends from extant forms, in the absence of adequate fossils, these seem to have little bearing or effect on the so-called phylogenetic or evolutionary systems of classification of the Angiosperms except to assist in working out the relative degree of primitiveness or advancement of individual higher taxa. They do, however, help give us an insight into the evolutionary processes involved. Because of the size and complexity of the problems involved, the methodologies of evolutionary extrapolation and reconstruction have not been consistently applied in the construction of systems of classification which owe much more to the traditional methods of taxonomy than to the principles of evolution. On the other hand, we should press for the publication of the detailed reasons for the form and content of the currently adopted systems of classification, such as those of CRONQUIST, TAKHTAJAN, and THORNE. Admittedly, the latter author has commenced publication of the detailed rationale for part of his system, for example, his phylogenetic classification of the *Annoniflorae* (THORNE 1974), a move greatly to be welcomed. Until this is more commonly done, we must remain somewhat sceptical about their systems, qua classifications, and modifications to them in the absence of a data-base. The horizontal, basically phenetic approach of DAHLGREN is a welcome development, especially as there is now becoming available an impressive body of comparative data (some of it presented in this volume) which can be used to test it.

Botanists can scarcely complain if the currently available systems of classification do not meet their needs unless they make these needs clearly known.

References

BENTHAM, G., 1873: Notes on the classification, history and geographical distribution of Compositae. J. Linn. Soc. (Bot.) **13**, 335—577.

CONSTANCE, L., 1964: Systematic botany—an unending synthesis. Taxon **13**, 257—273.

CRONQUIST, A., 1975: Some thoughts on angiosperm phylogeny and taxonomy. Ann. Missouri Bot. Gard. **62**, 517—520.

DAHLGREN, R., 1975: A system of classification of the angiosperms to be used to demonstrate the distribution of characters. Bot. Notiser **128**, 119—147.

EL-GAZZAR, A., and HAMZA, M. K., 1975: On the Monocots-Dicots distinction. Publ. Cairo Univ. Herb. **6**, 15—28.

HEYWOOD, V. H., 1966: How many taxonomies? Rev. Roum. Biol. sér. Bot. **11**, 101—106.

— HARBORNE, J. B., and TURNER, B. L. (Eds.), 1977: The Biology and Chemistry of the Compositae. London-New York: Academic Press.

HENNIG, W., 1966: Phylogenetic Systematics. Urbana, Chicago and London: Univ. Illinois Press.

HUGHES, N. F., 1976: Palaeobiology of Angiosperm Origins. Cambridge: University Press.

JACOBS, M., 1969: Large Families—not alone! Taxon **18**, 253—262.

STEBBINS, G. L., 1975: Deductions about transspecific evolution through extrapolation from processes at the population and species level. Ann. Missouri Bot. Gard. **62**, 825—834.

THORNE, R. F., 1974: A phylogenetic classification of the Annoniflorae. Aliso **8**, 117—209.

TIFFNEY, B. H., 1977: Dicotyledonous angiosperm flowers from the Upper Cretaceous of Martha's Vineyard, Massachusetts. Nature **265**, 136—137.

WALKER, J. W., 1974: Aperture evolution in the pollen of primitive angiosperms. Amer. J. Bot. **61**, 1112—1137.

— 1976: Comparative pollen morphology and phylogeny of the Ranalean complex. In: Origin and early evolution of angiosperms (BECK, C. B., Ed.), 241—299. New York: Columbia Univ. Press.

Address of the author: Professor V. H. HEYWOOD, Department of Botany, Plant Science Laboratories, The University, Whiteknights, Reading RG6 2AS, England.

Plant Syst. Evol., Suppl. 1, 13—19 (1977)

Hugo de Vries-Laboratorium, University of Amsterdam, The Netherlands

Delimitation of the Major Taxa of the Higher *Cycadophytina*: Theoretical Criteria Versus Taxonomic Practice

By

A. D. J. Meeuse, Amsterdam

Abstract: The application of other than strictly morphological criteria in attempts towards producing a "phylogenetic" classification of the Flowering Plants is clearly unsound as long as the fundamental morphological interpretation of certain magnoliophytic features remains unsettled. Such "established" notions as gymnospermy versus angiospermy, the nature of the functional reproductive units, the sporophyll concept, the assumed primacy of phaneranthy and the entomophilous syndrome (including the sex distribution), etc., which postulates primarily decide the basic framework for the taxonomic arrangement in a system, may well be antiquated. A greater heterogeneity among the major taxa of the magnoliophytic assembly is in better agreement with the available evidence and finds some support when additional taxonomic criteria are applied. It is to be expected that appreciable advances will result from such ancillary criteria if applied to a framework based on a neological approach of the phylogeny and morphology of the *Magnoliophyta* founded upon alternative assumptions concerning their original anthomorphology and palaeoecology. The present author is entirely in favour of the application of evidence from all sources potentially providing useful taxonomic pointers, but does not believe that such data can be decisive for the delimitation and placing of higher taxa among the Angiosperms until some agreement concerning several fundamental, morphological concepts has been attained. This inevitably involves phylogenetic speculations concerning the nature and morphology of the early magnoliophytes which already require the application of some additional criteria such as palaeoecological evidence and certain palaeobotanic clues.

Although various kinds of criteria are gradually being given more prominence in classificatory botany (cf., e.g., JOHRI 1963, MAHESHWARI 1964, KUBITZKI 1969, 1973, MEEUSE 1970, BATE-SMITH 1972, PHILIPSON 1974, BOUMAN 1974, DAHLGREN 1975, DICKISON 1975, EYDE 1975, JENSEN et al. 1975, SPORNE 1975), their application to the distinction or rearrangement of higher taxa of the advanced cycadophytinous spermatophytes has mostly remained restricted to the family level. Recently the *Thelygonaceae* have, for instance, been transferred to the *Rubiales* in the most recently published systems of classification, and the inclusion of the *Batidaceae* and *Tropaeolaceae* in the *Capparales* has been proposed by Dahlgren (1975). These transfers were mostly based on phytochemical evidence, but this is not the case in Behnke's studies of microplastids which clearly seem to provide an absolute criterion (viz., the presence of a special type of plastid) for the segregation of the *Centrospermae* (= *Caryophyllidae*) from all other *Angiospermae* (see Behnke 1972, and this issue). This is also the only example of which one can truly say that it served to distinguish a major taxon at, at least, ordinal level on the ground of a non-conventional differential characteristic, although one must bear in mind that the separate status of the centrospermous alliance had already been accepted much earlier on the basis of traditional criteria. Behnke's data confirm the caryophyllalean affinity of several taxa whilst they point to the exclusion of other families from the assembly.

Apart from this example, several other attempts have been made to use criteria other than the customary macromorphological features, e.g., characteristics of the ovule (Philipson 1974, and this issue; see also Bouman 1974) and the androecial ontogeny (Leins 1971, Merxmüller 1972), but if one takes stock of the systems of classification proposed in the last decade (cf. for example those of Ehrendorfer 1971, Takhtajan 1973, and Dahlgren 1974), it is quite clear that although the proposers of these systems aim at a taxonomic arrangement reflecting the phylogenetic origin of the various groups, which suggests a good deal of original thinking, it appears that the current systems are not basically different in approach and scope from several proposed by about the turn of the century or perhaps not even from some developed over a century ago (at least not in principle). It is the purpose of the present paper to confront the methodological basis of those systems of classification of the *Magnoliophyta* or Angiosperms which have at present the greatest number of adherents with evidence which may yield alternative criteria for the definition of primitive or basic taxonomic categories within the Flowering Plants. Certain "classical" typological fundamentals developed in phytomorphology and secondarily employed in taxonomic classification inevitably dictate certain sequential arrangements of taxonomic groups at the family or ordinal level and thus lay down an overall foundation for the ultimate classification of all Angiosperms. As we have seen, readjustments emanating from the application of additional criteria mostly result in some minor "shifts" and rearrangements without changing the existing major framework of the systems in question. There is absolutely no point in discussing the application of unconventional criteria in the present context without a previous analysis of the above-mentioned, essentially typological tenets.

Systematists professing to aim (or claiming to have arrived) at a phylogenetic system have subjectively made a principal, aprioristic assumption regarding the more basic (= primitive) versus the more secondary (= advanced) groups and the corresponding morphological characteristics, and they started their classification from there. The taxonomic features of the group accepted as the relatively most primitive of all were in turn used in phylogenetic speculations regarding the ancestry and the palaeoecology of the assumed basic taxon[1].

Gradually a number of primary and secondary postulates emerged which were partly suffused with other doctrines such as the notions of gymnospermy versus angiospermy, and the "sporophyll" concept. The present author has repeatedly criticised this narrow point of view (see MEEUSE 1974a, 1975b, 1976) and singled out the most relevant basic postulates:

1. Flowering Plants are monophyletic.

2. All Flowering Plants bear "flowers" (or rather: functional reproductive units) of one kind only; these conventional flowers are all supposed to be uniaxial structures, basically corresponding (homologous) with a leafy shoot.

3. Primitive magnoliophytic forms, both the extinct and the surviving recent representatives, were (are) phaneranthous and entomophilous.

4. All angiospermous flowers are derived from the phaneranthous

---

[1] This can be substantiated by two recent examples. In an introduction to a symposium WALKER (1975: 515) explained that the foundations of angiosperm phylogeny were to be discussed on the basis of the TAKHTAJAN and CRONQUIST systems of classification which systems were chosen "because they are the most thoroughly documented ... systems ... which have taken into consideration data from all bases of angiosperm phylogeny".

The other example is also in a paper by WALKER (1976: 291) in which he, when discussing early angiospermoid pollen types, rejects the conclusion drawn by KUPRIANOVA (1967) concerning the occurrence in the Lower to Middle Cretaceous of plants closely allied to the recent *Chloranthaceae* on the ground that at that time such an already "advanced" family could not yet have evolved ("if paleobotany is to tell us anything about the origin of angiosperms pollen similar to that of *Magnolia*, *Degeneria*, and *Anaxagorea*, not *Ascarina*, must be sought in the fossil record"). DOYLE et al. (1975: 451), on the other hand, do not reject a relation between Barremian or Aptian *Clavatipollenites* and pollen grains of recent taxa including *Chloranthaceae* (and *Myristicaceae*, also already advanced in the sense of WALKER), and thus (implicitly) accept a possible taxonomic (and morphological) affinity between the latter and certain Cretaceous taxa.

archetype and this tenet in turn serves as a yardstick for the assessment of the degree of evolutionary advancement of all other kinds of so-called flowers as more or less modified derivatives of that preselected, and supposedly archaic prototype of all "flowers"; the corresponding taxa are subsequently arranged according to this assessment as not, or hardly, to extremely advanced descendants of the group with the most basic floral morphology.

5. Flowering Plants have foliar carpels enclosing the ovules (this condition is called angiospermy or angioody), and they differ in this respect appreciably and fundamentally from the gymnosperms.

The ensuing system of classification is rigid in that only one-directional sequences can be construed, all based on, and radiating from the same primary group. It follows that taxa with simply constructed functional reproductive units (FRUs = conventional "flowers"), such as most of the so-called apetalous ones, are of necessity placed at the end of such sequences as much derived (or depauperated, advanced, or reduced but in any case secondarily originated) taxa, e.g., amentiferous, cyperaceous, and salicaceous forms. (It must be pointed out here that the opposite starting point forms the basis of the Englerian and Wettsteinian systems in which priority of place is given to monochlamydeous dicotyledons, and a more advanced status to phaneranthous orders, but, mutatis mutandis, the same reasoning is followed although in a more or less reverse order; however, the situation is more complicated owing to a discrepancy in ENGLER's views on Angiosperm phylogeny and his "system": see MEEUSE 1972.)

Each of the above-mentioned five postulates has been subjected to severe criticism by the present author. Alternative assumptions can be formulated as follows:

As opposed to (1): Flowering Plants are pleiophyletic in so far that several evolutionary lineages leading to recent magnoliophytic groups had already become segregated before their common progenitorial taxon, from which such a group has descended, had completely attained the level of advancement of a truly angiospermous plant form.

As opposed to (2): the FRUs of the Flowering Plants are all derived from an ancestral, complex and pluriaxial structure (anthocorm), but represent either a condensed, whole anthocorm, or only a subordinate part of it, so that the conventional category of the flower is heterogeneous (MEEUSE 1975a); anthocorms and subordinate parts of anthocorms are, moreover, sui generis in respect of a leafy shoot and its appendages.

As opposed to (3): primitive magnoliophytic groups (and their immediate precursors) were aphananthous and anemophilous or at best incipiently entomophilous; dicliny prevailed, incipient monocliny only originating in those lineages which ultimately developed into predominantly phaneranthous and zoophilous forms (MEEUSE 1976).

As opposed to (4): angiospermous FRUs reflect an early divergent evolution and can be used as a yardstick to trace "impossible" and unacceptable derivations, in conjunction with the assessment of the phylogenetic origin and the morphological status of the androecial members (MEEUSE 1974)—as an example, the derivation of hamamelidid taxa from a "ranalean" (= in practice a promagnolialean or even magnolialean) progenitor, or vice versa, must be rejected as highly improbable and rather be explained as the result of a parallel, independent evolution of fairly long duration as accepted sub (1).

As opposed to (5): the "carpels" (monogyna in a more neutral terminology) and most of the other floral appendages are considered to be sui generis in respect of the functional leaves or trophophylls; the genitalia (and most intrafloral semaphylls) are direct derivatives of ovuliferous cupules and androsynangia-bearing organs of the pteridospermous and more advanced cycadophytinous gymnosperms and thus link the *Magnoliophyta* with the latter, whilst still serving as a yardstick for the assessment of the relative degree of phylogenetic advancement of the sex organs of recent angiospermous groups.

It is quite clear that, as was already pointed out before, there is no reason to introduce other than morphological criteria before the fundamental controversies are understood, and before one has agreed upon the principles to be accepted as a starting point for the basic framework of a classification.

According to the present author the *Magnoliophyta* are much more heterogeneous than is generally assumed. Some recent summaries (see, e.g., BATE-SMITH 1972, MERXMÜLLER 1972, PHILIPSON 1974), some primarily based on morphological evidence, also suggest a more pronounced intranscendence of the major groups of Angiosperms than is generally accepted. This is in good agreement with certain indications from phytochemical, palynological, anthecological, embryological, and ontogenetic studies, which will not be discussed here in detail. Such surveys also emphasize the great error repeated since at least a hundred years ago of lumping (nearly) all sympetalous dicots in what is now mostly called the "*Asteridae*". This group is so heterogeneous, and the character of sympetaly has so frequently been mentioned as an example of convergent evolution, that it is absurd to retain the artificial assembly of the "*Asteridae*". This aggregate must be split up and referred to several of the principal subordinate taxa of the dicots: *Campanulales* and *Asterales* most probably belong, together with *Araliales* s.s. (*Apiales* s.s., i.e., without *Cornales* and related families containing iridoid compounds), to a lineage which also includes the ranalean and sapindalean-

rutalean orders; a rosoid-saxifragoid-cornoid affinity may be accepted
for tubiflorous (lamialean) groups; the *Ericales* may possibly have a
dilleniid origin, etc. Such a heterogeneity is also found among the
monocotyledons or *Liliatae* (compare, e.g., MEEUSE 1975c). Helobial
groups (*Alismatales*, etc.) appear to be rather isolated, and the lilialean
assembly is not necessarily basic in respect of *Poales, Cyperales, Pandanales, Arecales, Zingiberales*, and *Arales*. A long phylogenetic history
is more plausible than a monophyletic derivation from a single basic
group of protomonocots, which in its turn, as several workers have it,
is descended from a dicotyledonoid (nymphaeoid, ranalean-ranunculid,
or ranalean-piperid) progenitorial taxon.

It is quite clear that such views necessitate a reconsideration of
the starting point of a classification. One of the conclusions may be a
different dividing line between the principal subordinate major groups,
which segregates the non-ranalean orders from an aggregate of all
monocotyledonous and ranalean taxa, as strongly intimated by HUBER
in this symposium. For this reason the fundamental morphological
features and the associated evolutionary trends derivable from the
occurrence of such features ought to be re-studied and re-assessed more
thoroughly. It is possible that, especially as far as the phylogenetic
aspects are concerned, additional criteria may need to be used in this
connection. Certain fossil finds discussed by the present author (MEEUSE
1976: 92—94) may, for instance, point to the more probable, early floral
architecture in protangiospermous forms.

References

BATE-SMITH, E. C., 1972: Chemistry and phylogeny of the Angiosperms.
    Nature **236**, 353—354.
BEHNKE, H.-D., 1972: Sieve-tube plastids in relation to Angiosperm sys-
    tematics—an attempt towards a classification by ultrastructural ana-
    lysis. Bot. Rev. **38**, 155—197.
BOUMAN, F., 1974: Developmental studies of the ovule, integuments and
    seed in some Angiosperms. Thesis, Univ. of Amsterdam. Naarden.
DAHLGREN, R., 1974: Angiospermernes taxonomi. Vol. 1. Copenhagen.
— 1975: Current Topics. The distribution of characters within an Angio-
    sperm System. 1. Some embryological characters. Bot. Notiser **128**,
    181—197.
DICKISON, W. C., 1975: The bases of Angiosperm phylogeny: vegetative
    anatomy. Ann. Missouri Bot. Gdn. **62**, 590—620.
DOYLE, J. A., VAN CAMPO, M., and LUGARDON, B., 1975: Observations
    on exine structure of *Eucommiidites* and Lower Cretaceous Angiosperm
    pollen. Pollen & Spores **17**, 429—486.
EHRENDORFER, F., 1971: Systematik und Evolution. In: Lehrbuch der
    Botanik für Hochschulen (STRASBURGER, E., et al.), 30. Aufl., p. 379—741.
    Stuttgart: G. Fischer.
EYDE, R. H., 1975: The bases of Angiosperm phylogeny: floral anatomy.
    Ann. Missouri Bot. Gard. **62**, 521—537.

JENSEN, S. R., NIELSEN, B. J., and DAHLGREN, R., 1975: Iridoid compounds, their occurrence and systematic importance in the Angiosperms. Bot. Notiser **128**, 148—180.

JOHRI, B. M., 1963: Embryology and taxonomy. In: Recent advances in the embryology of Angiosperms (MAHESHWARI, P., Ed.), 395—444. Delhi.

KUBITZKI, K., 1969: Chemosystematische Betrachtungen zur Großgliederung der Dicotylen. Taxon **18**, 360—368.

— 1973: Probleme der Großsystematik der Blütenpflanzen. Ber. dtsch. bot. Ges. **85**, 259—277 (1972).

KUPRIANOVA, L. A., 1967: Palynological data for the history of the *Chloranthaceae*. Pollen & Spores **9**, 95—100.

LEINS, P., 1971: Das Androeceum der Dikotylen. Ber. dtsch. bot. Ges. **84**, 191—193.

MAHESHWARI, P., 1964: Embryology in relation to taxonomy. In: Vistas in Botany (TURRILL, W. B., Ed.), **4**, 55—97. Oxford: Pergamon Press.

MEEUSE, A. D. J., 1970: The descent of the Flowering Plants in the light of new evidence from phytochemistry and from other sources. Acta Bot. Neerl. **19**, 61—72, 133—140.

— 1972: Sixty-five years of theories of the multiaxial flower. Acta Biotheor. **21**, 167—202.

— 1974: Some fundamental principles in interpretative floral morphology. In: Vistas in Plant Sciences (VARGHESE, T. M., Ed.), Vol. **1**, 1—78. Hissar.

— 1975 a: Changing floral concepts: Anthocorms, flowers, and anthoids. Acta Bot. Neerl. **24**, 25—36.

— 1975 b (1974): Floral evolution and emended Anthocorm Theory. In: Intern. Bioscience Monogr. (VARGHESE, T. M., Ed.), Vol. **1**. Hissar.

— 1975 c: Aspects of the evolution of the Monocotyledons. Acta Bot. Neerl. **24**, 421—436.

— 1976: Fundamental aspects of evolution of the *Magnoliophyta*. In: Glimpses in Plant Research (NAIR, P. K. K., Ed.), **3**, 82—100. New Delhi: Vikas.

MERXMÜLLER, H., 1972: Systematic Botany: an unachieved synthesis. Biol. J. Linn. Soc. **4**, 311—322.

PHILIPSON, W. R., 1974: Ovular morphology and the major classification of the dicotyledons. Bot. J. Linn. Soc. **68**, 89—109.

SPORNE, K. R., 1975: A note on ellagitannins as indicators of evolutionary status in Dicotyledons. New Phytol. **75**, 613—618.

TAKHTAJAN, A. L., 1973: Evolution und Ausbreitung der Blütenpflanzen. Jena: G. Fischer.

WAGENITZ, G., 1975: Blütenreduktion als ein zentrales Problem der Angiospermen-Systematik. Bot. Jahrb. Syst. **96**, 448—470.

WALKER, J. W., 1975: The bases of angiosperm phylogeny: Introduction. Ann. Missouri Bot. Gdn. **62**, 515—516.

— 1976: Comparative pollen morphology and phylogeny of the Ranalean complex. In: Origin and early evolution of the Angiosperms (BECK, C. B., Ed.), 241—299. New York: Columbia Univ. Press.

Address of the author: Prof. Dr. A. D. J. MEEUSE, Plantage Middenlaan 2 A, Amsterdam, The Netherlands.

Plant Syst. Evol., Suppl. 1, 21—31 (1977)

Institut für Allgemeine Botanik und Botanischer Garten,
Universität Hamburg, Federal Republic of Germany

Some Aspects of the Classification and Evolution of Higher Taxa

By

K. Kubitzki, Hamburg

Abstract: It is shown that a fully phylogenetic classification of the flowering plants, though not impossible in principle, is hard to attain at present because of the absence of any precise cladistic information about this plant group. The unwarranted confusion of grades and clades imposes a strongly typological element upon current systems intended to be phylogenetic. A plea is made for a striving after an understanding of the functional significance of characters which is indispensable for aiming at a true understanding of evolutionary processes and pathways.

In the study of the taxonomy and the evolution of the flowering plants the higher levels of the taxonomic hierarchy have been relatively neglected for a long time. The activity of most taxonomists has been concentrated mainly around the production and revision of regional Floras, and though much effort has been devoted in the last century towards the production of phylogenetic systems of classification of flowering plants, these have often been no more than attempted phylogenetic rearrangements of established groups, rather than a continued reassessment of the taxa themselves in the light of new evidence (YOUNG & WATSON 1970). Nevertheless, botanists have always been ready to assimilate, and to integrate into their classifications, information from new sources, though mostly in a somewhat selective manner, with the main emphasis on information from—at any given time—more fashionable fields, as more recently from population biology, natural product chemistry, and ultrastructural research. It is an interesting feature that there have always been certain newer developments which, at the time of their appearance, have been overestimated, with the danger of becoming discredited early, the rise and fall of serology during the first half of this century being but one example. Today, however, it is hardly likely that any of our highly valued ancillary disciplines will suffer the same fate.

As to evolutionary studies, these have focused during the last three decades or so mainly around the study of populations, though in this field also, important conclusions have been reached as to evolutionary pathways and patterns at and above the level of the species. As a result of these efforts, the process of evolution at the population level is now so well understood that the leaders of this field, such as G. L. STEBBINS and E. MAYR, have already turned to the question of the evolutionary mechanisms involved in the differentiation of higher taxa, and the renewed interest in these aspects of flowering plant evolution and classification may well be due to this attitude. At any rate, we see the factual basis of current classifications being analyzed by a growing number of systematists, and these activities have recently led to symposia like that on "The monocotyledons: Their evolution and comparative biology" (in "Quarterly Review of Biology, Vol. 48, No. 3/4, 1973), "Chemistry in botanical classification" (edited by BENDZ and SANTESSON), or "The bases of angiosperm phylogeny (in "Ann. Missouri Bot. Gard. Vol. 62, No. 3, 1976), and now also to this symposium which, contrary to those just mentioned, has been planned by its organizers to embrace also the methodological problems involved in the study of higher taxa.

The latter is an important aspect since the consumers of classificatory systems like teachers and students, plant physiologists, chemists, etc., mostly remain ignorant about the processes of classification and system-making, being primarily interested in their results. Thus it is hardly appreciated outside taxonomic circles (and often not even there!) that the systematist when being faced with problematical choices all too often has to opt for one out of several equally tenable solutions concerning circumscription and arrangement of taxa; moreover, few people are really aware of just how vacillating the phylogenetic basis of recent systems of classification in fact is!

If one looks into the classificatory basis of flowering plant systematics, first of all the observer is impressed by the vast amount of information which has been accumulated by generations of botanists and other workers, without being readily available to, or at least not commonly used by, the taxonomist. There are various reasons why the data worked out by different disciplines often do not reach those concerned with the task of building up classifications. One must agree with HEYWOOD (1973a) that the main, but still unachieved task of systematics, besides of its identificatory role, is "to collate, synthesize and order these data in a systematic fashion so that conclusions of all sorts may be drawn". It is probably true that some sorts of data are easier to handle, in this respect, than others; this may apply, e.g., to embryological characteristics which always have been well summarized (see, e.g., DAHLGREN 1975a), and, because of their innate unambiguity, this seems to be especially true for chemical data, as shown by some recent compilations of the systematic distribution and importance of xanthones (REZENDE & GOTTLIEB 1973), of benzyltetrahydroisochinolines (REZENDE et al. 1975) and of iridoid compounds (JENSEN et al. 1975). There are few examples, however, in which for a larger group the obtainable evidence from all fields has been collated though numerical methods which are perhaps the most important methodological achievement of systematic biology in recent times, are available for analyzing the similarities and differences of organisms upon which most classifications are based. There is no doubt that numerical methods become the more indispensable, the larger is the size of the taxa. It must be mentioned, however, that in practice we are still very far from making use of the wealth of information already accumulated.

This has led to the impression that taxonomists, in building their classifications, proceed in a highly selective manner. With regard to the lower taxa, this is only seemingly true and is a result of the common practice among taxonomists, who regard the taxonomic decisions to be more important than the information on which the decisions have been based, so that, in communicating their results, they justify their decisions in the form of a key or a description, but do not worry about listing the information available about the organism (HEYWOOD 1973). As to the higher categories, taxonomists are often assumed to select deliberately those facts which fit best in their preconceived ideas, but I think there are also inherent difficulties which contribute to narrow the number of usable characters. These may be seen in how to treat characteristics of scattered occurrence, how to formulate an average picture of a taxonomic group, which, in itself, embraces different anagenetic grades, and, above all, how to evaluate and integrate features which undoubtly have come into being as the result of parallelism and convergence. It is because of this, as my colleague Dr. HUBER informs me, that for the building up of an evolutionary classification of the monocotyledons, e.g., only a surprisingly small number of characters is suitable.

Mentioning of parallelism and convergence leads us to an inspection of the evolutionary components of classifications since, if there were divergence only, but no convergence, a purely phenetically based classification would by itself be a phylogenetic one. As PHILIPSON (1961) has stressed the probabilities of the occurrence of convergence among the flowering plants have been still underestimated so that many taxa of any rank, at closer inspection, will be revealed as unnatural grade groups. The question arises then if an evolutionary system of classification is possible at all and if so, how to achieve it. If we are to construct a system which intends to reflect the development of the plant kingdom or a part of it in time and space, in other words if we are going to try to elucidate phylogeny (the origin and evolution of taxonomic groups), and if such an attempt is based on knowledge of present-day forms alone, then the answer must clearly be, that this is an virtually unachievable goal. In view of the still meagre palaeontological evidence, a recon-struction of the interrelationships between all kinds of flowering plants, fossil and Recent, is for the present an unattainable task, though HUGHES (1976), in his recent book, shows that it is possible, in principle, even if such an attempt would enforce a profound change in methodology. As STEBBINS (1974) in particular has pointed out, the common ancestors of all modern flowering plants are completely extinct, and from all modern groups of plants and animals that possess a reasonably good fossil record it can be learned that the nature of the ancestors of ancient and widespread modern classes can never be reconstructed with certainty by putting together traits recognized as primitive which occur scattered among modern forms[1]. Nevertheless, continued efforts have been made to deduce phylogeny in this strict sense by directly relying on modern forms, though these have been condemned as "pseudophy-logeny" by spokesmen of phylogenetic classification, as LAM and ZIMMERMANN.

Widely prevalent, however, is a much more loose concept of phylo-geny, as, for instance, is nicely expressed by JOHNSON (1972), who feels phylogeny to imply primarily "retrospective predictivity; that is, it predicts what we may hope to find out, in the future, about the past—and thus 'explain' the present!" Nevertheless, we have avoided, in the title of this symposium, the self-committing term "phylogeny" and have preferred, to speak instead, of "evolutionary systems of classifi-cation". I am aware that there are also strictly phenetic classificatory approaches, like the notable one of YOUNG & WATSON (1970); yet it is hard to see why at least an attempted evolutionary interpretation of an otherwise perhaps more or less strictly phenetically based clas-sification should not be allowed for.

I would now like to try to analyse the phylogenetic/evolutionary features which current systems of classification do display. As to the circum-scription of the families and, to some degree, also to the orders of the flowering plants, and to their primary division in di- and monocotyledons, a rather broad agreement has been reached (for a contrasting view, see HUBER in this volume). The mutual relationships of the groups, however,

---

[1] It should be noted here that reliance on primitive characters is strictly rejected by the practitioners of phylogenetic systematics in zoology; advanced characters are used instead, though this method does not aim so much for the recognition of the common ancestor but for establishing sister group relationships.

are judged to be rather controversial, and their arrangement relative to one another and their sequence have been dealt with quite differently by various authors. If, however, the sequence and position of contemporaneous groups relative to each other (without practically any knowledge of their cladistic relationship) really is the main phylogenetic/evolutionary content of such classifications, then HEYWOOD (1973b) is right in considering such a "construction of phylogenetic schemes of the angiosperm(s) . . . a much overrated pastime which has been pursued far too unscientifically for far too long". Furthermore, neither the rank of the taxa, which of course is mainly determined by their phenetic distance, nor the "box-in-box structure" of the system (formerly often considered to be a proof a phylogeny) are very meaningful phylogenetic components. The main evolutionary element which influences taxonomic arrangements may well be the significance of morphological characters (in their broadest sense) in terms of primitiveness and advancement, and the determination of their trends which lead to the establishment of so-called "Merkmalsphylogenien" or "semophyleses". Moreover, there has been the recognition of genetic and ecological trends (e.g., allogamy → autogamy) and of certain syndromes which also tend to follow each other in regular successions (e.g., different modes of zoogamy; zoogamy → anemogamy, etc.). In the last cases in which ecologically important traits are composed of several single features which must cooperate harmoniously, their functional importance cannot be overlooked; I doubt, however, if such always exists and can be indicated with regard to isolated features. This is worth mentioning because it is universally agreed that evolution is accompanied by a constant improvement of adaptations at all levels, from which follows that the primitive should be less well adapted than the advanced. One objection to this is the notion, put forward especially by STEBBINS (1974), that it is normally not the single character but rather an adaptive complex of several characters which is adaptive. In other words, we mostly ignore completely what a character which may be important to us (mainly for attempting to elucidate phylogeny) might signify for the plant (see CROWSON 1970, KUBITZKI 1975). Because of this, and because of the impossibility to test semophyleses along with palaeontological evidence, these character phylogenies tend to include a strongly typological element. I would like to conclude that functional improvement, as a concomitant of a certain evolutionary trend of a single feature, can only be taken for granted if it is either self-evident (as perhaps in the case of BAILEY's sequence from tracheids to vessels with simple perforation), if the anagenetic improvement has been tested experimentally, or if this can be concluded on the basis of experiments by analogy.

These limitations do not at all minimize the value of the functional point of view which has the advantage of leading away from typology towards an understanding of adaptations. In this context, it is rather con-fusing that flower characters, which have been considered for a long time to be especially conservative and hence have been used from the very beginnings of plant taxonomy as the major yardstick for tracing affinities, have now become known to be extremely plastic in the face of the plant's needs to maintain, or to improve, an effective pollination system. It is because of this that biologists like CARLQUIST (1969) have claimed a mainly, if not exclusively, functional interpretation of flower structures, and more recently (CARLQUIST 1975) also of wood structure, claims which, however, appear to me somewhat one-sided, because they overlook the fact that every organism, past and present, besides the necessity of having been or being adapted to its environment, has been or is also the carrier of a phylogenetic load.

If we now turn more specifically to the higher taxa of the angiosperms, it must be emphasized that delimiting and describing a taxon of higher rank is equally an hypothesis as is the description of, say, a species. In the latter case, the hypothesis is that there will be found in nature thousands of individuals similar to those forming the limited sample the description is based upon (HEYWOOD 1973). And while the testing of this hypothesis in the case of the species is done by countless checking and rechecking of other individuals in the field, herbarium and labora-tory by later observers, in the case of higher categories the test will be to find out how far the classification is compatible with evidence which had not originally been used in the construction of the group.

In this context it would be interesting to look into the different major subdivisions of the flowering plants which have been proposed more recently. While TAKHTAJAN (1959) divided the dicotyledons into 13 superorders, the monocotyledons into 5, in 1964 he proposed a scheme with 6 subclasses in the dicotyledons and 5 in the mono-cotyledons. This major classification has also been adopted, with slight modifications, by CRONQUIST (1968), while THORNE (1968) continued with 20 superorders in the dicotyledons and 5 in the monocotyledons, and DAHLGREN (1975) with not less than 27 and 7 superorders, respec-tively. To me there is no doubt that the lower number of taxa used for the primary subdivision of dicotyledons and monocotyledons, while not being beyond the grasp of the consumers, has, together with the high didactic standard of TAKHTAJAN's (1959, 1969, 1973) and CRON-QUIST's (1968) books, been one of the main reasons for the broad accep-tance of these schemes.

Yet it is generally recognized that taxa of lower rank can be defined much more precisely than those of higher rank; the level of superorder is probably the last one in which a somewhat meaningful diagnosis can be given; the subclasses, instead, can only be characterized by circumscribing them and by indicating some general tendencies. While some of these taxa, like the *Magnoliidae*, and to a lesser degree also the *Asteridae*, might represent fairly correctly circumscribed taxonomic and evolutionary units, this may be doubted for others. In another occasion (KUBITZKI 1969) I have pointed to the somewhat vacillating value of the *Rosidae* and the *Dilleniidae*; and the *Hamamelididae*, though constantly becoming smaller · and smaller (cf. THORNE 1973), still give the impression of a vanishing grade group.

Things may become clearer by asking what is known or, what can be inferred, about the mode of origin of the higher taxa of the flowering plants. The idea of special mechanisms of macroevolution being involved, as opposed to those of microevolution, has long been abandoned. More recently it has also become clear that every morphological feature that distinguishes families and orders can in some groups vary at the level of genera and species (STEBBINS 1974). In spite of this obstacle, STEBBINS (1951), in a ingenious approach, has been able to demonstrate that the combination of characteristics, found in the larger, more successful families of the flowering plants must have evolved through the guidance of natural selection because these families therewith acquired particularly successful methods of solving the problems of fertilization and seed dispersal. Less is known and can be expected to having been operative as a guiding principle in the evolution at still higher levels; so that the ideas presented about the respective importance of secondary plant substances (see CRONQUIST and GOTTLIEB in this volume) deserve especial attention.

A basic tenet of most of these considerations is the tacitly made assumption of a reasonable monophyly of the higher taxa which is inferred from the phenetic evidence. The future, however, will hold many surprises to us, and with regard to several subclasses, we are already becoming aware that monophylesis is questionable, at least. One example is the *Asteridae*, the homogeneity of which has been challenged on the basis of leaf-morphological studies (HICKEY & WOLFE 1976); another, where strong doubts as to its evolutionary coherence seem to be justified is the *Caryophyllidae*. While the core of this group, as characterized by its peculiar pigments, sieve-tube leucoplasts, its deviating basic chromosome number $x = 9$ (RAVEN 1976) and other features must have a common, and possibly ancient, origin, it may well be quite unrelated to *Polygonales* and *Plumbaginales*; and in DAHLGREN's (1975) scheme the latter orders appear quite distant from *Caryophyllales*. To give allowance for some pleiophyly is probably inevitable, rather than making hasty adjustments of our classifications to the stepwise progress of our knowledge, provided that resulting groupings can clearly be defined phenetically.

A question that deserves special mentioning is that for the monophyly of the angiosperms as a whole. The reasons for supporting this hypothesis (summarized, e.g., by TAKHTAJAN 1959) are so well-known that they need not be repeated here. The classical argument of the uniformity of the embryo sac and of double fertilization has begun to loose its persuasive power (Meeuse 1964); and the constancy of the position relative to each other of micro- and megasporangia in the flower may well be related to its function in animal pollination, instead of signalizing a common origin. On the other hand, it is clear that the older pleiophyletic hypotheses of Greguss (see, e.g., 1964) and Emberger (1960) do not convince us any longer, because the former was based on rather superficial traits, while the latter inferred pleiophyly from nothing but the basic heterogeneity of present-day angiosperms. In more recent times, several workers (see, e.g., Meeuse 1971, Philipson 1975) have collated new evidence which supports the acceptance of rather different evolutionary lines within the angiosperms. While Philipson is cautious about inferring pleiophyly from this, Meeuse is convinced that this heterogeneity clearly favours the acceptance of a pleiophyletic origin of the flowering plants. Nobody, however, is at present able to indicate where these different lines start, nor where they, if at all, converge, and it is merely in the absence of a better alternative that the origin of the flowering plants is postulated from one group of the seed ferns. In the absence of any cladistic information, a decision between mono- and pleiophyly seems to me at present virtually impossible.

The question of the mutual interrelationships between the major groups of contemporary angiosperms is equally one of the more traumatic matters in our field, and I shall only briefly comment on this here. The question which subclass may be the key group of, and ancestral to, the other subclasses of the flowering plants is becoming less important as we are going to become aware of the complete disappearance of the common ancestor(s) of that group. On the one hand, living groups of higher rank can neither be ancestral to other living ones, nor can the nature of this ancestor be inferred with certainty from comparisons among Recent groups, as mentioned above. On the other hand, the different subclasses of the angiosperms have attained different anagenetic grades; the *Magnoliidae* representing the lowest, the *Asteridae* the highest level of specialisation among the dicotyledons (with many exceptions in both groups); and in monocotyledons, the situation is similar, whereas Walker's (1975, Fig. 4) recent scheme is rather misleading in this respect since all monocotyledons are confined in it to the lowest grade. The question is then, are subsequent levels of specialisation by themselves indicative of a phylogenetic relationship between major groups? If we put the question in this way the answer can only be in the negative. I am aware that the *Magnoliidae* is the group with the greatest concentration of known primitive characters, to which constantly more cryptic new ones are added, as the archaic seed structure in *Myristica* and *Annonaceae* (CORNER 1976); the presumably primarily columella-less exine in forms like *Degeneria*, *Eupomatia*, certain *Nymphaeaceae* and *Annonaceae* (WALKER & SKVARLA 1975); and the occurrence of the diarylpropanoid viranolol in *Virola*, one of the simplest flavanoids so far discovered (BRAZ FILHO et al. 1973). Primitiveness by itself, however, cannot be the proof of the *Magnoliidae* being ancestral to other subclasses, inasmuch as they occupy, as I have pointed out earlier (KUBITZKI 1969, 1973), a phenetically isolated position[2]. The group for which a relationship with the *Magnoliidae* can be vizualized is in my mind, the monocotyledons (cf. also HUBER, this volume), though two of the most distinctive features of the Ranalean complex are even absent from them: namely multistaminate flowers with a spiral phyllotaxis ("primary polyandry"), and the occurrence of benzyltetrahydroisochinoline alkaloids whose distribution and systematic importance recently has been dealt with by REZENDE et al. (1975). If we are going to postulate links between major extant groups we should be anxious to ensure that they fulfill this role not only with regard to isolated traits (as, e.g., *Trochodendron* with regard to supposed preadaptation to anemogamy), but with regard to an overall similarity. Otherwise, in our arrangements typological elements will largely continue to prevail.

To overcome these difficulties there is one way for elucidating relationships between extant groups which, though it has been widely used is far from being exhausted. This is, neither to use average parameters for major groups which in themselves often include a marked gradation from primitive to advanced states, nor to consider evolutionary tendencies (often leading to convergent, or at best homoiologous developments), but, as MARKGRAF (1955) has pointed out, to search in each group for primitive characteristics often only present in one or a few out of many genera which may be the clue for the understanding of the whole complex. It is perhaps needless to say that this is possible only in connection with monographic studies which have become so unfashionable. By this way, however, the systematic arrangement of major groups may still be improved largely. One other important aspect of systematics is not so much concerned with classification itself but with the processes and mechanisms of evolution which have led to the present-day end products of evolution. If one asks which aspect should have priority in future research the answer can only be that we

---

[2] Though I have been rather misunderstood by several authors, since I have neither "contested on chemical grounds ... the basal position of the Ranalean complex" (HEYWOOD 1973b: 369), nor have I stated that "modern *Magnoliidae* are advanced rather than primitive in chemical features" (CRONQUIST 1976: 17)!

have an obligation to attempt both approaches: it appears that not much time is left for the study of diversity because the early destruction of several of the earth's most fascinating biomes seems inevitable.

References

BENDZ, G., and SANTESSON, J. (Eds.), 1973: Chemistry in Botanical Classification. Nobel Symposium 25. New York-London: Academic Press.

BRAZ FILHO, R., FROTA LEITE, M. F., and GOTTLIEB, O. R., 1973: Constitutions of diarylpropanoids from *Virola multinervia*. Phytochemistry 12, 417—419.

CARLQUIST, S., 1969: Toward acceptable evolutionary interpretation of floral anatomy. Phytomorphology 19, 332—362.

— 1975: Ecological Strategies of Xylem Evolution. Berkeley: University of California Press.

CORNER, E. J. H., 1976: The Seeds of Dicotyledons. Vol. I, II. Cambridge: Cambridge University Press.

CRONQUIST, A., 1968: The Evolution and Classification of Flowering Plants. London: Nelson.

— 1976: The taxonomic significance of the structure of plant proteins: A classical taxonomist's view. Brittonia 28, 1—27.

CROWSON, R. A., 1970: Classification and Biology. London: Heinemann.

DAHLGREN, R., 1975: A system of classification of the angiosperms to be used to demonstrate the distribution of characters. Bot. Notiser 128, 119—197.

— 1975a: The distribution of characters within an angiosperm system. I. Some embryological characters. Bot. Notiser 128, 181—197.

DAVIS, P. H., and HEYWOOD, V. H., 1963: Principles of Angiosperm Taxonomy. Edinburgh and London: Oliver and Boyd.

EMBERGER, L., 1960: Les Végéteaux Vasculaires. In: Traité de Botanique (Systématique), Tome II, Fascicule I, II. (CHADEFAUD, M. et EMBERGER, L.), Paris: Masson et Cie.

GREGUSS, P., 1964: The phylogeny of sexuality and triphyletic evolution of the land plants. Acta Biol. Szeged. 10, 3—51.

HEYWOOD, V. H., 1973: Ecological data in practical taxonomy. In: Taxonomy and Ecology (HEYWOOD, V. H., Ed.), 329—347. London-New York: Academic Press.

— 1973a: Taxonomy in crisis? Or taxonomy is the digestive system in biology. Acta Bot. Acad. Sc. Hung. 19, 139—146.

— 1973b: The role of chemistry in plant systematics. In: Chemistry in Evolution and Systematics (SWAIN, T., Ed.), 355—375. London: Butterworths.

HICKEY, L. H., and WOLFE, J. A., 1976: The bases of angiosperm phylogeny: Vegetative morphology. Ann. Missouri Bot. Gard. 62, 538—589 (1975).

HUGHES, N. F., 1976: Palaeobiology of Angiosperm Origins. Cambridge, London, etc.: Cambridge University Press.

JENSEN, S. R., NIELSEN, B. J., and DAHLGREN, R., 1975: Iridoid compounds, their occurrence, and systematic importance in angiosperms. Bot. Notiser 128, 148—180.

JOHNSON, L. A. S., 1972: Evolution and classification in *Eucalyptus*. Proc. Linn. Soc. N.S. Wales 97, 11—29.

KUBITZKI, K., 1969: Chemosystematische Betrachtungen zur Groß-gliederung der Dicotylen. Taxon **18**, 360—368.
— 1973: Probleme der Großgliederung der Blütenpflanzen. Ber. dtsch. bot. Ges. **85**, 259—277 (1972).
— 1975: Relationships between distribution and evolution in some hetero-bathmic tropical groups. Bot. Jahrb. **96**, 212—230.
MARKGRAF, F., 1955: Über neuere Pflanzensysteme. Ber. dtsch. bot. Ges. **67**, 20—22 (1954).
MEEUSE, A. D. J., 1964: Some phylogenetic aspects of the process of double fertilization. Phytomorphology **13**, 237—244.
— 1970: The descent of the flowering plants in the light of new evidence from phytochemistry and from other sources. Acta Bot. Neerl. **19**, 61—72, 133—140.
PHILIPSON, W. R., 1961: Relationship and convergence in angiosperms. Phytomorphology **10**, 367—376.
— 1975: Evolutionary lines within the dicotyledons. New Zealand J. Bot. **13**, 73—91.
RAVEN, P. H., 1976: The bases of angiosperm phylogeny: Cytology. Ann. Missouri Bot. Gard. **62**, 724—764 (1975).
REZENDE, C. M. A. da M., and GOTTLIEB, O. R., 1973: Xanthones as sys-tematic markers. Biochem. Syst. **1**, 111—118.
—, — and MARX, M. C., 1975: Benzyltetrahydroisochinoline-derived al-kaloids as systematic markers. Biochem. Syst. Ecol. **3**, 63—70.
STEBBINS, G. L., 1951: Natural selection and the differentiation of angio-sperm families. Evolution **5**, 299—324.
— 1974: Flowering Plants. Evolution Above the Species Level. London: E. Arnold.
TAKHTAJAN, A., 1959: Die Evolution der Angiospermen. Jena: Fischer.
— 1964: The taxa of the higher plants above the rank of order. Taxon **13**, 160—164.
— 1969: Flowering Plants. Origin and Dispersal. Edinburgh: Oliver and Boyd.
— 1973: Evolution und Ausbreitung der Blütenpflanzen. Stuttgart: Fischer.
THORNE, R. F., 1968: Synopsis of a putatively phylogenetic classification of the flowering plants. Aliso **6**, 57—66.
— 1973: The "*Amentiferae*" or *Hamamelidae* as an artificial group: A summary statement. Brittonia **25**, 395—405.
WALKER, J. W., 1976: Evolutionary significance of the exine in the pollen of primitive angiosperms. In: The Evolutionary Significance of the Exine (FERGUSON, I. K., and MULLER, J., Eds.), 251—308. Linn. Soc. Symp. Ser. No. 1.
— and SKVARLA, J. J., 1975: Primitively columellaless pollen: a new concept in the evolutionary morphology of angiosperms. Science **187**, 445—447.
YOUNG, D. J., and WATSON, L., 1970: The classification of the dicotyledons: A study of the upper levels of the hierarchy. Aust. J. Bot. **18**, 387—433.

Address of the author: Prof. Dr. KLAUS KUBITZKI, Jungiusstraße 6–8, D-2000 Hamburg 36, Federal Republic of Germany.

Plant Syst. Evol., Suppl. 1, 33—51 (1977)

The Botany School, University of Cambridge,
Cambridge, England

Some Problems Associated With Character Correlations

By

K. R. Sporne, Cambridge, England

Abstract: Among dicotyledons, 122 positive correlations at the 50 : 1 level of significance (and 170 at the 20 : 1 level) occur between twenty-six characters, some of which are morphological, some anatomical and some biochemical. Most of these characters are more abundant among families known to have appeared early in the fossil record than they are among those which appeared later. These facts have been used to discover which families are primitive and which advanced, in a way which minimizes subjective judgements. However, such judgements have not been completely eliminated; they have to be applied at all stages.

Decisions are necessary as to: which taxonomic scheme to use; which level of taxon to take as the statistical unit; how to treat incomplete data and "mixed taxa"; whether correlated characters are functionally associated and concerned in relative efficiency, or whether they are indicators of relative advancement, having been involved in evolutionary trends; whether all characters should be given equal weight.

At least 125 years have elapsed since arguments first started as to the course of evolution among flowering plants; and they continue still. The discussion has mainly concerned the nature of the most primitive members of the group, one school of thought favouring the catkin-bearing plants, while others favour ranalian types. Yet others, believing that the angiosperms are polyphyletic, hold that several quite different kinds of flowering plants could be equally primitive.

Until the early years of the present century, there seemed to be little hope of settling such arguments, for the fossil record of flowering plants was thought to give little or no help, and the various protagonists relied almost entirely on subjective judgements and circular arguments. However, BAILEY and his colleagues (whose work has been summarized elsewhere, by SPORNE 1976a) showed that among dicotyledons certain characters are associated (i.e. they occur together more frequently than they would if they were merely randomly distributed within the group).

Furthermore, since some of these characters were more abundant in plants from Cretaceous and Tertiary deposits than among present-day dicotyledons, they concluded that the associated characters are primitive ones. Further development of these ideas, and their extension to include wood characters, led to the introduction by CHALK (1937) of simple statistical tests of significance. By means of 2×2 contingency tests, values of χ^2 were calculated, from which could be judged the probability that the observed frequencies of association might have occurred by chance. It was thus possible to discover which characters are statistically associated (i.e. "correlated").

This important step forward encouraged the hope that future discussions of angiosperm phylogeny would be based on objective reasoning and that the subjective element might be considerably reduced. Starting in 1939, I have been applying essentially the same techniques as those of CHALK to the widest possible range of dicotyledon characters; and the work is still in progress. Data have been analysed concerning some forty characters, of which twenty-six are significant indicators of evolutionary status.

These characters are numbered from 1 to 26 in Table 1, where the degree of correlation between them is indicated by various symbols. Positive correlations are shown by "plus" signs, and negative ones by "minus" signs. Bold symbols indicate correlations that are significant at the 50:1 level, and faint symbols those that are significant at a level between 20:1 and 50:1. "Crosses" indicate associations that are meaningless (e.g. to show a positive correlation between the woody habit and the possession of particular wood characters would be absurd since, by definition, wood characters can only be present in woody plants). A "dot" implies absence of correlation. Lines 27 and 28 indicate the extent to which the twenty-six characters occur more (or less) often than one would expect among those families of flowering plants that have been identified in the fossil record. Line 29 indicates the extent to which they occur more (or less) frequently in rain-forest floras than in the rest of the world.

The information contained in Table 1 includes, and also supplements, that which has been published elsewhere (SPORNE 1974, 1975, 1976a, CHENERY & SPORNE 1976). The purpose of this article is to examine the procedures that have been adopted during the assembling, analysis and interpretation of the data, the problems that have arisen and the extent to which subjective judgements have had to be made in their solution.

Bearing in mind the lack of agreement among botanists as to the way in which monocotyledons stand in relation to dicotyledons, it was decided to treat the two groups separately, beginning with the dicotyledons. The monocotyledons were subsequently studied by a colleague, LOWE

(1961). However, STEBBINS (1951) in a similar survey, but of a restricted range of characters, chose to deal with the angiosperms as a whole.

One then had to decide whether to work with species, genera or families as the statistical units. Theoretically, species should be the obvious choice, for their delimitation is traditionally supposed to involve much less subjective judgement than does that of larger taxa. Furthermore, there is much less variation in the expression of characters within a species than there is within a genus or a family. However, there are probably some 200,000 species of dicotyledons in the world. To take the species as the statistical unit would, therefore, be quite impracticable; and this was even more true in 1939 than it is nowadays, when computers are widely available. But, even today, the species is still impracticable because of inadequate data. Thus, embryology, the study of which is tedious and time-consuming, is known for only a minute proportion of the total number of species. The same is true for biochemical characters. It is most improbable, therefore, that an embryologist working in India, for example, would choose to investigate the very same species as a plant biochemist working in Denmark. This being so, it would be impossible to discover whether embryological and biochemical characters are correlated or not, so long as the species was taken as the statistical unit.

YOUNG & WATSON (1970), interested more in taxonomy than in phylogeny, carried out an analysis of the occurrence of some eighty-three attributes among genera of dicotyledons but, in spite of using computerized methods, they felt obliged to restrict their survey to 543 genera. Their selection was made, first of all, by choosing those families with the greatest number of genera and species. From these they selected the largest genera, the number of genera chosen to represent each family depending on its size. The resulting list of 600 genera was then reduced to 543 by removing those for which the data proved to be either inadequate or of doubtful accuracy. While such selection methods are, doubtless, justifiable in a taxonomic investigation, they would be inappropriate in a phylogenetic one, where small genera might be just as important as large ones for showing the course of evolution.

I decided to use the family as the statistical unit, but not without considerable misgivings, for this procedure demanded that one should treat as equal families as different in size as *Compositae* and *Adoxaceae*. The former contains some 19,000 species, while the latter consists of a single species. In deciding which taxonomic scheme to use, the main criterion was availability of data. Differing views as to the relationships between families were, at this stage of the procedure, irrelevant. Bearing in mind the vast amount of factual information contained in the two editions of ENGLER & PRANTL (1887—1915 and 1924—), ENGLER's system, as presented by DIELS (1936), was an obvious choice.

The 259 families recognized by Diels were, accordingly, listed in alphabetical order on recording sheets, (alphabetical order being chosen so as to avoid, as far as possible, any personal bias during the process of recording data). On one sheet would be recorded presence or absence of character "X" (for example) and, on another, presence or absence of character "Y". Then, by superimposing the two sheets and viewing them by transmitted light, those families with both characters X and Y would be immediately obvious and could be recorded on a third sheet. Suppose that the number of families with character X is "x" and that the number with character Y is "y". Then, on a "null hypothesis", the expected number (m') with both would be $\dfrac{x \cdot y}{259}$. If the actual occurrence (m) is greater than m', then the characters X and Y are said to be positively correlated; and, if less, then negatively correlated. The greater the difference between m and m', the more highly significant is the correlation. By calculating the value of χ^2, the value of "p" can be determined from tables, where p represents the probability that the difference between m and m' has occurred purely by chance. In most of my published work, the value of $p = 0.02$ was set as the criterion of significance. This is more rigorous than the value of $p = 0.05$, which is normally set in biological work, but it was felt that greater caution was necessary, because so many of the data are in an unsatisfactory state. In Table 1, the "bold" symbols represent those correlations which meet this criterion. However, many of my colleagues have asked to see the effect of taking the less rigorous criterion of $p = 0.05$. The result has been to introduce all the "faint" symbols in Table 1.

The procedure outlined above seems quite straightforward and simple but, in fact, frequently proves otherwise. Perhaps the most disturbing complication arises from the existence of "mixed" families, i.e. those families in which some genera, or species, may exhibit a particular character, while others do not. The recording sheets for such a character will include three symbols: "$+$" against those families in which it is constantly present, "$-$" against those from which it is constantly absent, and "$\pm$" against those which are mixed. One procedure is to treat those marked "$\pm$" as if they had been marked "$+$", thereby producing a list of families which possess the character non-exclusively ("n.e."). Another procedure is to treat those marked "$\pm$" as if they had been marked "$-$", since some of their members lack the character; those marked "$+$" would then constitute a list of families possessing the character exclusively ("e."). For any pair of characters, X and Y, therefore, it might be necessary to carry out four estimates of χ^2, as follows: X (e.) with Y (e.); X (n.e.) with Y (n.e.), X (e.) with Y (n.e.), X (n.e.) with Y (e.).

With few exceptions, when this is done, the first gives the highest value of χ^2, and the last two give the lowest. Doubtless, this is a reflection of the fact that mixed families tend to have something in common. Thus, they could be mixed in respect of both characters X and Y (and also in respect of many other characters) because they are large families that ought to be split. A good example of such a family is provided by the *Leguminosae*, as defined by DIELS, which proves to be mixed in respect of a large number of characters. Thus, even the number of carpels must be recorded as variable, because of the few genera, like *Archidendron*, with up to fifteen carpels. Most taxonomists now split the *Leguminosae* into three families of which only one, *Mimosaceae*, has to be recorded as variable in respect of carpel number.

Even so, it would still be true to say of the flowers of *Mimosaceae* that it is very rare for them to have more than one carpel. The temptation is strong, when reading such a description to ignore the rare exceptions and to record a " $+$ " for the character state "carpels one". However, this temptation must be firmly resisted, for there are continuous series of intermediates between "very rarely present" and "almost always present". These are variously described as "occasionally present", "sometimes present", "often present", "usually present", etc. The only way of avoiding the accusation of personal bias is to record a " $\pm$ " for all such statements. A single genus (or even species) is no less important phylogenetically when it belongs to a large family than when it belongs to a small one.

The meaning to be attached to the statement that character "X" is present in a family, or that character "Y" is absent will depend, not only on the kind of character, but also on the kind of publication in which the statement appears. The most reliable are likely to be statements about taxonomic characters, such as those used in ENGLER & PRANTL, for they have been checked (presumably) in every species within the family, and can usually be verified by reference to herbarium sheets. Characters whose recognition requires long and tedious techniques may, however, raise problems. It is common to find statements such as "The following families exhibit character X", the families then being listed without any clear indication as to the scope of the investigation. In this form, the information is almost useless for statistical purposes, and the only safe procedure is to record a " $\pm$ " for each of the listed families, recognizing that all one really knows is that the character is present in at least some species belonging to those families. Families that are not included in the list cannot, however, be assumed to lack character "X", unless there is a clear statement that this is so. Omission of a family might mean no more than that it has not been examined.

Ideally, one hopes to find the data presented in such a way as to show: (a) those families in which all the species examined exhibit character "X", (b) those in which some species exhibit it but others do not, (c) those in which none of the species examined exhibit the character, and (d) those families which have not been examined. If there is a clear statement as to which families have not been examined, then these families must be excluded during any search for possible correlations between character "X" and other characters. Thus, where information is known to be lacking on two characters, the total "population" for statistical analysis may be much smaller than the 259 families recognized by DIELS. However, it is better to work with small, and reliable, numbers than to pretend that the data are complete. There will also, of course, be reductions in the total population by the exclusion of families to which certain characters, by definition, cannot apply. Thus, when analysing wood characters, herbaceous families must be excluded and, when analysing fusion of petals, apetalous families must be excluded.

For one reason or another, therefore, the size of the statistical population may be very much reduced. Thus, the correlation between "apotracheal parenchyma" (character 4, in Table 1) and "leuco-anthocyanins" (character 9), the population is as low as 145, not only because herbaceous families have to be excluded, but also those which BATE-SMITH (1962) had been unable to examine biochemically and those whose wood METCALFE & CHALK (1950) had not described.

Sometimes, the decision as to whether to exclude a particular family or not is made more difficult by the problem of homology. This is clearly illustrated by the *Aizoaceae* (or *Ficoidaceae*). LAWRENCE (1951), without hesitation, states that the perianth is uniseriate, composed of a calyx of five to eight sepals, and that there are no petals (the apparent petals being petaloid stamens). In this he agrees with DIELS (1936). HUTCHINSON (1959), however, is equally confident that both petals and sepals are present. The decision is a vital one because, in the search for possible correlations with petal number, petal fusion, corolla symmetry, etc., the *Aizoaceae* would have to be excluded if they truly lack petals, but should be included if they truly possess petals. Since my work has been based on DIELS, I excluded this family from such calculations. However, if I were now to repeat all my investigations using HUTCHINSON's scheme, the *Aizoaceae* ought to be included.

Similar problems arise in connection with the aril which, strictly speaking, is an extra envelope outside the integuments, which grows up around the ovule from the point of attachment of the funicle. There are various so-called "arillodes", which may resemble arils and may perform similar functions, but whose homologies are obscure. They include "strophioles" (swellings on the raphe), and "caruncles" (swellings near the micropyle). Some morphologists have argued that such arillodes are merely analogous structures. Subjective judgements cannot be avoided in such circumstances.

Some critics of the work summarized in Table 1 have argued that subjective judgements must have played a part in the choice of charac-

Table 1. Correlations among twenty-six characters in dicotyledons; also, correlations between these characters and their occurrence in pre-Oligocene families, pre-Tertiary families and present-day rain-forest families. Bold symbols indicate correlations of significance greater than 50 : 1 and faint symbols those of significance between 50 : 1 and 20 : 1. A "dot" indicates that the observed figure is not significantly different from that calculated on a random basis. A "cross" indicates a meaningless correlation

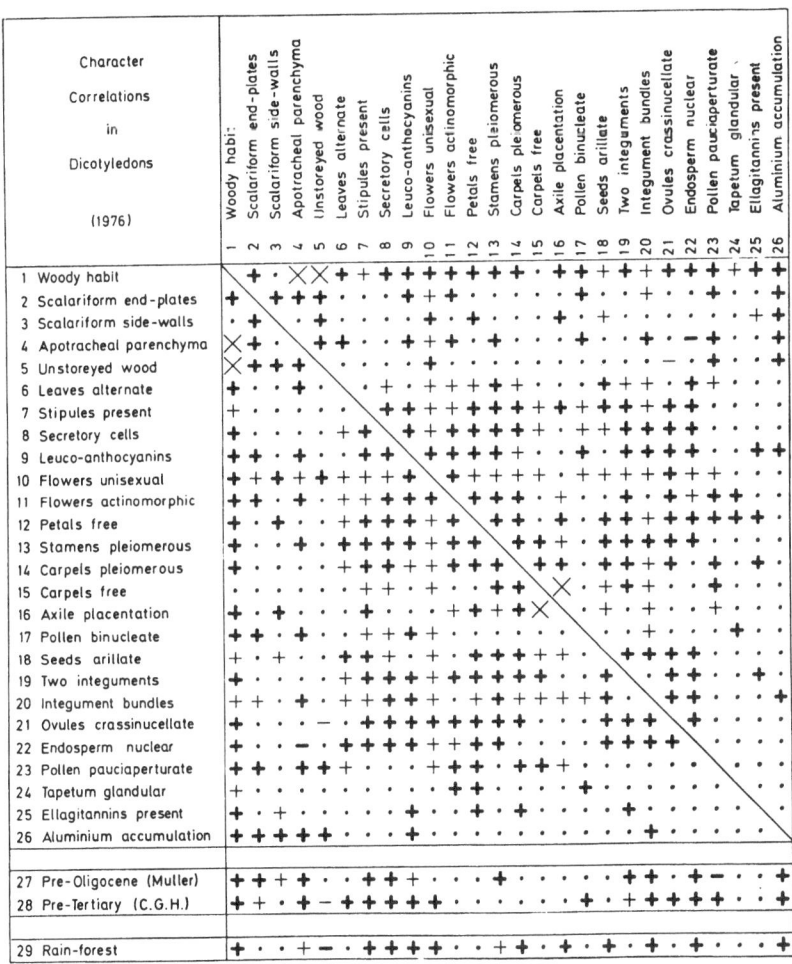

ters to be studied, but this is not so. In fact, far more characters have been considered than the twenty-six listed in Table 1, but any which gave non-significant results were merely not mentioned in any publications. Some characters, indeed, had to be abandoned at the outset, for lack of data. Among those with adequate data, but which proved to be non-significant, were several involving facts provided by Davis (1966), e.g. patterns of embryogenesis (Onagrad, Asterad, Solanad, etc.), cytokinesis of pollen mother cells (i.e. successive or simultaneous), pollen tetrad types (i.e. tetrahedral, isobilateral, decussate, etc.), number of nuclei in the tapetal cells, pollen grains united when dispersed, and endosperm ruminate. Data concerning girdling vascular bundles in the flower receptacle, derived from Saunders (1937, 1939) and supplemented by Sporne (1976b), gave no significant correlations; neither did data derived from Yakovlev & Zhukova (1973) concerning the colour of the embryo (i.e. whether green or not). Data derived from Shunji Imai et al. (1936) concerning the presence of phyto-ecdysones gave no significant results; neither did those derived from Behnke (1972), concerning the type of plastids in the sieve-tubes; nor did those derived from Yampolsky & Yampolsky (1922) concerning dioecism. The compound leaf, toothed leaf margin, inferior ovary, anemophilous pollination, copious endosperm and starchy endosperm, likewise, gave no significant correlations. Ovule shape was shown to be involved in only one significant correlation, anatropous ovules being positively correlated with unisexual flowers (Sporne 1969).

Other critics apparently fail to understand the meaning of the statement that two characters, A and B, (or their alternative states, A' and B') are correlated, and point to the existence of "exceptions" in which A' occurs with B (or A with B'), as if this destroys the correlation. It must be emphasized that the statement is a statistical generalization. It is not a statement that A and B always occur together. If they did, then statistical tests of significance would not be necessary.

It is perhaps surprising, however, that even when characters A and B are correlated (and also, of course, the associated characters A' and B'), the number of families exhibiting the combinations AB and A'B' can be in a minority within the "population". For example, suppose that, in a population of 100 families, 75 possess character A and 25 character A'; suppose, also, that 20 possess character B and 80 character B'. Furthermore, suppose that all those with character B also possess character A. Clearly, the majority (55 families) possess the combination AB', yet a 2×2 contingency test shows that A is positively correlated with B (and A' with B'), for the expected number of families with both A and B, on a random basis, is $\dfrac{20 \times 75}{100} = 15$. This is significantly less than the observed number, 20. The value of $\chi^2 = 8.34$ shows that the correlation is highly significant, for it corresponds to a value of "p" which is less than 0.01.

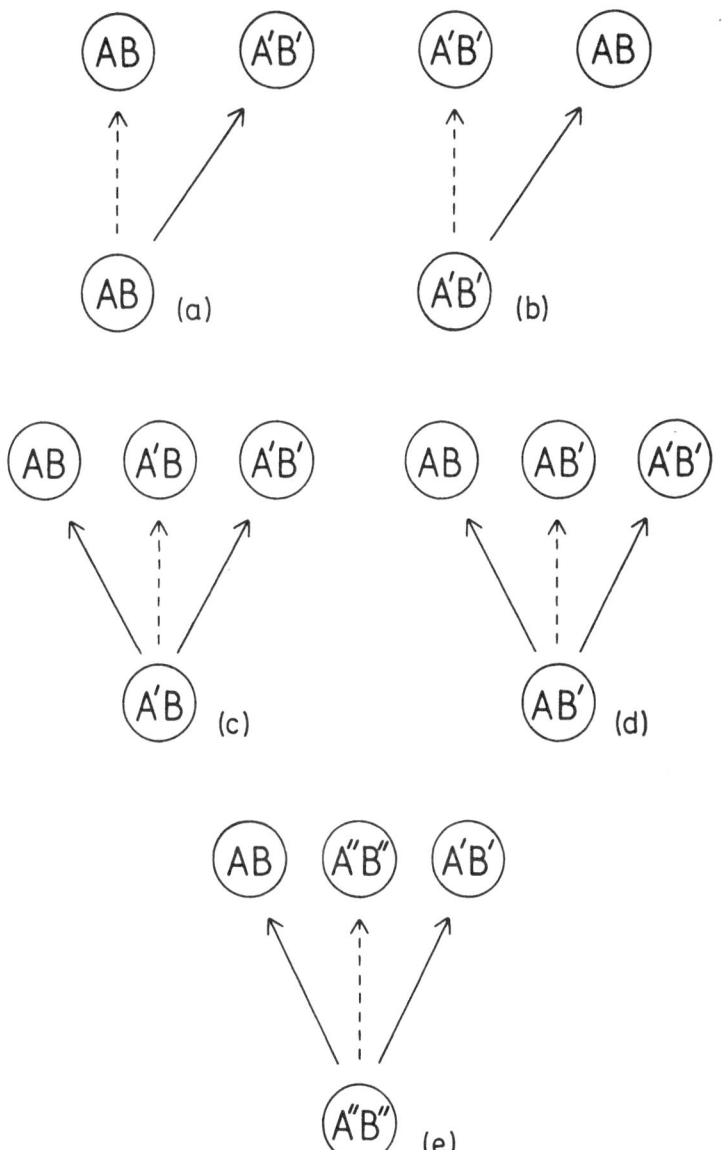

Fig. 1. Five model evolutionary processes, each of which, if repeated many times in different taxa, could have led to correlations today of A with B and A′ with B′. In each, the broken lines indicate the survival of ancestral conditions, while the continuous lines indicate evolutionary changes

It has often been claimed, in discussions on evolution, that the commonest combination of characters within a group represents the primitive type for that group. However, the example just quoted shows clearly that this is not necessarily true. Character A could be a primitive one which has evolved so slowly that only 25% of the families have achieved the advanced state, A', while character B could be a primitive one which evolved so rapidly that only 20% have retained it. Alternatively, the whole evolutionary scheme could be viewed upside down, with A' as a primitive character evolving rapidly into A, and B' a primitive character evolving slowly into B.

That one should expect primitive characters to show positive correlations was suggested many years ago (Sporne 1948). It follows from the definition of a primitive taxon as one which has retained a relatively large number of primitive (i.e. ancestral) characters. Primitive characters, instead of being randomly distributed among present-day families, therefore, tend to be concentrated among those which are primitive. This, in itself, is sufficient to cause positive correlations between primitive characters (e.g. A and B) and, so long as the corresponding advanced states (A' and B') are defined merely as the opposites of A and B, they too must show positive correlation. However, if divergent evolution has occurred, A diverging into C and D, while B has diverged into E and F, then it does not follow automatically that any of the advanced combinations, CE, CF, DE or DF, will be positively correlated.

Thus, although one should expect primitive characters to be positively correlated, advanced ones may or may not be so, according to the way in which they are described. But the very fact that advanced characters, as well as primitive ones, may be correlated creates a problem in that, without other evidence (e.g. from the fossil record), it is impossible objectively to decide whether A and B are the primitive states, or A' and B'. However, even in the absence of such other evidence, one can at least suggest that characters have been involved in evolutionary trends, and that if one pair is primitive the other is advanced.

Stebbins (1951) questions this conclusion and puts forward an alternative explanation for character correlations. He suggests that characters which are correlated are those which, together, contribute to outstanding biological success, and he describes organisms possessing such characters as having attained "adaptive peaks". Thus, A + B + C + D might represent one adaptive peak, while the opposite states, A' + B' + C' + D' might represent another.

In certain circumstances, this explanation is completely compatible with the idea that such characters represent the beginning and end of evolutionary trends but, in other circumstances it is not. Fig. 1 illustrates five model evolutionary processes, in all of which the broken lines represent the survival of ancestral character-combinations, so as to give the primitive condition today, while the unbroken lines represent evolutionary changes, leading to advanced states. Each, if repeated many times in different taxa, could have resulted in positive correlations, not only between A and B, but also between A' and B'. The first two models, 1a and 1b, represent evolutionary trends. Thus, in 1a, A and B are primitive characters which have survived together from the ancestral taxon, while A' and B' represent advanced characters. It is, of course, unlikely that A' and B' would have appeared simultaneously, as the model seems to imply but, for the sake of simplicity, the intermediate combinations A'B and AB' have been omitted. Model 1b represents trends in the opposite direction, from A' to A, and from B' to B, yet the statistical result is the same, viz. that A is positively correlated with B, and A' with B'. In these models, the two possible explanations of correlations are compatible with one another for, in order to have survived at all, any character combinations that occur today must be relatively efficient, whether they be primitive or advanced. Thus, the combination AB may be specially efficient in those regions of the present-day world whose climate resembles that in which fossil ancestors lived, while A'B' may be specially efficient in other regions; and the concept of adaptive peaks is, therefore, not incompatible with that of evolutionary trends.

However, models 1c, 1d, and 1e, represent evolutionary schemes in which the correlations of A with B and A' with B' cannot be explained on the basis of trends. Here, the concept of adaptive peaks provides the only acceptable explanation. However, such schemes are applicable only when the correlated characters are relatively few, and when they are likely to be functionally associated with each other. Thus, they might explain correlations between floral characters, which are intimately concerned in pollination mechanisms, or between seed characters, which are intimately concerned in dispersal and seedling establishment. They are unlikely to explain all the 122 correlations (at the 50:1 level of significance) between the twenty-six characters listed in Table 1.

Fig. 2 summarizes much of the information in Table 1, twenty-two of the characters having been re-arranged in seven groups, as follows: growth habit (one character); vessel characters (two); leaf characters (three); flower characters (five); pollen characters (three); seed characters (five); biochemical characters (three). Characters 4, 5, and 16 have been omitted because, by definition, they are subordinate ones. The number of intra-group correlations (at the 50:1 level of significance) is shown in brackets, as a proportion of the total possible number. Thus, among the five flower characters, there are seven correlations, out of a possible ten. Such intra-group correlations could, indeed, be ex-

plained on the grounds of functional association. The lines crossing
Fig. 2 indicate inter-group correlations. Thus, between the three leaf
characters and the five seed characters, there are twelve correlations,
out of a possible fifteen. It is unlikely that these can be explained

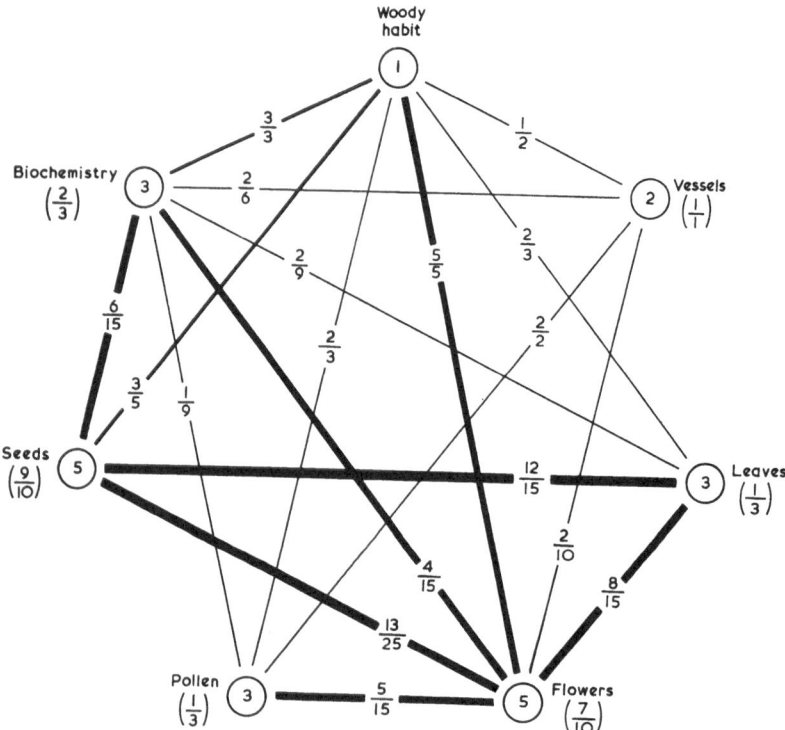

Fig. 2. Correlations among twenty-two characters, summarized from Table 1.
The characters are re-arranged into seven groups, the number in each
being shown in the circles. Lines indicate inter-group correlations, whose
number in each case is shown as a fraction of the total possible number.
Alongside each group, the number of intra-group correlations is shown
in brackets

on the grounds of functional association; and it is much more likely that
they result from evolutionary trends. The same is true of most of the
inter-group correlations, for they are between characters which occur in
widely different regions of the plant, at widely different times in its
development, and even in different phases of its life-cycle. However,
if these correlations are, indeed, the result of evolutionary trends,

some knowledge of the fossil record is vital in establishing the direction in which they have gone.

As explained elsewhere (SPORNE 1976a), the fossil record has been used in two quite different ways, one of which is regarded by HUGHES (1976) as improper, and the other less so. The first involves attempts to assign fossil remains to modern taxa. Thus, fossil wood is compared with known modern timbers, in the hope of finding an exact match. Similarly, attempts have been made to match fossil leaves with those of living species, but the results should be treated with extreme scepticism, not only because of the improbability of locating the right species out of a total of some 200,000, but also because of the variability of leaf shape from plant to plant, and even from branch to branch on one and the same plant. To this should be added the remarkable variability shown by some heteroblastic species, as they pass through successive developmental phases. Fossil pollen grains have been widely studied in recent years, in the confident expectation that they would give more reliable identifications than either fossil wood or leaves, but even so, they are not without problems. Thus, although the pollen grains of many families can be assigned to a particular genus, or even species, those of some others cannot easily be assigned to their family.

The results of such attempts to identify fossil remains (be they of wood, leaves or pollen grains) are often presented in the form of lists of families, or genera, that are claimed to have existed in past ages. However, as HUGHES points out, such claims involve the assumption that, because pollen grains like those of (say) *Ilex* existed in Senonian times, the plant producing them had the totality of features which characterize present-day *Ilex*. Such an assumption is unwarranted, for it implicitly denies that evolution can have occurred since the Senonian stage.

Nevertheless, if those families whose pollen has allegedly been identified in pre-Oligocene deposits (MULLER 1970) are compared with the whole present-day flora of the world, significant differences are found. These are summarized in Line 27 of Table 1. Here, existence in pre-Oligocene deposits has been treated, for statistical purposes, as if it were a morphological character with which others are correlated; and the symbols have the same meaning as in the main part of the Table. Thus, a "+" sign indicates a positive "correlation", showing that the relevant character occurs more frequently among those families that have been identified by MULLER in pre-Pleistocene deposits than it does in the present-day flora of the world. CHESTERS et al. (1967) give details of the first recorded appearance of various families of dicotyledons, identified on the basis of both pollen and other morphological characters. If those which were in existence by the end of the Cretaceous are compared with the present-day world flora, significant differences are shown to occur; and they are summarized in line 28 of Table 1.

It is important to realise that, in all these comparisons with "fossil" families, we are concerned, not with the characters that they might have shown in the past, but with those that they exhibit now. In view of the cautionary words of HUGHES, therefore, it is perhaps surprising that any significant results are obtained at all. Even more surprising is that the very characters which are most significant in such comparisons are also those which appear in the body of Table 1. In view of this, it must be fairly certain that the twenty-six characters are primitive, rather than advanced. Equally, it is fairly certain that those families which appeared first in the fossil record have retained, to this day, a large number of primitive characters and that their rate of evolutionary progress must have been slower than that of the more advanced families which appeared later.

The second way in which the fossil record can be used involves no more than the accurate description of the fossils themselves (i.e. no assumptions need to be made about the missing parts of the plant). Successive horizons are then compared, in order to discover any evolutionary trends that may have occurred. This has been done, with great success, by MULLER (1970), who has demonstrated a trend from few to many in the number of apertures in the grain. Here, at last, are characters which the student of phylogeny can label with confidence as primitive and advanced, respectively. "Pollen pauci-aperturate" represents the primitive state (as opposed to "pollen multi-aperturate", the advanced state), and takes its place, as character 23 in Table 1. The fact that it is correlated with more than ten of the other twenty-five characters leaves little doubt that they, too, are primitive. The negative correlation in line 27 may seem surprising, but there is a simple reason, as explained elsewhere (SPORNE 1975). It arises from the fact that it is difficult to assign sulcate pollen grains to a particular family. As a result, families with such pollen tend not to be listed as having occurred in pre-Oligocene times; yet modern families with sulcate have pollen been treated statistically as having pauci-aperturate pollen.

Recent work by HICKEY (1973) and by DOYLE & HICKEY (1976) holds out great hope for the future use of leaf characters in a similar way. HICKEY has devised a system of terminology for the various types of leaf architecture in living dicotyledons; DOYLE & HICKEY have now applied this system to fossil leaves from successive mid-Cretaceous deposits in the U.S.A. They have demonstrated clear trends, not only in overall shape and size, but also in venation. As a result, yet more primitive characters can be added to the list; and it will be interesting to discover how far they are correlated with other primitive characters, among present-day dicotyledons. However, before this will become possible, much work still remains to be done in assembling the necessary data.

When making comparisons that involve fossil floras, one is very much aware that some plants were more likely than others to become fossilized. Thus, many of the best localities for fossil angiosperms are those which were estuarine, where macroscopic remains of leaves, twigs or fruits were likely to have come from a circumscribed region, or even from particular types of vegetation. For example, mangrove or rain-forest plants might be over-represented in estuarine localities. Such deposits, therefore, represent, not merely a minute sample of the world's vegetation, but also one which is biassed.

Fossil pollen introduces less of this kind of bias, because pollen grains can be transported much further than macroscopic remains, for they can be carried in air currents, as well as in rivers and on the surface of the sea. They can, therefore, have come from a much wider range of habitats. However, the pollen record suffers from other kinds of bias. Thus, pollen grains of some species, for a variety of reasons are unlikely ever to have reached the site of deposition; others may be so delicate that they never became fossilized or, if they did, then they fail to withstand the chemical treatment to which they are subjected by palynologists. The effects of this kind of bias can be minimized if a particular fossil flora is compared, not with the present-day flora, but with other fossil floras. Such a comparison has been made (SPORNE 1973) between the pre-Oligocene pollen record of MULLER (1970) and his complete list of pollen identifications from all deposits up to, but excluding, the present day; and a similar comparison was made between the pre-Tertiary families recorded by CHESTERS et al. (1967) and their complete list of families identified from all deposits up to, but excluding, the present day. Although the number of significant differences is lower than for comparisons with the present-day world flora, nevertheless, the results strongly support the conclusion that the characters listed in Table 1 are primitive.

The knowledge that certain characters are primitive and others advanced can be used to make an assessment of the relative advancement of the various families of dicotyledons for, by definition, primitive families are those with a relatively large number of primitive characters, and advanced families are those with relatively few primitive characters. The first attempts (SPORNE 1949) to calculate an "advancement index" were based on the twelve characters which, at that time, had been shown to be correlated. These included "the unisexual flower". The latest calculations (SPORNE 1969) are based on twenty-one characters, which do not include "the unisexual flower". This character, more than any other, is likely to have been involved in evolutionary reversals, the recognition of which would have been a matter entirely of subjective judgement. For this reason, if for no other, it was decided to omit this controversial character from the calculations. It should, however, be noted that its inclusion would have made scarcely any difference to the final assessments.

The methods used for evaluating the advancement index of a family are open to criticism for a number of reasons, of which the most serious is that, yet again, subjective judgements are unavoidable. Basically, the procedure is to assign "marks" to each family, according to the number of advanced characters that it shows. However, the question immediately arises as to whether all the twenty-one characters are equally important as indicators of advancement. Because it is impossible to discover how important they might be to the plant, it may be argued that they should be treated as equivalent. Yet it is not possible to do this where the expression of one character is dependent on the presence of another. Thus, "apotracheal parenchyma" and "storeyed wood" are both subsidiary to "woody habit". Illogical though it may be, the decision was made to give one "mark" for the possession of "storeyed wood" and one to "herbaceous habit", as if they were equally deserving. Similar problems arise in relation to the corolla, where "petals fused" and "corolla zygomorphic" are both subsidiary to "petals present"; and a similar decision was taken.

Another problem is that caused by "mixed" families, where it may be argued that a family most of whose members possess a particular character should receive a higher award than one in which the character is only rarely present. Again, a decision has had to be made. All mixed families have been treated alike, and have been awarded a "half-mark" for each character in respect of which they are mixed. The effect of such a procedure, when applied to a family like the Leguminosae, which is mixed in respect of most characters, is to produce an advancement index with relatively little meaning. The figure obtained is merely an average for a wide range of component taxa which, in modern classifications, have been elevated to the status of families, of which one would now have a low advancement index, while the others would have relatively high ones.

Clearly, the time is ripe to change from the taxonomic scheme of Diels (1936) to a more recent one, especially now that computer facilities are readily available. Even a superficial glance at recent classifications, e.g. Melchior (1964), Takhtajan (1966, 1969) and Cronquist (1968), shows that many of the old families have been split into new ones whose characters are much less mixed.

In the meantime, the advancement index should be taken only as a very rough guide to relative advancement. Nevertheless, it has already proved its predictive value in relation to those characters which have been investigated since the latest re-calculation of advancement indices. Thus, the average advancement index for the fifteen families with exclusively multi-aperturate pollen grains was found to be 77.2 (Sporne 1972), whereas that for the 155 families with exclusively pauci-aperturate pollen grains was 54.4. These figures gave a strong indication that further investigation was justified; and the correlations listed in line 23 of

Table 1 were subsequently found to exist. Likewise, the fifty families with ellagitannins in their tissues were found to have a low average advancement index of 47.3, as compared with an average for the world flora of 56.6 (SPORNE 1975), and the thirty-seven families in which aluminium accumulators occur have an average of 49.0 (CHENERY & SPORNE 1976). The correlations that were subsequently found are listed in lines 25 and 26, respectively.

The most important conclusion to have come from calculating the advancement indices of families has been to show that several families, in quite different orders, are almost equally primitive. No single family that is alive today can, therefore, be regarded as the ancestor of all other dicotyledons. This should surprise no one who is aware of the long time that has elapsed since angiosperms first appeared, but it may help to silence the rather sterile arguments that still occur from time to time.

Among the most advanced families, there are seen to be at least two quite different kinds of flower. On the one hand, there is the zygomorphic, gamopetalous type, found in *Phrymaceae*, *Pedaliaceae*, *Martyniaceae*, etc.; on the other, there is the extremely reduced type found in *Callitrichaceae* and *Hippuridaceae*.

In the past, conclusions such as this would have involved a circular argument which, when reduced to its essentials, would have run as follows: "These families are primitive because they possess primitive characters, and primitive characters are those which are possessed by these primitive families." At last this circular argument has been broken, for we have discovered which characters are primitive without having had to ask the question: "Which families are primitive?" Indeed, the asking of this question was postponed until an answer, which was not entirely based on personal beliefs, could be expected with some degree of confidence. However, it has not been possible to eliminate subjective judgements completely. Indeed, subjective judgements have had to be made at every stage.

References

BATE-SMITH, E. C., 1962: The phenolic constituents of plants and their taxonomic significance. J. Linn. Soc. (Bot.) **58**, 95—173.

BEHNKE, H.-D., 1972: Sieve-tube plastids in relation to angiosperm systematics—an attempt towards a classification by ultrastructural analysis. Bot. Rev. **38**, 155—197.

CHALK, L., 1937: The phylogenetic value of certain anatomical features of dicotyledonous woods. Ann. Bot. (London) **1**, 409—427.

CHENERY, E. M., and SPORNE, K. R., 1976: A note on the evolutionary status of aluminium-accumulation among dicotyledons. New Phytol. **76**, 551—554.

4

CHESTERS, K. I. M., GNAUCK, F. R., and HUGHES, N. F., 1967: *Angio-spermae*. In: The fossil record (HARLAND, W. B., et al., Eds.), 269—288. London: Geological Society.

CRONQUIST, A., 1968: The evolution and classification of flowering plants. London: Nelson.

DAVIS, G. L., 1966: Systematic embryology of the angiosperms. New York: Wiley.

DIELS, L., 1936: A. ENGLER's Syllabus der Pflanzenfamilien, 11. Aufl. Berlin: Borntraeger.

DOYLE, J. A., and HICKEY, L. J., 1976: Pollen and leaves from the mid-Cretaceous Potomac Group and their bearing on early angiosperm evolution. In: Origin and early evolution of angiosperms (BECK, C. B., Ed.), 139—206. New York: Columbia University Press.

ENGLER, A., und PRANTL, K., 1887—1915: Die natürlichen Pflanzenfamilien. (23 Vols.). Leipzig: Engelmann.

—, — (1924—): Die natürlichen Pflanzenfamilien, 2. Aufl. Leipzig: Engelmann.

HICKEY, L. J., 1973: Classification of the architecture of dicotyledonous leaves. Amer. J. Bot. **60**, 17—33.

HUGHES, N. F., 1976: Palaeobiology of angiosperm origins. Cambridge: University Press.

HUTCHINSON, J., 1959: The families of flowering plants, Second edition. Vol. 1: Dicotyledons. Oxford: Clarendon Press.

LAWRENCE, G. H. M., 1951: Taxonomy of vascular plants. New York: MacMillan.

LOWE, J., 1961: The phylogeny of monocotyledons. New Phytol. **60**, 355—387.

MELCHIOR, H., 1964: A. ENGLER's Syllabus der Pflanzenfamilien, 12. Aufl. 2. Band: Angiospermen. Berlin: Borntraeger.

METCALFE, C. R., and CHALK, L., 1950: Anatomy of the dicotyledons. (2 Vols.) Oxford: Clarendon Press.

MULLER, J., 1970: Palynological evidence on early differentiation of angiosperms. Biol. Rev. **45**, 417—450.

SAUNDERS, E. R., 1937 and 1939: Floral morphology: a new outlook, with special reference to the interpretation of the gynaeceum. (2 Vols.) Cambridge: Heffer.

SHUNJI IMAI, TOMOYOSHI TOYOSATO, MICHIHIKO SAKAI, YASUO SATO, SHOJI FUJIOKA, EIKO MURATA, and MINORU GOTO, 1969: Screening results of plants for phytoecdysones. Chem. pharm. Bull., Tokyo **17**, 335—339.

SPORNE, K. R., 1948: Correlation and classification in dicotyledons. Proc. Linn. Soc., London **160**, 40—47.

— 1949: A new approach to the problem of the primitive flower. New Phytol. **48**, 259—276.

— 1969: The ovule as an indicator of evolutionary status in angiosperms. New Phytol. **68**, 555—566.

— 1972: Some observations on the evolution of pollen types in dicotyledons. New Phytol. **71**, 181—185.

— 1973: The survival of archaic dicotyledons in tropical rain-forests. New Phytol. **72**, 1175—1184.

— 1974: The morphology of angiosperms. London: Hutchinson.

SPORNE, K. R., 1975: A note on ellagitannins as indicators of evolutionary status in dicotyledons. New Phytol. **75**, 613—618.

— 1976a: Character correlations among angiosperms and the importance of fossil evidence in assessing their significance. In: Origin and early evolution of angiosperms (BECK, C. B., Ed.), 312—329. New York: Columbia University Press.

— 1976b: Girdling vascular bundles in dicotyledon flowers. Gdns' Bull., Singapore.

STEBBINS, G. L., 1951: Natural selection and the differentiation of angiosperm families. Evolution **5**, 299—324.

TAKHTAJAN, A. L., 1966: Systema et phylogenia magnoliophytorum. Moscow: U.S.S.R. Acad. Sci.

— 1969: Flowering plants—origin and dispersal (Transl. JEFFREY, C.). Edinburgh: Oliver and Boyd.

YAKOVLEV, M., and ZHUKOVA, G., 1973: Angiosperms with green and colourless embryo. (In Russian) Leningrad: V. L. Komarov Botanical Institute, Academy of Sciences of the U.S.S.R.

YAMPOLSKY, C., and YAMPOLSKY, H., 1922: Distribution of sex forms in the phanerogamic flora. In: Bibliotheca Genetica (BAUR, E., Ed.), Vol. 3. Leipzig: Borntraeger.

YOUNG, D. J., and WATSON, L., 1970: The classification of dicotyledons: a study of the upper levels of the hierarchy. Aust. J. Bot. **18**, 387—433.

Address of the author: Dr. K. R. SPORNE, Botany School, Downing Street, Cambridge CB 2 3 EA, England.

Plant Syst. Evol., Suppl. 1, 53—76 (1977)

Instituto de Química, Universidade de São Paulo, Brazil

Biochemical Systematics: Methods and Principles

By

M. Aparecida H. Cagnin, Ceres M. R. Gomes, Otto R. Gottlieb,
M. Claudia Marx, A. Imbiriba da Rocha, M. Fátima das G. F. da Silva,
and J. Aparício Temperini *, São Paulo

Abstract: Chemosystematics will not evolve to become a scientific discipline in absence of accepted chemotaxonomic procedures. Our approach to this problem includes, initially, the mapping of metabolites by consideration of their biosynthetic relationship, indicated by skeletons, substitution patterns and frequency of occurrence in nature. The maps form the basis for three-dimensional graphic presentations of the chemical constitution (relative to a given metabolic class) of plant groups, which aid in the deduction of basic chemosystematic principles, as well as for two-dimensional graphic presentations of taxonomic distances among subunits of these groups, which further the understanding of their evolutionary relations. Progress of this type of work with respect to several phenylalanine (coumarins, benzylisoquinolines, *Amaryllidaceae* alkaloids), phenylalanineacetate (isoflavonoids) and tryptophane-mevalonate (indole alkaloids) derived biogenetic classes of compounds are described, and its potentiality illustrated by reference to several taxa of flowering plants.

Introduction

The existence of a relation between chemical composition and systematic affinity of plants was sensed ages ago (ROCHLEDER 1854). Why are we, inspite of considerable effort (HEYWOOD 1973), still only sensing this relationship, why are we still able to express it by nothing better than this dangerous presence/absence criterion?

The answer: lack of taxonomic procedures applicable to chemical data. In absence of scientific methodology, deduction of basic principles is impossible and biochemical systematics will never attain status of a scientific discipline.

---

* Read by OTTO R. GOTTLIEB to whom also enquiries should be addressed.

Work sponsored by Fundação de Amparo à Pesquisa do Estado de São Paulo.

Chemotaxonomic Methods

Biogenetic Group Maps

Secondary metabolites can be classified into biogenetic groups, **BGs**. At the head of each BG stand precursor units (**P**) which generate all other representatives by a sequence of biosynthetic pathways. Such sequences can be mapped for skeletons, as shown for the isoflavonoid BG (Fig. 1). Each skeleton is represented in nature by a number of derivatives, through substitution by hydroxy, alkoxy and other groups at specific C-atoms. In the BG-maps diversity of derivatives diminishes, generally from left to right, with increasing number of reaction steps separating the particular skeleton from the precursor; and from top to bottom, the reaction sequences being written in such a way as to lead progressively to smaller collections of derivatives. The lower the frequency of substitutional derivatives of a skeleton (or of a sequence of skeletons), the less widespread, the more specialized, must be the reaction step (or steps) leading to that skeleton (or sequence of skeletons).

The complete substitution pattern of a particular derivative probably already exists in a particular P of the BG, and as many BG-maps can be written as there are substitution patterns. Again, the lower the frequency of skeletal variants endowed with a certain substitution pattern, the less widespread, the more specialized, must be the reaction step leading to that substitution pattern.

Diagrams of Chemical Constitution

The drawing of BG-maps is based on the critical revision of published work and thus to a great extent independent of personal talent or opinion. On such maps each compound is represented by a certain position relative to each other. This information can be transferred to tridimensional diagrams. The simplification involved in the transfer entails a certain amount of judgement as to significance by the operator who will be required to ask himself: what are the structurally most relevant attributes of a natural compound with respect to its value as systematic marker of the taxon under scrutiny, skeletons, oxygenation patterns, or additional parameters such as secondary modifications by oxidation, reduction, O- or C-alkylation? A decision having been reached concerning this matter, the attributes are marked on coordinate axes, in order of decreasing frequency of occurrence.

One diagram is reserved for each subunit of a taxon. All compounds of the chosen marker BG of the subunit are represented on this diagram by points whose coordinates are the appropriate values, x, y and z, of the chosen attributes. In the present case, *Leguminosae* genera,

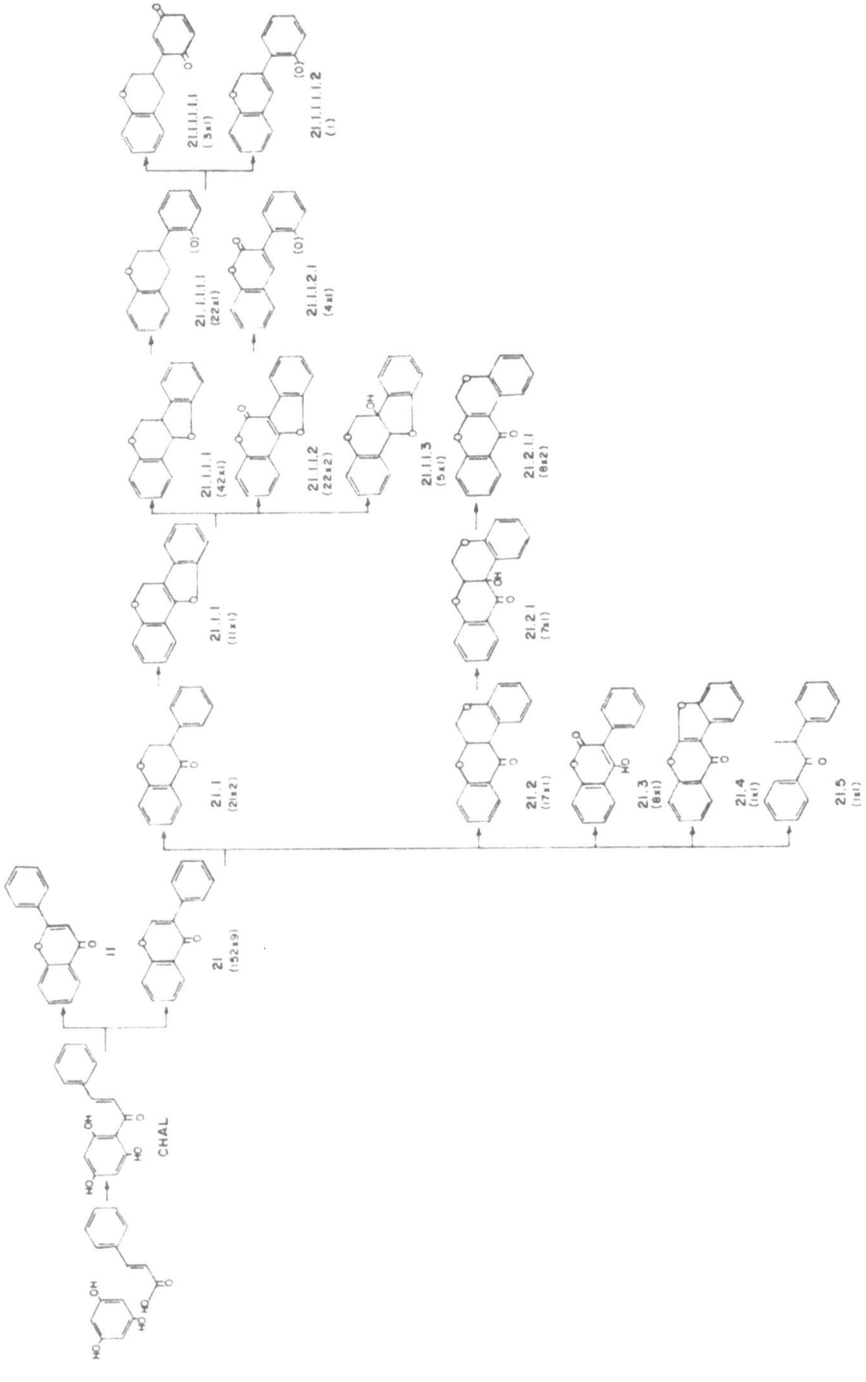

Fig. 1. Biosynthetic relationship and code of the isoflavonoid biogenetic group of skeletons. The biosynthesis involves, as PPs, one cinnamic acid and three acetic (malonic) acid units which form a chalcone (CHAL), the precursor of two BGs, the flavonoids which derive from flavone (code 11), and the isoflavonoids (code 21). In ( ): numbers of derivatives × numbers of families in which they occur (REZENDE et al. 1975)

the attributes are oxygenation of the shikimate-derived B-rings without (initial part of the x axis) or with (final part of the x axis) a 2'-oxy group, 5,7- or 7-oxygenation (y axis) and eventual additional oxygenation of their acetate derived A rings (z axis) (Fig. 2). The skeletons of the derivatives are marked by a code on the points of intersection. The diagrams reveal similarities and dissimilarities of chemical constitution of the genera, and the examples featured in Fig. 3 give an idea

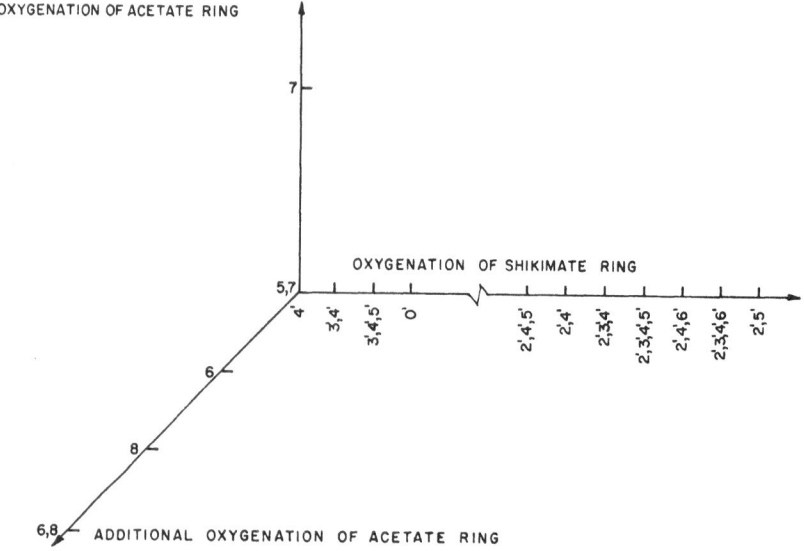

Fig. 2. Key to diagrams of isoflavonoid constitution of *Leguminosae* genera

concerning development along an isoflavone line (1st line), a pterocarpanoid line (2nd line) and a rotenoid line (3rd line). Such similarities and dissimilarities are visual guides in both aspects of the taxonomist's activities (HEYWOOD 1973): (a) circumscription and description of taxa; (b) formulation of phylogenetic relationships between taxa. Once these relationships have been traced through lines (p. 56–62), the diagrams also allow one to predict the chemical composition of extinct ancestral forms and of unanalyzed existing groups.

Affinity Plots

The diagrams of chemical constitution of taxonomic units refer to compounds and, as we shall see (p. 65–74), are useful in the deduction of basic principles of chemical evolution. What, however, is also needed, and the major goal of our effort, is the evolutionary mapping, not of

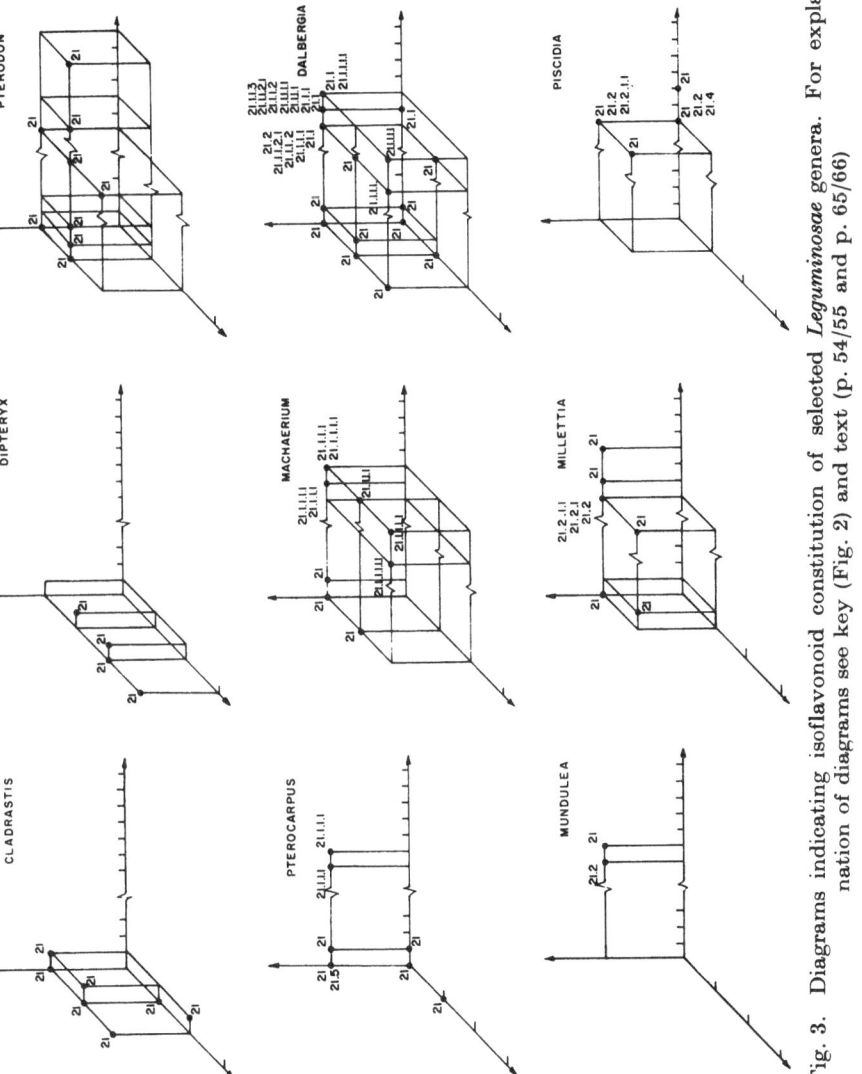

Fig. 3. Diagrams indicating isoflavonoid constitution of selected *Leguminosae* genera. For explanation of diagrams see key (Fig. 2) and text (p. 54/55 and p. 65/66)

compounds, but of their sources, the organisms or groups of organisms by which they are produced. The information contained in BG-maps can also be used in ways which make this objective possible, in the construction of phylogenetic trees.

The initial step in this direction involves quantification of the data of BG-maps by the estimation, for each compound, of a relative probability of occurrence (**RPO**). This probability diminishes with increasing number of required reaction steps and increasing com-

Table 1. Relative probability of occurrence (RPO) of isoflavonoid skeleta (cf. Table 2) given by distance from chalcone (in number of reaction steps)/number of substitutional derivatives

Code CHAL	Number of reaction steps					
	1	2	3	4	5	6
21	1/135 = 0.0074	2/20 = 0.1000	3/11 = 0.2727	4/42 = 0.0952	5/22 = 0.2273	6/3 = 2.0000
				4/19 = 0.2105	5/4 = 1.2500	6/1 = 6.0000
				4/5 = 0.8000		
21.2		2/17 = 0.1176	3/7 = 0.4286	4/5 = 0.8000		
21.3		2/8 = 0.2500				
21.4		2/1 = 2.0000				
21.5		2/1 = 2.0000				

plexity of reaction mechanism. The number of reaction steps can clearly be estimated with relative ease. The specialization of a reaction, however, can also be gauged considering its inverse proportionality with the frequency of occurrence of the product (or sequence of related products). The RPO for each compound can thus be estimated with respect to skeleton and substitution.

Skeletons: An example of the calculations is given in Table 1 for the isoflavonoid BG, the pertinent figures being superimposed on the position occupied by each isoflavonoid skeleton on the BG-map. P, in column 1, is alloted 1 point and skeletons in successive colums, 2, 3, 4, ... points. The point value of each skeleton is now divided by the frequency of occurrence, as given by the number of its derivatives.

Substitutions: The precursor substitution pattern, here 5,7,4'-oxygenation (Table 2), is alloted 1 point, and successive patterns, formally

Table 2. Relative probability of occurrence (RPO) of isoflavonoid substitution given by distance from 5,7,4'-oxygenation (each additional oxygenation or deoxygenation = 1 step)/number of skeletal derivatives

Position of oxygenation on rings A	B 2',4',5'	4'	2',4'	3',4'	2',3',4'	3',4',5'	2',3',4',5'	2',4',6'	2',3',4',6'	—	2',5'
7	4/55 = 0.0727	2/22 = 0.0909	3/38 = 0.0789	3/11 = 0.2727	4/10 = 0.4000	4/2 = 2.0000	5/3 = 1.6666	4/3 = 1.3333	5/3 = 1.6666	3/1 = 3.0000	
5,7	3/7 = 0.4286	1/28 = 0.0357	2/18 = 0.1111	2/19 = 0.1053	3/2 = 1.5000						6/1 = 6.0000
6,7	5/10 = 0.5000	3/9 = 0.3333	4/1 = 4.0000	4/8 = 0.5000	5/3 = 1.6666	5/2 = 2.5000					
7,8	5/10 = 0.5000	3/3 = 1.0000	4/3 = 1.3333	4/2 = 2.0000	5/3 = 1.6666		6/1 = 6.0000			4/1 = 4.0000	
5,6,7	4/4 = 1.0000	2/10 = 0.2000		3/1 = 3.0000							
5,7,8				3/2 = 1.5000							
5,6,7,8		4/1 = 4.0000		4/1 = 4.0000							
6,7,8	4/1 = 4.0000										

derived by introductions or eliminations of substituents, receive additional points according to the number of operations involved. The point value (REZENDE & GOTTLIEB 1973) for each substitution pattern is now divided by the frequency of occurrence, as given by the number of its representatives within the BG, irrespective of skeletons.

Table 3

Isoflavonoid constituent		RPO of isoflavonoid	
Structure	Name	skeleton	substitution

R = H	(3R)-laxifloran	0.2273	0.4000
R = H	(3R)-lonchocarpan	0.2273	1.6666

R = H	(6aR,11aR)-philenopteran	0.0952	1.6666
R = Me	(6aR,11aR)-9-OMe "	0.0952	1.6666

R = H	(6aS,12aS)-rotenone	0.1176	0.0727
R = H	(6aS,12aS)-deguelin	0.1176	0.0727
R = OH	(±)-tephrosin	0.4286	0.0727

Mean of skeletal RPO values
$$x = [(0.2273 \times 2) + (0.0952 \times 2) + (0.1176 \times 2) + (0.4286 \times 1)]/16.5593$$
$$= 0.0790$$
Mean of substitutional RPO values
$$y = [(0.4000 \times 1) + (1.6666 \times 3) + (0.0727 \times 3)]/58.4622$$
$$= 0.0961$$

For each compound a skeletal and a substitutional RPO-value can thus be determined. As an example, Table 3 lists RPO-values for the 7 isoflavonoid constituents of the genus *Lonchocarpus*. Which of the compounds represents the evolutionary state of the unit? None in isolation, of course, but all together as represented by the weighted means of their RPOs with respect to skeletons and substitutions.

These means, marked on the x (skeleton) and y (substitution) axes, give a point which represents the biological unit. Once all units of a taxon are plotted on the same graph, their taxonomic distances can

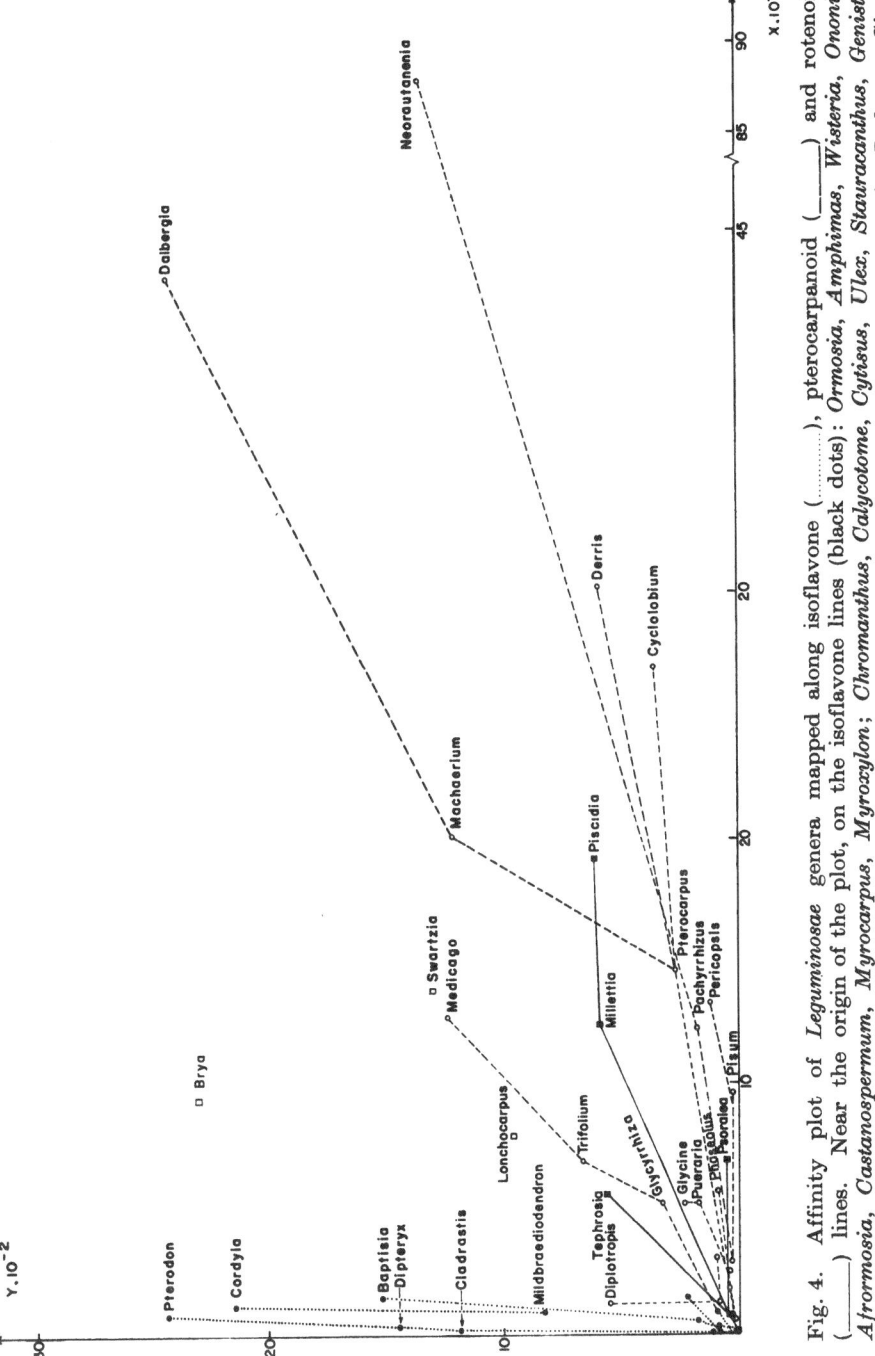

Fig. 4. Affinity plot of *Leguminosae* genera mapped along isoflavone (........), pterocarpanoid (———) and rotenoid (– – –) lines. Near the origin of the plot, on the isoflavone lines (black dots): *Ormosia, Amphimas, Wisteria, Ononis, Afrormosia, Castanospermum, Myrocarpus, Myrocxylon; Chromanthus, Calycotome, Cytisus, Ulex, Stauracanthus, Genista, Erinacea, Laburnum, Adenocarpus, Chamaecitysus, Teline, Chamaespartium, Lygos, Lupinus, Thermopsis; Lathyrus, Cicer;* on the pterocarpanoid line (circles): *Baphia, Maackia, Ougenia, Desmodia, Aldina, Sophora, Canavalia, Vigna; Lotus, Flemingia, Platymiscium; Andira;* on the rotenoid line (black squares): *Mundulea, Amorpha*

be read directly. It is, additionally, possible to connect the plotted
points by reasonable lines, thus placing the units on to the branches
of a genealogical tree, as exemplified for *Leguminosae* genera in Fig. 4.
How can the viability of an affinity line be tested? By consideration
of the chemical composition of each unit. For this purpose, first, the
centres of the coordinate axes of the tridimensional diagrams (p. 56, 57)
are superimposed on the pertinent plotted points. Next, each diagram X
is considered with respect to its neighbours. If a neighbour, Y, farther
away from the origin of the plot, is characterized by compounds which
are identical or derivable from compounds of X, the units possess
chemical affinity and their connection, $X \to Y$, is a reasonable postulate.
Only the BG precursors of X will frequently be missing from Y, possibly
because of a more efficient turnover in the more specialized unit.

Similarity Indexes for Biogenetic Groups

The assessment of the relative vigour with which a biological unit
develops certain pathways may also be based on BG-maps, comparing
number of derivatives (Du) and of skeletons (Su) produced by the unit

Table 4. Percent isoflavonoid similarity indexes for families of *Embryobionta*

Class	Subclass	Family	$100 \times SI/9261.54$
Polypodiopsida	*Leptosporangiatae*	*Polypodiaceae*	0.04
Pinopsida		*Podocarpaceae*	0.04
Magnoliopsida	*Magnoliidae*	*Myristicaceae*	0.04
	Hamamelididae	*Moraceae*	0.04
	Caryophyllidae	*Chenopodiaceae*	0.02
		Amaranthaceae	0.02
	Rosidae	*Rosaceae*	0.25
		Leguminosae	100.00
	Asteridae	*Compositae*	0.06
Liliopsida	*Liliidae*	*Iridaceae*	0.29
		Stemonaceae	0.06

with number of derivatives (Dt) and of skeletons (St) produced by a super-
ordinate unit. Percentual similarity indexes (**SI**) relative to a given BG
may be tabulated for each unit under scrutiny, as exemplified for the
isoflavonoid BG (Table 4).

$$SI_{BG} = \frac{D_u}{D_t} \times \frac{S_u}{S_t} \times 10^4$$

Chemosystematic Principles

Direction of Chemical Evolution

Any evolutionary series based on secondary chemical constituents can theoretically be read in either direction (HARBORNE et al. 1976). In other words, consider a sequence of metabolites, $A \to B \to C \to D$, whose biosynthetic relationship we are able to predict. Now, if a taxon contains C, one will not be able to say if it evolved from a taxon accumulating B (by expansion of the reaction sequence) or from a taxon accumulating D (by reduction of the reaction sequence). This, however, is essential knowledge and it will never be possible to use chemical characters as basis for phyletic classification until it becomes clear when and where to expect expansion or reduction of a reaction sequence, until some basic principles predict the direction in which an evolutionary series has to be read.

Prior to examining natural reaction sequences, it is possible to distinguish two types, according to their association with primary or secondary metabolism. Most, if not all of the primary metabolites (PPs) are, of course, synthesized by all plant species as precursors of biopolymers. In the present context we are, however, interested only in their function as heads of biogenetic groups of secondary metabolites, which they do or do not produce according to the individual plant, and it is thus permissible to discuss expansion or retraction of a reaction sequence involving such primary precursors. The second type of reaction sequence operates within each biogenetic group of secondary metabolites. Having, thus, separated reaction sequences into types, let us examine these in order.

First Principle

Results of SI-calculations are given in Table 5 for several biogenetic groups in correlation with plant taxa, named according to CRONQUIST (1961, 1968). In its present, far from complete version, the picture leads to a working hypothesis for a first principle of biochemical systematics: among taxa of *Embryobionta* of high hierarchical level, biochemical evolution of secondary metabolites involves the gradual substitution of biogenetic groups formed from primary precursors of the shikimate pathway by biogenetic groups formed from primary precursors of the acetate pathway.

Indeed, reduction in importance of the lignan BG, seems to have led to accumulation of cinnamic acid, inhibitor of its synthesis from phenylalanine (HASLAM 1974), which thus became available for the production of the benzylisoquinoline BG. The introduction of nitrogen into shikimate metabolites conferred probably an important competitive advantage which primitive angiosperms had over most gymno-

Table 5. Percentual similarity indexes of several biogenetic groups for taxa of *Embryobionta*

Pathways BGs	Shikimate				Shikimate + Mevalonate			Mevalonate
	Ligmans	Benzyliso-quinolines	Amar. Alk. Colchicins	Cactus Alkaloids	Coumarins	Anthran. Ac. Deriv.	Indole Alkaloids	Steroid Alkaloids
Pinophyta	53.35	0.00	0.00		0.00			
Magnoliophyta								
Magnoliopsida								
Magnoliidae	33.05	100.00	2.29		0.04			
Hamamelididae	3.00	0.01	0.00		0.61			
Caryophyllidae	0.00	0.00	0.00	100.00	0.02			
Dilleniidae	4.00	0.01	0.00		0.09			
Rosidae	72.78	5.42	0.00		100.00			
Asteridae	100.00	0.03	0.00		2.92	100.00	100.00	100.00
Liliopsida								
Alismatidae	0.00	0.00	0.00		0.00			
Arecidae	0.00	0.02	0.00		0.00			
Commelinidae	0.26	0.00	0.00		0.09			
Liliidae	0.00	0.01	100.00		0.04			100.00

sperms. Chemical evolution continued by further contraction of the shikimate pathway and use of the consequently available chorismic acid in the production of anthranilic acid and tryptophan. Coincidental with this contraction was the introduction of a progressively higher proportion of terpenoid chains into shikimate-derived compounds. The ultimate consequence of this evolutionary trend is the suppression of shikimates and the restoration of terpenoids as the dominant group of biologically active compounds.

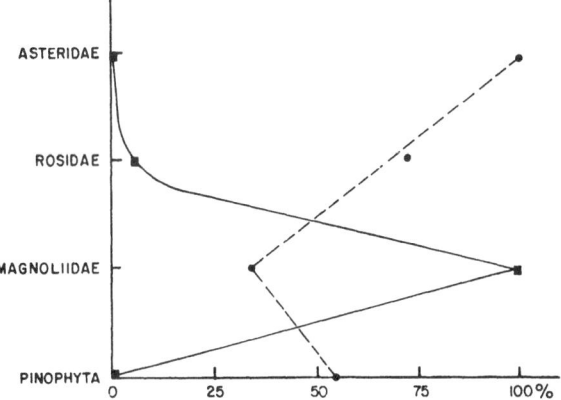

Fig. 5. Percent similarity indexes for lignans (●) and benzylisoquinolines (■) in *Pinophyta* and *Magnoliophyta*

Partial reversal of this major trend is a concomitant possibility. A plot of percentual SIs not only shows renewed vigour of diversification of the lignan BG in connection with the *Rosidae-Asteridae*, but also points to the competitive, substitutive development of the lignan vs. the benzylisoquinoline BGs (Fig. 5).

The first principle cannot be viewed with unreserved confidence. This is not due only to lack of completeness of our analysis, which we hope to remedy in the near future, but chiefly to the fact that an element of circularity was involved in its deduction. Chemical evolution was correlated with a morphological classification, itself based on inference of significance.

Secondary Metabolites as Systematic Markers

Isoflavonoid Lines in the *Leguminosae*. Isoflavonoids occur widely scattered throughout the *Polypodiophyta*, *Pinophyta* and *Magnoliophyta*. As shown by SIs (Table 4), however, only in the family *Leguminosae* does structural diversity justify their use as systematic markers.

As may be gauged from the isoflavonoid chemistry of the *Legumi-nosae* available as at December 1975, several parallel evolutionary lines can be distinguished: three isoflavone lines, a pterocarpanoid line and a rotenoid line (Fig. 4). Although it seems clear that all three isoflavonoid classes were already represented by biosynthetically simple derivatives at an early stage of the evolutionary history of the family, the order in which they are given indicates, nevertheless, progressive specialization in relation to their environment. Thus, whereas only the isoflavone skeleton occurs in genera along lines 1 and 2, the peripheral genera of the isoflavone line 3 show additionally the most widespread skeletal variation: isoflavone → pterocarpanoid; and the peripheral species of the more highly developed pterocarpanoid lines 4.2 and 4.4 contain additionally rotenoids. Finally, the genera of the rotenoid line 5 seem to be substantially lacking in the less specialized pterocarpanoids.

The time is now ripe to ask how the tribal classification, based on chemical characteristics, agrees with the morphological evidence. Although this is an interesting topic, its importance must not be overestimated. The iso-flavonoid data refer only to 232 species belonging to 70 genera. Worse still, it is impossible to state how many of these species have been examined with adequate thoroughness.

Given these cautionary arguments, let us compare the location of tribes along the chemical lines based on isoflavonoid data (Fig. 6*a* from top to bottom: 3 isoflavone one rotenoid and 3 pterocarpanoid lines) with the relevant parts of a very recent reappraisal of morphological rela-tionships within the *Lotoideae* (Fig. 6*b*, Polhill 1976). If, indeed, chemical evolution within the isoflavonoid BG proceeds by gradual diversification of substitutions and skeletons, practically no serious discrepancy between the two schemes is apparent. The slight differentiation of the primaeval forms makes the determination of origins of lines frequently difficult. Thus, e.g., although the lines leading to *Trifolieae* and *Tephrosieae* run intriguingly close on the affinity plot, suggesting the existence of a common ancestor, this is chemically unknown. The isoflavonoid data suggest, furthermore, that the *Dalbergieae*, as well as the *Phaseoleae* are polyphyletic, a fact which will hardly be contested on morphological grounds.

In conclusion, it is evident that considerable insight into the evolu-tion of the *Leguminosae* results, if tribal divisions are extended back along chemical lines of development until they become blurred in the *Sophoreae*. It is intellectually pleasing to acknowledge that the con-cluding part of this sentence paraphrases Polhill (1976) who wrote regional where we write chemical.

Indole Alkaloid Lines in the *Apocynaceae*. Monoterpene-indole alka-loids occur in three related plant families. The present section is concerned, nevertheless, only with their use as systematic markers in the *Apocy-naceae*, subfamily *Plumerioideae*. Biosynthetically, tryptophan and

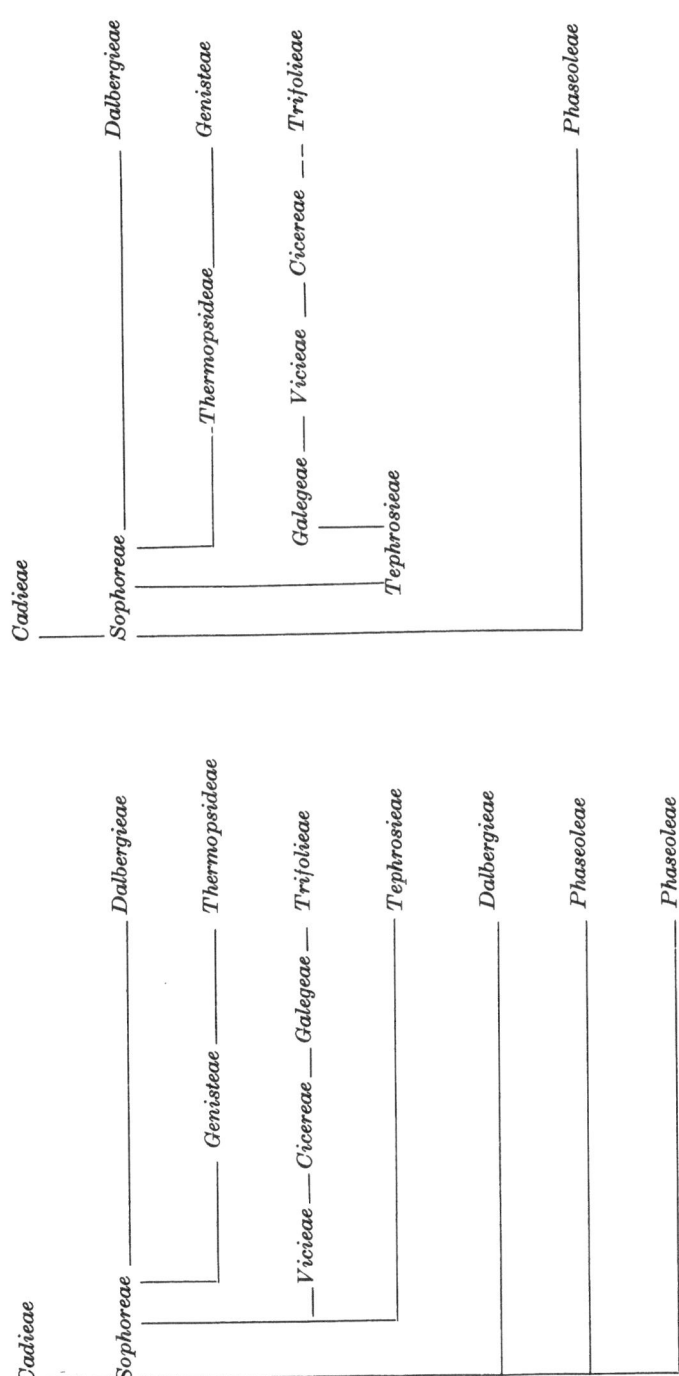

Fig. 6. Evolutionary relationship of *Lotoideae* tribes, based on a) chemical evidence; b) morphological evidence

mevalonate, the PPs of the indole alkaloids, produce such a large number of skeletons, that it is impossible to present our BG map in reasonable format for publication. Fortunately, however, this profusion of skeleta can be grouped into so-called corynanthe, aspidosperma and iboga types (Geissman & Crout 1969), according to the disposition of C-atoms in the mevalonate moiety, which may have preserved the

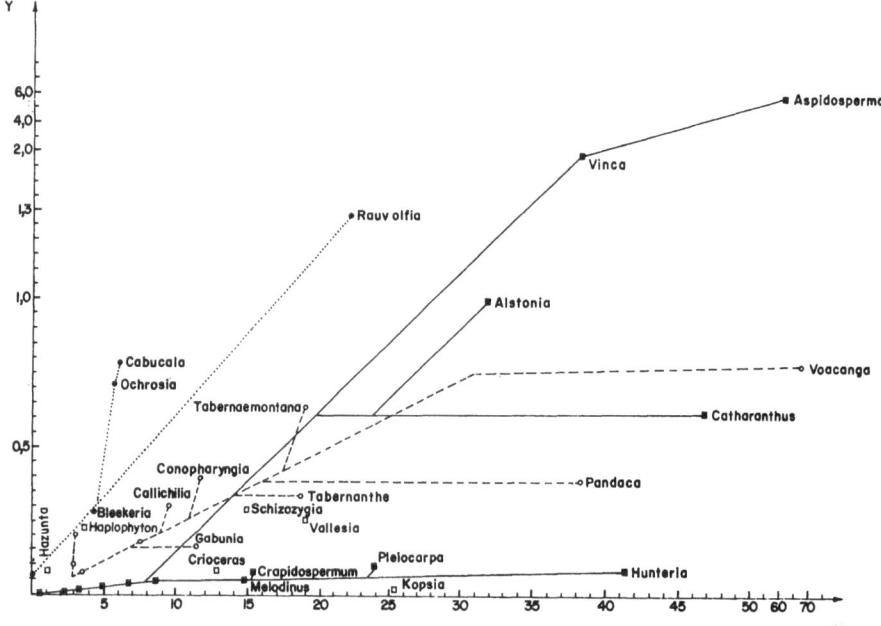

Fig. 7. Affinity plot of *Apocynaceae* genera mapped along corynanthe (............), aspidosperma (_____) and iboga (_____) lines. Genera of uncertain affinity are marked □. Near the origin of the plot, on the corynanthe line (black dots): *Excavatia*; on the aspidosperma line (black squares): *Geissospermum, Diplorrhynchus, Gonioma, Rhazya, Picralima, Amsonia*; on the iboga line (circles): *Peschiera, Stemmadenia, Rejoua, Ervatamia*

original arrangement (corynanthe) or suffered two kinds of rearrangement (aspidosperma, iboga).

The alkaloid chemistry, available as at December 1975, of 284 species belonging to 36 genera of the *Plumerioideae*, suggests the existence of three lines which, in order, indicate presumably progressive specialization (Fig. 7). (1) The corynanthe line which develops by oxygenation of corynanthe type alkaloids. The peripheral genus *Cabucala* contains already the biosynthetically simplest of the aspidosperma type alkaloids. (2) The aspidosperma line, characterized, in addition to corynanthe alkaloids, by aspidosperma type alkaloids. After an initial extension (2.1) formed by genera showing little differentiation, the line branches into 2.2, which develops by skeletal diversification, and 2.3 which develops by skeletal and oxidative diversification. (3) The iboga line, characterized, in addition to corynanthe and aspidosperma alkaloids, by iboga type alkaloids. Oxidative diversification is restricted to iboga (and aspidosperma) constituents.

Now let us compare again chemical lines and morphological tribes. The picture here is very clear. The genera of *Rauvolfieae* fall along line 1, with the exception of *Vallesia* and *Kopsia* which seem closer to line 2. The genera of *Alstonieae* form line 2 which sustains a branch occupied by the genera of *Carisseae* (line 2.2). Although the peripheral genera *Alstonia* and *Catharanthus* produce already inclusively iboga type alkaloids, they should still belong to line 2 in view of the typical oxidative diversification of their corynanthe alkaloids, absent from the genera of line 3. The connection of *Haplophyton* to line 2 is unclear. The genera of *Tabernaemontaneae* form line 3.

Benzylisoquinoline Lines in the *Magnoliidae*. Benzylisoquinoline alkaloids occur widely scattered throughout the *Magnoliophyta*. As shown by SIs, however, only in the orders of the subclass *Magnoliidae* does structural diversity and natural distribution justify their use as systematic markers. The biosynthesis of benzylisoquinolines involves, as PPs, two tyrosine units which form a benzyltetrahydroisoquinoline (BTIQ), the precursor of their BG.

The alkaloid chemistry of 132 genera, available as at December 1975, suggests the evolution of the families of *Magnoliidae* along two interconnected lines (Fig. 8). (1) The aporphine line which develops by skeletal and oxidative diversification of aporphine type alkaloids. The peripheral pairs *Magnoliaceae—Annonaceae* and *Monimiaceae—Lauraceae* produce already sporadic examples respectively of berberine and morphine alkaloids of simple type and oxygenation. (2) The berberine line which develops by skeletal and, in the case of subline 2.1 including the peripheral *Papaveraceae* and *Fumariaceae*, also by oxidative diversification of berberine type alkaloids. These families are differentiated by the capacity of diversifying skeletons and oxygenation of aporphine type alkaloids, peculiarity of the *Papaveraceae*. Both produce only simple morphines, in contrast to *Menispermaceae* on subline 2.2, characterized by skeletal and oxidative variants of the aporphine and berberine, as well as of morphine-type alkaloids.

The chemical lines of the affinity plot (Fig. 8) and the morphological relations proposed by CRONQUIST (1968) for the orders *Magnoliales-Ranunculales-Papaverales*, agree down to details, such as the placement of *Magnolia-*

ceae-Annonaceae and *Monimiaceae-Lauraceae* in separate clusters, and the opinion that the *Papaveraceae* and the *Fumariaceae* are to be regarded as parallel groups which show different individual specializations. Even the position of *Ranunculaceae* on the affinity plot and Cronquist's statement that there is no obvious reason why the *Ranunculaceae* might not be ancestral to all other families of the order are contradictory only in appearance. Indeed, if *Thalictrum*, chemically as well as morphologically (Tamura

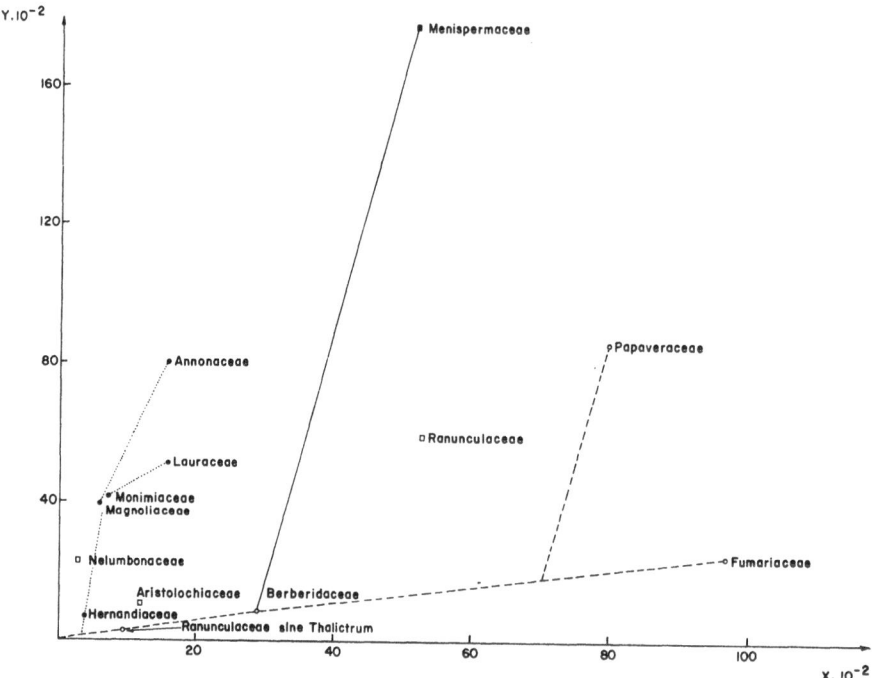

Fig. 8. Affinity plot of *Magnoliidae* families mapped along aporphine (.............), berberine (_____) and berberine/morphine (_____) lines

1962) the most advanced genus, is omitted from our calculation of average chemical composition, the point representing *Ranunculaceae* falls precisely on the line leading to *Berberidaceae* (Fig. 8).

Alkaloid Lines in the *Amaryllidaceae*. The *Amaryllidaceae* contain a special type of alkaloids which has been isolated but very rarely from other sources. Their biosynthesis involves, as PPs, phenylalanine and tyrosin which form a benzylphenylethylamin, the precursor of their BG.

The alkaloid chemistry of 145 species belonging to 29 genera, available as at December 1975, suggests the existence of two distinct lines within the *Amaryllidaceae* (Fig. 9). Development of the primitive stock,

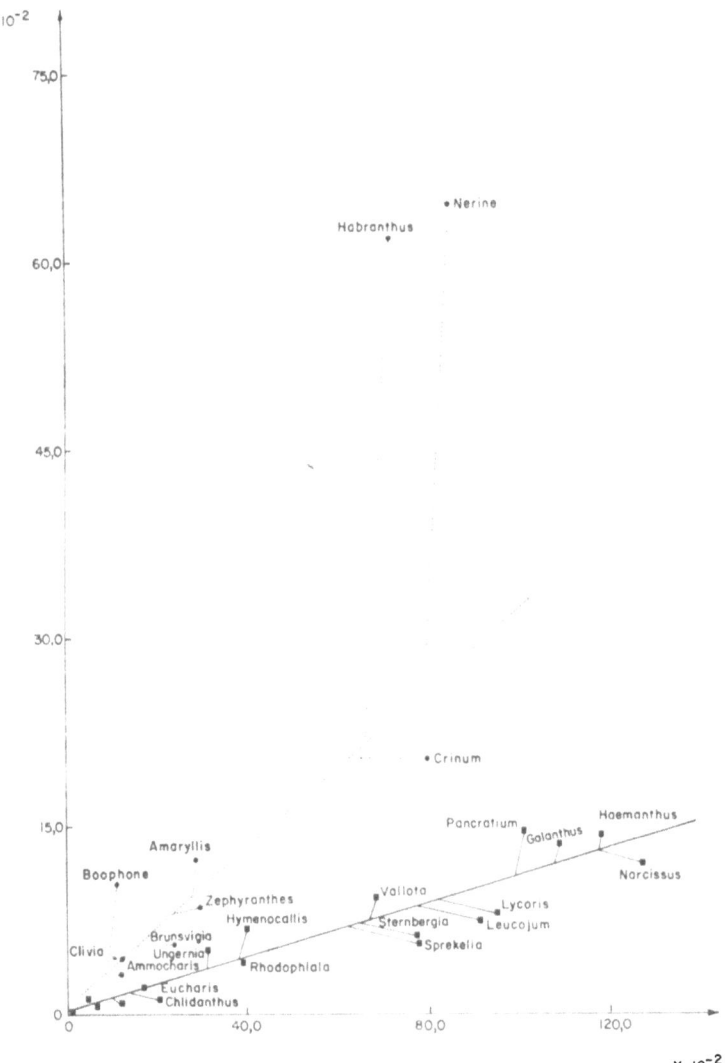

Fig. 9. Affinity plot of *Amaryllidaceae* genera mapped along crinine (..........)
(line 1) and lycoricidine lines (_____) (line 2). Near the origin of the plot,
on the lycoricidine line (black squares): *Eurycles, Cyrtanthus, Calostemma,
Eustephia, Elisena*

which already contained representatives of both, was accompanied
either by modification predominantly of the phenylethyl moiety of
crinine type alkaloids, or by modification of the benzyl moiety of lycoricidine type alkaloids.

Now again, how does the chemical classification compare with morphological evidence? The *Euchareae* appear as basic group which leads either to genera pertaining to *Crineae* and to the genus *Amaryllis* (line 1) or to a series of tribes (line 2). The most conspicuous fact here concerns only the separation of *Crineae* and it seems worthwhile to examine their relatives for possible morphological affinity. This suggestion gains momentum through the observations of TRAUB (1957) concerning the occurrence of hybrids *Amaryllis* × *Crinum*, *Amaryllis* × *Brunsvigia*.

On presently available evidence, however, it must appear to the morphologist that each tribe is equipped to perform its own parallel chemical evolution, while it does appear to the chemist that each line is equipped to perform its own parallel morphological evolution. The analysis looks, nevertheless, at least partly meaningful.

Coumarin Lines in the *Umbelliferae*. Coumarins occur widely scattered throughout the *Magnoliophyta*. The present section is concerned, nevertheless, only with their use as systematic markers in *Umbelliferae*, one of the two families in which they possess adequate structural diversity. Biogenetically, all coumarins of *Umbelliferae* seem to be derived from umbelliferone. Unsubstituted coumarin is synthesized by a different route and hence does not belong to the same BG.

The coumarin chemistry, available as at December 1975, of 245 species belonging to 47 genera of the *Umbelliferae* suggests the existence of 3 lines (Fig. 10). (1) The coumarin line, which, as the other lines, develops by progressive oxygenation of the coumarin skeleton. The peripheral genus *Bupleurum* contains an angular furocoumarin. (2) The linear furocoumarin line, divided into several sublines according to co-occurring types of skeletons. The peripheral genera of subline 2.1, *Anethum* and *Apium*, contain in addition two pyranocoumarins. (3) The pyranocoumarin line, subdivided according to the angular or linear nature of the skeletons.

Here, the points representing genera of particular tribes mostly do not occupy particular lines. They are confined to circular sections, which we nickname "annual rings" since they suggest evolutionary development. Thus the genera of *Peucedaneae* occupy all sections, *Apieae* all sections minus the outermost, *Smyrnieae* all sections minus the two outermost, etc. The present situation, in which representatives of several tribes contain compounds derived through several lines, is considered to be a sign of phylogenetic advancement for a family.

This recalls one of the recurring themes of our results: the whole basic chemistry is imprinted on organisms at the beginning of the taxon's evolutionary history. The possibilities of variation along one

Fig. 10. Affinity plot of Umbelliferae genera mapped along coumarin (··········), furanocoumarin (———) and pyranocoumarin (—·—·—) lines. Near the origin of the plot, on the coumarin line: *Astrodiscos, Daucus, Scandix, Falcaria, Thapsia, Astydamia*; on the furanocoumarin line: *Trachyspermum, Petroselinum, Foeniculum, Coriandrum, Cicuta, Smyrniopsis, Levisticum, Laser, Phellopterus, Ferulago, Malabaila, Leptotaenia, Zozimia, Apium, Cymopterus, Conium, Anethum, Cachrys*; on the pyranocoumarin line: *Capnophyllum, Pteryxia, Laserpitium*. 10. *Peucedaneae*, 9. *Apieae*, 8. *Smyrnieae*, 11. *Laserpitieae*, 7. *Coriandreae*, 6. *Scandiceae*, 12. *Dauceae*

particular biochemical pathway being exhausted, the line initiates variation along another. In time, units along all lines will produce compounds through several pathways of a BG. Thus, for morphologically differentiated groups, clearcut chemical differentiation (see *Magnoliidae, Leguminosae, Apocynaceae*) indicates primitiveness, whereas blurred chemical differentiation (see *Umbelliferae, Amaryllidaceae*) indicates advancement.

Second Principle

While the first principle was based, at least partly, on a correlation with current morphological evolutionary thought, no such bias influenced the deduction of a working hypothesis for a second principle: among taxa of *Embryobionta* of low hierarchical level, biochemical evolution of secondary metabolites involves gradual specialization by substitutional or/and skeletal diversification of compounds within biogenetic groups.

Indeed, to summarize the proposed procedure, initially the evolutionary sequences of compounds are mapped on biosynthetic (relative closeness of structure to ubiquitous PPs indicates relative primitiveness of a compound) and distributional (relative dispersion of occurrence indicates relative primitiveness of a compound) grounds. Next, taxa (presently we work with families or genera, but the method is, of course, equally well suited for species or orders) are considered as average conditions of the evolutionary stages of the compounds contained. This results in the sequencing of taxa which are, finally, examined in the light of existing morphological schemes. The operations are thus performed in the order suggested by Heywood (1973), to avoid elements of circularity in the argument.

It is hoped, therefore, that the independence of our chemical method and accepted morphological evidence is such as to warrant mutual comparison of information. The examples cited in which such a comparison was performed show that, in a general way, within a biogenetic group, given a sequence of metabolites, A → B → C → D (see p. 63), if a taxon contains C, one is able to state that it may have evolved from a taxon accumulating B. Neither A nor B need be accumulated by the more advanced taxon; their transitory existence is assured by the biosynthetic mechanism.

Clearly, it is again proper to enquire about the possibility of reversal. Although this may occur, we have come across so few cases which may function as possible examples, that it must be considered a minor evolutionary trend.

Third Principle

The precursors of all BGs are of course present in all plants where they provide essential macromolecules. It is thus understandable, or even plausible, that chemical characters are polyphyletic, i.e. that compounds pertaining to any BG may appear widely scattered through-out the plant kingdom. Clearly, however, only if secondary metabolites are accumulated in certain plant groups will they serve a useful purpose in biochemical systematics. It may appear, a priori, that the distinction between haphazardly distributed metabolites and specially accumulated metabolites would be difficult or even impossible. This is not the case. Ubiquitous or widespread compounds usually remain structurally close to the primary precursors, while useful markers undergo structural variation through chemical evolution. As already noted, evolutionary vigour of a BG may be ascertained by SIs.

One and the same species usually contains, against a background of chemosystematically irrelevant metabolites, only a limited number of relevant BGs. Different BGs with analogous functions may be produced by closely related plant taxa, according to environmental conditions, resulting in chemical vicariance. Enhancement of products along one biosynthetic route seems to trigger a regulating mechanism which suppresses the formation of compounds along another.

Morphological vicariance, due to differential environmental conditions, is a well know phenomenon. We have ascertained that the chemical rela-tionship of morphologically vicariant species can be closer than the chemical relationship of congeneric species of the same habitat (GOTTLIEB & STE-FANI 1970). Analogously, the morphological relationship of chemically vicariant taxa may be closer than the morphological relationship of taxa of similar chemical composition. An understanding of the interaction of environment and secondary plant metabolism, incipient at the present time, will have to be reached before the reasons for chemical vicariance will become clear.

All this leads to a series of statements which we embody presently in a third principle of biochemical systematics: in view of the ubiquity of BG-precursors, presence of certain compounds in selected taxa gives no assurance as to their affinity. Only if metabolites are considered with respect to structural variation within their biogenetic route or group may the analysis lead to a significant result. Even then, however, due to chemical vicariance, difference in chemical composition of two taxa says nothing about their lack of affinity.

Conclusion

Taxonomic methods (p. 54–62), applied to chemosystematic prob-lems (p. 63–75), indicate that secondary metabolism is only moderately useful, if employed as an auxiliary criterion in the construction of an evolutionary classification. Indeed, at present, the interplay or con-nection of form and chemistry of an organism is, at best, debatable.

Secondary metabolism, just as comparative morphology is an independent, primary criterion. It is considered highly probable that phyletic classifications of plants will be, before long, proposed by chemists, not least because they promise to be more useful, than purely morphological ones.

References

CRONQUIST, A., 1961: Basic Botany. New York: Harper & Row.
— 1968: The Evolution and Classification of Flowering Plants. London: Nelson.
GEISSMAN, T. A., and CROUT, D. H. C., 1969: Organic Chemistry of Secondary Plant Metabolism, p. 533. San Francisco: Freeman, Cooper & Co.
GOTTLIEB, O. R., and STEFANI, G. M., 1970: Xanthones from *Kielmeyera excelsa*. Phytochemistry 9, 453—454.
HARBORNE, J. B., HEYWOOD, V. H., and KING, L., 1976: Evolution of yellow flavonols in flowers of *Anthemideae*. Biochem. Syst. Ecol. 4, 1—4.
HASLAM, E., 1974: The Shikimate Pathway, p. 198. New York: Wiley.
HEYWOOD, V. H., 1973: The role of chemistry in plant systematics. Pure Appl. Chem. 34, 355—375.
POLHILL, R. M., 1976: personal communication.
REZENDE, C. M. A. da M., and GOTTLIEB, O. R., 1973: Xanthones as systematic markers. Biochem. Syst. 1, 111—118.
—, — and MARX, M. C., 1975: Benzyltetrahydroisoquinoline-derived alkaloids as systematic markers. Biochem. Syst. Ecol. 3, 63—70.
ROCHLEDER, F., 1854: Phytochemie, p. 260. In: Pflanzenchemie und Pflanzenverwandtschaft (MOLISCH, H., 1933), p. 3. Jena: G. Fischer.
TAMURA, M., 1962: Taxonomical and phylogenetical consideration of the *Ranunculaceae*. Acta Phytotax. Geobot. 20, 71—81.
TRAUB, H. P., 1957: Classification of *Amaryllidaceae*. Subfamilies, tribes and genera. Plant Life (Herbertia) 13, 76—83.

Address of the author: Prof. Dr. OTTO GOTTLIEB, Instituto de Química, Universidade de São Paulo, Caixa Postal 20780, 01000 São Paulo, Brazil.

Plant Syst. Evol., Suppl. 1, 77—95 (1977)

Botany Department, University of Queensland, Australia

Quantitative Studies of Inter-relationships Amongst the *Liliatae*

By

H. T. Clifford, St. Lucia

Abstract: There is an immense amount of taxonomic data and new information is accumulating rapidly, thereby making it increasingly difficult for taxonomists to comprehend more than a small amount of what is available. Hence it is essential that advantage be taken of modern computing facilities for storing and retrieving information but more especially for generating classifications. The introduction of such technology should lead to a full exposure of the data and methodology thereby enabling potential users of a classification to assess its value for the problem in hand.

As an example of the methodology a sample of eighty-eight liliate families has been classified using data pertaining to fifty-one attributes and using an intensely clustering sorting strategy. Of the four major groups of families emerging from the analysis two, the *Zingiberales* and a group of water plants corresponding closely with the *Alismidae* are well established in traditional classifications. The two remaining groups did not agree closely with any recognized grouping in established classifications but nevertheless showed a strong internal homogeneity in that one is comprised largely of wind-pollinated and the other mainly of insect-pollinated families.

Introduction

Taxonomy has long been investigated with the mystique of personality playing a dominant role in decision making. Such a viewpoint was clearly expressed by HUTCHINSON (1960) when he stated "the delimitation of families, of genera and of species is sometimes very much a matter of taste and personal idiosyncracy but I would also add of judgement and experience". A similar sentiment is echoed by AIRY-SHAW (in WILLIS 1973) in the preface to his influential Dictionary where he states that taxonomy is a "personal matter".

In like vein HEYWOOD (1974: 44) has written as follows in connection with what he sees as the failure to reconcile traditional and numerical taxonomists: "Part of the answer, at least, lies in the extent to which traditional methods of classification have involved such subtle processes as Gestalt perception and a related phenomenon, typology, which bring into play complex human mental processes which it is difficult to appreciate and analyse, let alone machine copy." The suggestion that perception is at the basis of the taxonomies of all cultures has also been advocated recently by Dwyer (1976).

As perception is an intensely personal matter it is not surprising that taxonomists sometimes expound their opinions with an almost fanatical fervour. To avoid misunderstanding it is therefore essential that any taxonomic work should include a statement as to the reason for its production. The reasons may well differ from person to person and from time to time with the same person. For example, it may be intended that the classification reflect the probable phylogeny of the group in question, the breeding relationship of its members, provide well defined morphological groups useful for making subsequent identifications or generate groups of maximum predictive value, to mention but four possibilities. How the classification was generated should also be recorded together with an account of the attributes used and those neglected or rejected. The declaration of all three of these matters will enable the users of taxonomic systems to assess their reliability for the tasks to which they may be applied. For as MacKay (1969) has observed, pattern is meaningful only to the agent who seeks it and a classification of value in one situation may be valueless in another.

Because taxonomic studies are fundamental to so much biology it is important that they should be as objective as possible. It would seem more important that a monographer or reviser of a group should discuss all the attributes studied rather than to list those regarded as diagnostic, and to state how much data was available rather than to list all the specimens examined. The attitude adopted to the formation of taxonomic groups should also be stated. Were they defined in terms of internal consistency or in terms of discontinuities between groups of equal rank? Those following after would then be able to build on rather than repeat much of the work. To this end it is essential that adequate records be kept and since many of these would be unacceptable to journal editors means of establishing and maintaining data banks must be sought.

Such data banks would also reveal the considerable extent to which extrapolation inevitably creeps into classifications. For example, it seems odd that cotyledons should serve as a basis for classification or nomenclature when the seedlings of so few of the members of the class have been described, especially as the attribute is neither constant within nor confined to the class.

Most classifications are based on a wide range of attributes and then one or few diagnostic attributes are chosen to circumscribe the groups recognized. The tradition of extrapolating diagnostic attributes is of

long standing, for example, JUSSIEU (1789) has algae, liliates and magnoliates in his sixth order of the *Acotyledons*, the *Naiades*, indicating both
that he had seen few seedlings of the members of the order, and that
he based his grouping on other and undefined attributes.   Likewise,
DE CANDOLLE (1813) included the cycads amongst his *Endogens* or
Monocotyledoneae possibly because of their palm-like habit.

One means of implementing data banks would be to store the accumulated information on magnetic tape and to use the computer as the
storage-retrieval system.   Programs for these purposes are available
but before they can be used efficiently, considerable thought must be
given to procedures for updating the records, keeping track of their
reliabilities and for both adding to and deleting information with the
passage of time.   Data recorded as appropriate at one time may be
inappropriate at a later date because the purpose for which the information was required may change.   Thus seeds possessed of an aril have
been described for many species but the nature of the aril is not the
same for all species.

For some users, including ecologists, the possession or otherwise of
an aril by a seed may alone be significant, but for a phylogenetically
interested taxonomist the ontogeny of the aril will be important.   Hence to
a considerable extent the kinds of data accumulated will depend upon the
likely requirements of its users.   The development of data banks has only
just begun and whether they are all to be computer based or published
more conventionally remains to be seen.   To date the only large published
taxonomic data bank at generic level appears to be that prepared by
CLIFFORD & WATSON (1977) for the *Poaceae* of Australasia.   It is clear
that patience and co-operation are required for the establishing of satisfactory definitions and that the coding problem is still the "Achilles heel"
of numerical approaches to taxonomy (EL-GAZZAR & WATSON 1970).

In addition to its functioning as a data storage-retrieval system
the computer has an important role in the generating of classifications.
It is with respect to this function that there has been most apprehension
amongst practising taxonomists.   Such distrust is difficult to appreciate
especially when considering higher level taxonomies where it seems
impossible for the human mind to assess all the available information.
To this end witness the few large families that have been monographed
this century.

Indeed JACOBS (1969) has maintained that in a single lifetime a
taxonomist is unlikely to be able to deal critically with more than about
1000 species.   Hence, if his work is to be related to that of others some
assistance or co-operation is required.   In this regard some chores are
more easily delegated than others.   For example, it is not necessary for
monographers to write descriptions of their own taxa; nor, might it be

added, their own keys, for such tasks could be readily undertaken by a well trained plant morphologist. As was long ago pointed out by DIELS (1924: 68), ,,Begrifflich und inhaltlich sind Phytographie und Systematik zwei ganz verschiedene Dinge. Wer sie vermengt, kennt beide nicht."

However assistance in these ways leads only to amplification of the descriptions of recognized taxa and doesn't really help with the general problem of establishing classifications in which relationships may be expressed. It is here that the computer has a central role to play in that the user may specify the bases on which the taxa are to be grouped or divided, the amount of similarity or dissimilarity to be allowed within and between groups and the relative importance to be assigned to each attribute. Such objectivity is to be welcomed especially as the right to criticize and amend the results still rests in the hands of the taxonomist. It is merely that the data and methodology are fully exposed.

Further advantages are that the process of classification may be made interactive with the computer thereby enabling the taxonomist to interrupt the analysis at any stage. For all but the largest data matrices the cost of producing a classification or ordination is slight, and in any event much of this cost involves the printing of the results. Hence if these are displayed visually the consequences of changing the attributes or taxa can be investigated almost instantaneously. Such a facility is particularly useful for the study of higher level taxa where, for example, some of the groups generated may possess only one state of a two- or more-state attribute. To be able to so control the classificatory program as to be able to experiment, in a matter of seconds, with different strategies is an exciting prospect.

A less exciting, but none-the-less important, aspect of computer aided taxonomy stems from the ease with which descriptions of groups may be generated and contrasted with one another. Such comparisons may direct the attention of the user to errors in the data or to the discriminatory value of attributes not previously appreciated as being taxonomically significant.

In what follows, the results of subjecting a sample of liliates to a number of numerical classificatory procedures are compared with their relationships as expressed in two recently proposed classifications. The higher level taxonomy of the *Liliatae* is by no means stable and unfortunately it is impossible to decide with certainty which if any of these classifications is correct. Furthermore, in the absence of a reliable fossil record it is not possible to speculate with confidence on the phylogeny of the class. Finally, due to a lack of sound theories of morphogenesis it is, for the majority of attributes, rarely possible to predict with certainty which are their primitive and which their advanced states.

Hence, there are difficulties in deciding which structures are alike because of their sharing a common ancestry and which are alike because though differing in origin they are responding in a similar manner to the same type of selection pressure. Lack of agreement as to which are advanced and which primitive characters is frequently the basis of the differences observed between many of the proposed phylogenetic classifications. Such a lack of agreement is aggravated in that determination of relationships depends largely on the framework within which the comparisons are made and so reflects the interests of the individual taxonomist. Thus an authority on sedges might regard the carices as a family (*Kobresiaceae*) quite separate from the *Cyperaceae* as did GILLY (1958) whereas a reviewer of the flowering plants would probably regard them as different only at subfamilial level. That is the nature of the classification cannot be divorced from the viewpoint of the observer and depends upon whether the system is being viewed globally or locally.

The choice of a sampling system for selecting both taxa and attributes poses problems. Here, in order to ensure that the total variability within the class has been encompassed, most families recognized during this century have been accepted as valid and each has been scored using data from any or all of their species. Alternative schemes of sampling are to choose a subset of species at random from the total set of liliates or to choose of species in proportion to the numbers of species per family. The former alternative is unrealistic in demanding a list of all species of *Liliatae* and the latter is unrealistic in that small families are oversampled or completely sampled in comparison with large families. The families selected for study are listed in Table 2.

In circumscribing the class *Liliatae* a rather conservative viewpoint has been adopted and disputing contenders for membership such as the *Nymphaeaceae* and *Petrosaviaceae* have been neglected. Their consideration would be more appropriate to the problem of classifying the *Magnoliophyta* than to the problem in hand.

The choice of attributes has been determined by two principal considerations, availability of information and the degree of agreement or otherwise as to which structures and organs are homologous. Comparisons between taxa are most usefully made in terms of structures known to be homologous, but, as has already been noted there are difficulties in determining these homologies. Because there is dispute as to the nature of the perianth in different families few such data have been recorded. Furthermore, even when structures appear to be homologous they may originate in quite different ways as has been discussed by TOMLINSON (1969) for the paracytic stomates of the *Bromeliaceae* and *Pandanaceae*.

No conscious attempt has been made to select attributes on the basis of their supposed taxonomic usefulness for such judgements cannot be properly made until the classification has been established or according to CRONQUIST (1968: 10) "perceived". The classification in turn

Table 1. The attributes and their states

Attribute	States
1. Endosperm	nuclear, helobial, cellular
2. Embryo sac	crassinucellar, tenuinucellar, pseudocrassinucellar
3. Tapetum	glandular, amoeboid
4. Embryogeny	caryophyllad, asterad, onagrad, chenopodiad
5. Ovule form	orthotropous, anatropous, hemianatropous, campylotropous, anacampylotropuos
6. Origin of micropyle	inner integ., both integ., neither integ., outer integ.
7. Secondary antipodal nuclei	reported, not reported
8. Sporangia per anther	2, 4, more than 4
9. Pollen nuclei	2, 3
10. Anthers	sessile or basifixed, dorsifixed
11. Pollen grains	single, tetrads
12. Ovules per placenta	1, more than 1
13. Ovary	Superior, subinferior, inferior
14. Placentation	basal, axile, parietal, pendulous, marginal, scattered
16. Anther dehiscence	extrorse, lateral, introrse
17. Pollen	aperturate, non-aperturate
30. Pollen aperture	distal, zonal, global
18. Pollen mother-cell tetrad formation	simultaneous, successive
19. Pollen shape	thread-like, otherwise
20. Pollen aperture	porate, colpate, spiroaperturate
21. Seeds	arillate, nonarillate
22. Seeds	operculate, non-operculate
23. Anthesis	protandrous, protogynous, homogamous
15. Endosperm	scarce, abundant
24. Endosperm	ruminate, non-ruminate
28. Endosperm	starchy, otherwise
25. Placentae per pistil	1, 2, 3, 4, more than 4
26. Germination	remotive epigeal, remotive hypogeal, admotive
27. Seedling axis	bearing cataphylls, lacking cataphylls
29. Perisperm	present, absent
31. Embryo	macropodous, non-macropodous
32. Leaves	2-ranked, more than 2-ranked
33. Ligule	present, absent
34. Leaf-base	sheathing, not sheathing stem
35. Venation	reticulate, convergent, pinnate
36. Guard-cells	present, absent
37. Guard-cell shape	graminoid, otherwise
38. Subsidiary cells	0, 2, more than 2
39. Vessels (roots)	present, absent
40. Vessels (stems)	present, absent
41. Vessels (leaves)	present, absent
42. Velamen	present, absent

Table 1 (continued)

Attribute	States
43. Hypodermis (leaf)	present, absent
44. Raphide bundles	present, absent
45. Embryo in seed	basal, axile
46. Silica	present in epidermis, absent from epidermis
47. Secondary growth	occurring, not occurring
48. Squamulae	present, absent
49. Root endodermis	as in *Iris*, otherwise
50. Vernation	conduplicate, convolute, plicate, involute, flat
51. Fruit	achene or nut, berry, capsule, drupe, schizocarp

depends upon the sample of families investigated. For example, attributes constant within groups have little taxonomic value. Hence the possession of two cotyledons is of little significance amongst a sample of magnoliates but would be of considerable significance if the sample included liliates as well. It ought to be noted that amongst a wider sample of vascular plants two cotyledons would not necessarily be a diagnostic character.

Aside from such theoretical considerations, comparisons between families are also severely restricted by the lack of comparative family descriptions. Unfortunately, these descriptions often contain much extrapolated data and mistakes that have long passed unchallenged. To expose such misinformation demands recourse to the study of actual plant material. Whilst this is in itself a pleasurable task it is at best time consuming and at worst frustrating when the plant does not grow locally. Of the fifty-one attributes scored for numerical analysis, thirty-two refer to reproductive and nineteen to vegetative attributes. These attributes and their states are defined in Table 1.

Of the total possible records for the chosen attributes about 75% have been obtained. To quote all sources from which the data were acquired would result in an immense reference list and so only major references will be cited. Wherever possible the data have been cross-checked either with other references or with plant material. The principal reference works consulted were as follows: embryology (DAVIS 1966); anatomy (TOMLINSON 1969, CUTLER 1969); seedlings (BOYD 1932); vessels (CARLQUIST 1975); biochemistry (GIBBS 1974, HEGNAUER 1963); seeds (MARTIN 1946); morphology (ARBER 1921).

Classificatory Procedures

As a first step in the analysis the dissimilarities between the families, considered in pairs, were calculated using an information-gain statistic. Clustering was then undertaken on the basis that at each stage in the fusion cycle there was a minimum increase in diversity on fusion. The procedure is highly group-sized dependent (CLIFFORD & WILLIAMS 1973) but has an advantage in that it results in sharply defined groups. The relationships between the groups cannot be read directly from the dendrogram and so the inter-family dissimilarities were subjected to a principal co-ordinate analysis (GOWER 1967) which is an ordination based on the inter-taxa dissimilarities. From the results of the two analyses a set of four major sub-groups within the *Liliatae* was recognized.

Having established the groups their properties were investigated in two ways. For each division of the dendrogram the frequencies of the attribute states on either side of the division were determined; for each of the first three principal co-ordinate axes the attributes most closely correlated therewith were determined.

For most numerical taxonomic studies these analyses are adequate in that they are the minimum and yet shed light not only on the classification of the taxa but also indicate the role played by the attributes in generating that classification (CLIFFORD, WILLIAMS & LANCE 1967).

It is unfortunate that the comparisons between families could not be made with complete data as the classification may differ according to the attribute set employed (CLIFFORD & LAVARACK 1974, CLIFFORD 1976). However, to restrict the analysis to attributes known over all families would have so reduced the attribute set as to make any discussion of results of little value. For most families a wide variety of data was available so the bias likely to be introduced through the use of special subsets of data has been reduced.

Discussion of Results

The results arising from the application of the clustering strategy are summarized at the ten-group level in Fig. 1, and the families constituting each group are listed in Table 2. From the dendrogram it is clear that four major groups may be recognized and that their members differ markedly with respect to their habitat preferences and modes of pollination. Those attributes by which the groups may be distinguished vary from branch to branch of the dendrogram and since the classification is polythetic the groups are defined mostly by their joint possession or lack of several attribute states.

No detailed discussion will be undertaken of the family assemblages within the ten groups. However, the inclusion of the *Corsiaceae* amongst the water-plants calls for special comment. There were relatively few data available for the family and these were mostly attributes common to the class and so its presence in group 1 is subject to dispute. Early in the analysis the *Corsiaceae* became associated with the *Aponogetonaceae* and the somewhat variable *Hydrocharitaceae* with which it remained.

That it is anomalous in the group is also supported by its position relative to the other members of group 1 as shown in the ordination diagram (Fig. 2). Even so, it is interesting to note that DEYL (1955) included the *Corsiaceae* in the same order as the *Triuridaceae*, a family CRONQUIST (1968) considered as related to the water plants.

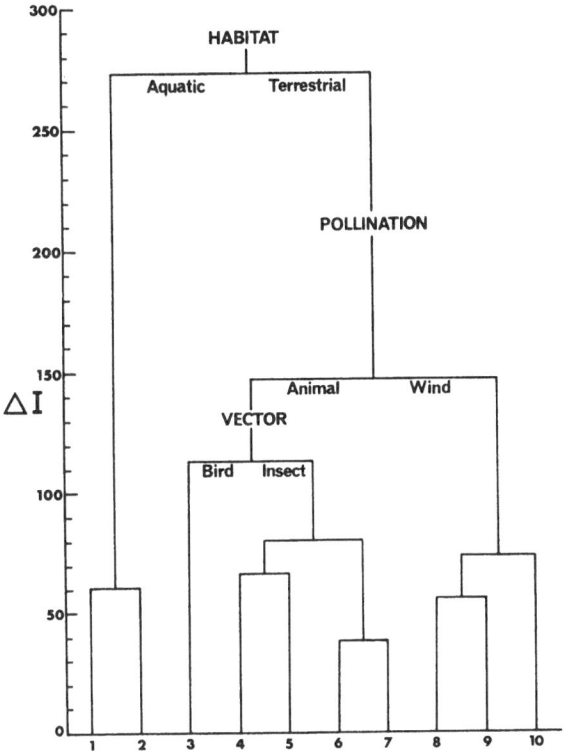

Fig. 1. Dendrogram resulting from clustering the families of *Liliatae*, truncated at the 10-group level with the principal habitats and pollination mechanisms of the groups superimposed

The prime division within the *Liliatae* separates the majority of the aquatic members (marine and freshwater) from the remainder. Four attributes are strongly associated with this separation and these are summarized in Table 3 where the numbers accompanying them indicate in percentages the frequencies of the attribute states in each group. That the percentages sometimes exceed 100 reflects the fact that some families include individuals possessing more than one state of the attribute.

Within the aquatic plant families two groups may be recognized in

Table 2. The family compositions of the 10 groups defined by the dendrogram in Fig. 1. Following each family name are two numbers, the first is that indicating its position in the alphabetical list of family names, the second the number of attributes for which the family has been scored

I. Mostly aquatic families (*Alismidae*)

1.	Alismataceae	3, 46	2.	Cymodociaceae	26, 25
	Aponogetonaceae	10, 39		Juncaginaceae	44, 38
	Butomaceae	17, 45		Lilaeaceae	46, 26
	Corsiaceae	22, 22		Naiadaceae	53, 42
	Halophilaceae	35, 28		Posidoniaceae	62, 24
	Hydrocharitaceae	38, 38		Potamogetonaceae	63, 43
	Limnocharitaceae	48, 24		Ruppiaceae	67, 39
	Thalassiaceae	76, 23		Scheuchzeriaceae	69, 38
				Zannichelliaceae	86, 32
				Zosteraceae	88, 40

II. Mostly bird-pollinated, tropical, terrestrial families (*Zingiberales*)

3.	Cannaceae	18, 47
	Costaceae	23, 42
	Heliconiaceae	37, 30
	Lowiaceae	49, 28
	Marantaceae	50, 39
	Musaceae	52, 41
	Strelitziaceae	73, 45
	Zingiberaceae	87, 46

III. Mostly insect-pollinated, temperate, terrestrial families

4.	Agavaceae	2, 48	5.	Asparagaceae	14, 41
	Alliaceae	4, 35		Bromeliaceae	15, 47
	Aloeaceae	5, 37		Burmanniaceae	16, 35
	Alstroemeriaceae	6, 42		Commelinaceae	21, 49
	Amaryllidaceae	7, 49		Dioscoreaceae	28, 45
	Aphyllanthaceae	9, 33		Geosiridaceae	32, 15
	Apostasiaceae	11, 29		Haemodoraceae	34, 43
	Gilliesiaceae	33, 30		Isophysidaceae	41, 21
	Hypoxidaceae	39, 40		Liliaceae	47, 42
	Iridaceae	40, 45		Ruscaceae	68, 35
	Orchidaceae	54, 43		Smilacaceae	70, 32
	Petrosaviaceae	57, 29		Tecophilaceae	75, 33
	Philesiaceae	58, 37		Trichopodaceae	79, 35
	Phylidraceae	59, 38		Velloziaceae	73, 37
	Thismiaceae	77, 21			
	Xanthorrhoeaceae	84, 44			

6.	Araceae	12, 49	7.	Croomiaceae	24, 27
	Lemnaceae	45, 38		Roxburghiaceae	66, 20
	Petermanniaceae	56, 33		Stemonaceae	72, 33
	Taccaceae	74, 40			
	Trilliaceae	80, 37			

Table 2 (continued)

IV. Mostly wind-pollinated terrestrial families

8.	*Abolbodaceae*	1, 31	9.	*Anarthriaceae*	8, 33
	Cartonemataceae	19, 35		*Centrolepidaceae*	20, 41
	Thurniaceae	78, 30		*Ecdeiocoleaceae*	29, 23
	Eriocaulaceae	30, 43		*Flagellariaceae*	31, 39
	Xyridaceae	85, 45		*Joinvilleaceae*	42, 30
	Hanguanaceae	36, 31		*Rapataceae*	64, 33
	Juncaceae	43, 46		*Restionaceae*	65, 37
	Mayacaceae	51, 36			
	Pontederiaceae	61, 47			
	Triuridaceae	81, 28			
10.	*Arecaceae*	13, 49			
	Cyclanthaceae	25, 39			
	Cyperaceae	27, 48			
	Pandanaceae	55, 42			
	Poaceae	60, 50			
	Sparganiaceae	71, 43			
	Typhaceae	82, 44			

terms of the number of ovules per placenta in each group. All members of group 2 have one and all members of group 1 have several ovules per placenta.

Amongst the terrestrial liliates there are two major groups which are behaviourly well separated in that one (8, 9, 10) has largely wind-pollinated and the other (3, 4, 5, 6, 7) largely animal-pollinated members. The four attributes by which these two groups may be distinguished are summarized in Table 4.

The animal-pollinated liliates in turn comprise two distinct groups of families one of which (3) contains a high proportion of bird-pollinated and the other (4, 5, 6, 7) a high proportion of insect-pollinated species. The former is also largely tropical in contrast to the latter being largely temperate. The principal distinctions between the two groups are summarized in Table 5.

Because the *Zingiberales* and *Alismidae* are clearly defined in the cluster analysis and also stand out from the remaining families in the principal co-ordinate analysis (Fig. 2) it was decided to reanalyse the data excluding these two groups of families. A clustering analysis only was undertaken and this confirmed the lack of well defined groups in the "insect-pollinated" section of the liliates.

Within the largely insect pollinated families it is inappropriate to recognize any further groupings above those resulting from truncating the dendrogram at the ten-group level. That is because amongst the families of the four groups 4, 5, 6 and 7 there are no attributes which show any marked correlations. For the same reason no attempt has been made to recognize major groupings between the three groups (8, 9 and 10) of largely wind-pollinated plants.

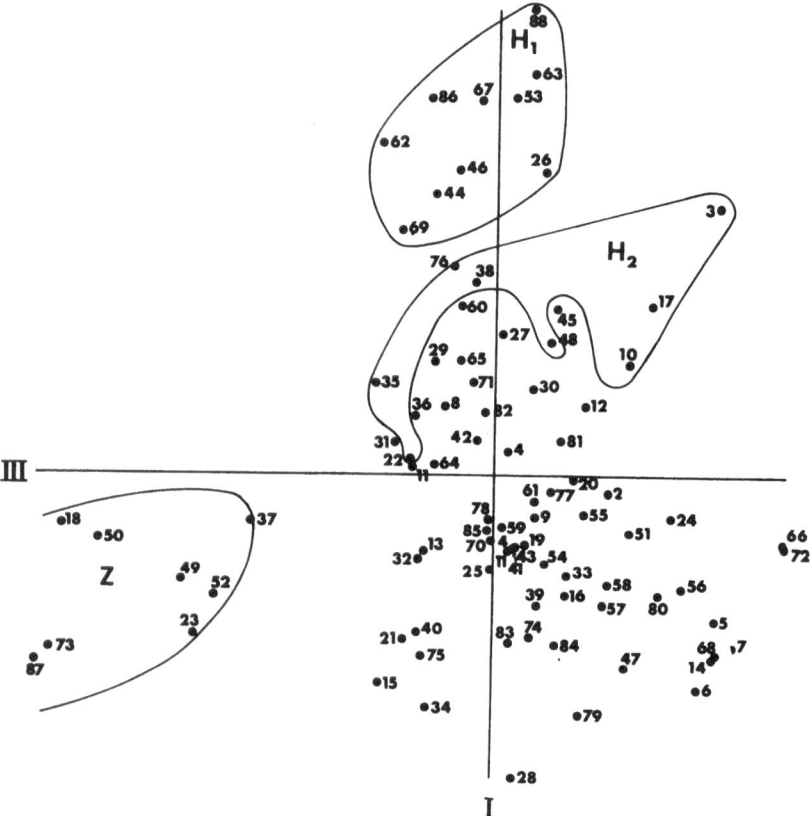

Fig. 2. The positions of the families of *Liliatae* relative to axes I and III of a principal co-ordinate analysis. The numbers refer to the families of Table 2; Z, *Zingiberales*; H₁ (Group 1) and H₂ (Group 2), *Alismidae*

A consideration of the attributes most closely correlated with the three principal co-ordinate axes revealed that the three attributes most closely correlated with the first principal co-ordinate axis were presence or absence of squamulae, the nature of the anther tapetum and the presence or absence of endosperm in the seed. Of these three, squamulae and endosperm attributes were earlier noted (Table 3) as being highly diagnostic of the primary division within the *Liliatae*. The three attributes most closely correlated with axis two were the presence or otherwise of vessels in the stems and

Table 3. The four attributes most diagnostic of the primary subdivision of the *Liliatae* into largely "Aquatic" and largely "Terrestrial" families

Attribute	Percentage Frequency	
	Aquatic	Terrestrial
Endosperm		
present	6	94
absent	94	8
Squamulae		
present	87	2
absent	13	100
Stomates		
present	44	100
absent	56	0
Embryo		
macropodous	64	0
otherwise	36	100

Table 4. The four attributes most diagnostic for separating the "Wind-pollinated" group of families from the "Animal-pollinated" groups

Attribute	Percentage Frequency	
	Animal	Wind
Ovary		
superior	42	100
subinferior	4	0
inferior	56	0
Endosperm		
starchy	20	87
otherwise	80	13
Subsidiary cells		
0	49	0
2	42	96
more than 2	12	4
Fruit		
achene, nut	4	25
berry	22	0
capsule	78	58
drupe	0	21

leaves and the shape of the embryo, basal or axile, none of which was diagnostic of any of the four groups originating from the classificatory study. The three attributes most closely correlated with the third axis were the presence or absence of perisperm, the type of leaf venation and the occurrence or otherwise of a leaf hypodermis, all three of which are diagnostic for distinguishing between "bird-pollinated" *Zingiberales* and the "insect-pollinated" *Liliales*.

Table 5. The five attributes that serve to separate the *Zingiberales* from the remaining families of animal pollinated families ("*Liliales*")

Attribute	Percentage Frequencies	
	Zingiberales	"*Liliales*"
Perisperm		
present	100	0
absent	0	100
Venation		
reticulate	0	22
convergent	0	78
pinnate	100	2
Leaf hypodermis		
present	100	14
absent	0	89
Subsidiary cells		
0	0	60
2	75	34
more than 2	38	5
Ovary		
superior	0	51
subinferior	0	5
inferior	100	46

General Discussion

The emergence from the analysis of four major groups is of interest in that two recent classifications, namely those of CRONQUIST (1968) and TAKHTAJAN (1969) recognize four subclasses in the *Liliatae* (Table 6). Though there is an overall similarity between the two classifications there are sufficient differences to justify comparing them with each other and with the computer-generated groupings of families.

In all three systems the water families (*Alismidae, Alismatidae*) stand out clearly from the remaining liliates and are widely recognized as being a distinctive group of families. The inclusion of the *Triuridaceae* in the subclass is problematical in that CRONQUIST includes them, but TAKHTAJAN does not, placing the family instead amongst the "lilies". The numerical classification has placed them in yet a further place amongst the group of largely wind-pollinated families.

Next to the water-families the wind-pollinated palms and their allies are probably the most consistently recognized high level liliate

Table 6. The classifications of TAKHTAJAN (1969) and CRONQUIST (1968) compared

TAKHTAJAN	CRONQUIST
Alismidae	**Alismatidae**
Alismanae	
Alismales	*Alismatales*
Hydrocharitales	*Hydrocharitales*
Najadales	*Najadales*
	Triuridales
Liliidae	**Liliidae**
Lilianae	
Triuridales	
Liliales	*Liliales* (inc. *Iridales*)
Iridales	
Zingiberales	
Orchidales	*Orchidales*
Commelinidae	**Commelinidae**
Juncanae	
Juncales	*Juncales*
Cyperales	*Cyperales* (inc. *Poales*)
Commelinanae	
Bromeliales	*Bromeliales*
Commelinales	*Commelinales*
Eriocaulales	*Eriocaulales*
Restionales	*Restionales*
Poales	
	Typhales
	Zingiberales
Arecidae	**Arecidae**
Arecanae	
Arecales	*Arecales*
Cyclanthales	*Cyclanthales*
Arales	*Arales*
Pandanales	*Pandanales*
Typhales	

assemblage. However, there is less certainty as to the orders to be included in the subclass (*Arecidae*). The *Typhales* are included by TAKHTAJAN and excluded by CRONQUIST. The numerical classification again supports TAKHTAJAN but disagrees with them both in that they also include the *Arales* in the same major grouping as the palms.

Although the *Arales* (*Lemnaceae*, *Araceae*) appear amongst the "lilies" in Fig. 1 a further ordination analysis of the data excluding the *Alismidae* and *Zingiberales* indicated that even amongst the "lilies" they stand out clearly from the remainder. Thus whilst the *Arales* are not to be associated closely with the palms they sit uneasily amongst the "lilies".

Whereas the grasses and the sedges are included in the *Commelinidae* by both TAKHTAJAN and CRONQUIST neither of these families is associated with the *Commelinaceae* in the numerical analysis. Instead they are associated with the palms and their allies. The reality of this association is supported by its recurrence for it also appeared in an earlier analysis (CLIFFORD 1967) based upon much less data. Indeed, the *Commelinidae* has been largely disrupted by the numerical analysis and the *Commelinaceae* have become associated with the lilies, a relationship long ago suggested by WARMING (1920).

Two differences of opinion between TAKHTAJAN and CRONQUIST with respect to the familial compositions of the *Commelinidae* and the *Liliidae* have been discussed. A third difference of opinion concerns their treatments of the *Zingiberales*. These TAKHTAJAN places with such confidence in the *Liliales* that he relates them to the family *Liliaceae*, subfamily *Asphodeloideae*. With equal confidence CRONQUIST places the gingers in the *Commelinidae* stating that in his opinion "The only respect in which the *Zingiberales* are more like the *Liliidae* than the *Commeliniidae* is that the vessels are confined to the roots instead of occurring in all vegetative organs".

Given such divergent opinions it would seem that different attributes are being regarded as more or less significant by each taxonomist. In terms of the numerical analyses which give no weight to any particular attribute the *Zingiberales* emerge as a single homogeneous group with no close links to any other group. The resolution of such disparate viewpoints depends upon the solution of two quite separate problems.

The first of these is whether evolutionary and morphogenetic theories can be devised to provide better guides to phylogenetic trends and the, second is, given a series of groups, which criteria are to be employed to determine their rank. At present means of distinguishing between parallel and convergent evolution appear to pose irresolvable difficulties for predicting phyletic relationships and so attention might be more profitably directed towards the determination of rank.

Earlier taxonomists such as ENGLER (1883) regarded the *Zingiberales* (*Scitamineae*) as a group equivalent in rank to the *Alismidae*. Such a rank is supported in particular by the principal co-ordinate analysis (Fig. 2) which, unlike the dendrogram, gives some indication of the separateness of taxa independently of the way in which they cluster. Ordination procedures are none-the-less not always a guide to rank. It may be argued that rank is itself unimportant, and that like other aspects of taxonomy "is a matter of judgement and not susceptible to precise and rigid decision" (HEYWOOD 1974: 280). However, this view would be supported neither by all taxonomists nor all ecologists, some of whom such as MAAREL (1972) have been using taxa of different ranks as a basis for the classification of vegetation.

The application of numerical methods to the determination of rank is bedevilled by a basic principle of taxonomy, namely that taxa are described in terms of their attributes. Consider the following common situation. For a series of individuals or taxa a 2-way table is drawn up in which the rows are taxa and the columns their attributes (or vice versa). Given such a table with entries in the cells it is customary to classify the taxa in terms of their attributes. It could equally be of interest to classify the attributes in terms of the taxa.

Indeed host-parasite relationships are often investigated both ways. Taxonomists concerned with the hosts often accept the parasites as indicators of relationships amongst the hosts. Likewise parasite taxonomists often look to the hosts for evidence of relationships amongst the parasites. In so doing each accepts the classification of the other and so the argument becomes somewhat circular.

To avoid this impasse methods are being sought for classifying two-way tables using both the row and column information (CLIFFORD 1975). To date little success has been achieved largely because most of the methods available demand quantitative scores. One basis for fixing such scores would be to define the states of attributes in terms of their degree of phylogenetic advancement or otherwise. Regrettably too few attributes lend themselves to being scored this way. Those interested in pursuing this matter from a classificatory viewpoint should consult MacNAUGHTON-SMITH (1965) and those interested in ordination approaches, HILL (1973).

The principal advantage of investigating rank from a combined consideration of both taxa and attributes is that it avoids undue emphasis being placed upon either a particular taxon or a particular attribute. It would also eliminate the necessity of working with dissimilarity measures for which the basis of dissimilarity was neglected once the measure had been calculated. To lose sight of the attributes at any stage is unfortunate for "any useful system of classification must concern itself both with the objects being classified and the characters used to differentiate the groups" (CRONQUIST 1968: 26).

Until joint consideration of attributes and taxa is possible by other methods resort must be had to ordination procedures such as principal co-ordinate analysis, the power of which has been illustrated with respect to the manner in which the *Zingiberales* were segregated from the liliate families in Fig. 2.

Finally it should be noted that the numerical approaches to the

classification of higher taxonomic categories differ in no ways from those
appropriate to lower categories. At all ranks it is essential that taxonomic procedures be made as explicit as possible. It is here that quantitative methods have a role to play in removing the mystique from taxonomy which should be seen not be an esoteric art practised by the few
(Kubitzki 1975) but as a cornerstone of biology practised by the majority.

The writer takes pleasure in acknowledging the receipt of much friendly
advice from Mr. David Hassall and Drs. M. B. Dale and V. H. Boughton
during the preparation of this paper.

References

Arber, A., 1925: Monocotyledons: a morphological study. Cambridge:
    Cambridge University Press.
Boyd, L., 1932: Monocotyledonous seedlings. Trans. and Proc. Bot. Soc.
    Edinburgh **31**, 5—223.
Candolle, A. P. de, 1813: Théorie élémentaire de la Botanique. Paris.
    In: The families of flowering plants (Hutchinson, J., 1960), Vol. I.
    Oxford: Clarendon Press.
Carlquist, S. J., 1975: Ecological Strategies of Xylem Evolution. Berkeley:
    Univ. California Press.
Clifford, H. T., 1975: Host-parasite relationships. In: Lecture Notes in
    Mathematics (Street, A. P., and Wallis, W. D., Eds.), 452. Combi-
    natorial Mathematics III. Berlin-Heidelberg-New York: Springer.
— 1976: The influence of attributes in the classification of the grasses
    (*Poaceae*). In: Pattern Analysis in Agricultural Science (Williams,
    W. T., Ed).. Melbourne and Amsterdam: C.S.I.R.O. and Elsevier
    Scientific Publishing Coy.
— and Watson, L., 1977: Identifying Grasses. Brisbane: Queensland
    University Press.
— Williams, W. T., and Lance, G. N., 1969: A further numerical con-
    tribution to the classification of the *Poaceae*. Aust. J. Bot. **17**, 119—311.
— Lavarack, P. S., 1974: The role of vegetative and reproductive at-
    tributes in the classification of the *Orchidaceae*. Biol. J. Linn. Soc. **6**,
    97—110.
— and Williams, W. T., 1973: Classificatory dendrograms and their
    interpretation. Aust. J. Bot. **21**, 151—162.
Cronquist, A., 1968: The evolution and classification of flowering plants.
    London and Edinburgh: Nelson.
Cutler, D. F., 1969: *Juncales*. In: Anatomy of the Monocotyledons. Vol. IV
    (Metcalfe, C. R., Ed.). Oxford: Clarendon Press.
Davis, G., 1966: Systematic Embryology of the Angiosperms. New York:
    John Wiley and Sons.
Deyl, M., 1955: The evolution of plants and the taxonomy of the mono-
    cotyledons. Acta Musei Nationalis, Pragae XI B No. 6, Botanica No. 3,
    1—143.
Diels, L., 1924: Aufgaben der Phytographie und der Systematik. In:
    Handbuch der biologischen Arbeitsmethoden. Abt. XI, Teil 1 (Ab-
    derhalden, E., Ed.). Berlin: Urban und Schwarzenberg.

DWYER, P. D., 1976: Systematics, Ecology and Biological Resources. Search **7**, 294—298.

EL-GAZZAR, A., and WATSON, L., 1970: Taxonomy of *Labiatae*. New Phyt. **69**, 451—486.

ENGLER, A., 1883: Syllabus der Pflanzenfamilien. Auf. 3. Berlin: Borntraeger.

GIBBS, R. D., 1974: Chemotaxonomy of Flowering Plants. 4 Vols. Montreal and London: McGill-Queen's University Press.

GILLY, C. L., 1952: Phylogenetic development of the inflorescence and generic relationships in the *Kobresiaceae*. Iowa J. Science **26**, 210—212.

GOWER, J. C., 1967: Multivariate analysis and multi-dimensional geometry. Statistician **17**, 13—28.

HEGNAUER, R., 1963: Chemotaxonomie der Pflanzen. Bd. 2. Basel und Stuttgart: Birkhäuser.

HEYWOOD, V. H., 1974: Chemosystematics—an artificial discipline. In: Chemistry in Botanical Classification, Nobel Symposium **25** (BENDZ, G., and SANTESSON, J., Eds.). New York and London: Academic Press.

HILL, M. O., 1973: Reciprocal averaging: an eigen-vector method for ordination. J. Ecol. **61**, 237—250.

HUTCHINSON, J., 1960: Families of Flowering Plants. 2 Vols. Oxford: Clarendon Press.

JACOBS, M., 1969: Large families—not alone! Taxon **18**, 253—262.

JUSSIEU, A. L. DE, 1789: Genera plantarum secundum ordines naturales disposita. Paris. In: Historia Naturales Classica Tome 35 (STAFLEU, E. A., Ed., 1964). Weinheim: J. Cramer.

KUBITZKI, K., 1975: Systematics and evolution of seed plants. In: Progress in Botany **35** (ELLENBERG, H., et al., Eds.). Berlin-Heidelberg-New York: Springer.

MAAREL, E. VAN DER, 1972: Ordination of plant communities on the basis of their plant genus family and order. In: Grundfragen und Methoden in der Pflanzensoziologie (MAAREL, E., VAN DER, and TÜXEN, R., Eds.). The Hague: Junk.

MACKAY, D. M., 1969: Recognition and action. In: Methodologies of Pattern Recognition (WATANABE, S., Ed.). London: Academic Press.

MACNAUGHTON-SMITH, P., 1965: Some statistical and other techniques for classifying individuals. Home Office Res. Unit Rep. No. 6. London: H.M.S.O.

MARTIN, A., 1946: The comparative internal morphology of seeds. Amer. Midl. Nat. **36**, 513—560.

TAKHTAJAN, A., 1969: Flowering plants, origin and dispersal. Washington: Smithsonian Institution Press.

TOMLINSON, P. B., 1969: *Commelinales-Zingiberales*. In: Anatomy of the Monocotyledons. Vol. III (METCALFE, C. R., Ed.). Oxford: Clarendon Press.

WARMING, E., 1920: A Handbook of Systematic Botany. London: Allan and Unwin Ltd.

WILLIS, J. C., 1973: A dictionary of the flowering plants and ferns, 8th ed. Revised by H. K. AIRY-SHAW. Cambridge: Cambridge University Press.

Address of the author: Dr. H. T. CLIFFORD, Botany Department, University of Queensland, St. Lucia, Brisbane 4067, Australia.

Plant Syst. Evol., Suppl. 1, 97—109 (1977)

Royal Botanic Garden, Edinburgh, U.K.

Classification Above the Genus, as Exemplified by *Gesneriaceae*, With Parallels From Other Groups

By

B. L. Burtt, Edinburgh

Abstract: *Gesneriaceae* is used as the starting point for a consideration of some aspects of classification between family and genus. *Gesneriaceae* may be split into two subfamilies, one with normal cotyledons, the other with one cotyledon becoming enlarged after germination: a parallel is drawn with the use of the plane of foliar distichy in *Zingiberaceae*. In both cases the resulting improvements are shown to be vital to the clear recognition and statement of evolutionary problems.

The status of *Klugieae*, one of the 4 tribes of *Gesneriaceae-Cyrtandroideae* is re-examined and its elevation to subfamily is discussed in relation to the system of the family. There is a rather general trend against simple dichotomy at subfamily level, but its retention in *Gesneriaceae* is recommended. The complexity of various classifications between family and genus is assessed by use of the ratio between number of genera and the rank immediately above (terminal suprageneric taxa) and the number of the latter that are monogeneric. When these ratios are too low a system looses much of its practical value.

The treatment of possible linking genera between major groups is discussed in relation to *Jerdonia* (*Scrophulariaceae/Gesneriaceae*) and *Triplostegia* (*Dipsacaceae/Valerianaceae*). The existence of such links does not justify the union of the families. In contrast, new evidence confirms the very close relationship of *Selaginaceae* to *Scrophulariaceae* tribe *Manuleeae*. This is not dependent on a small linking genus: the affinity is much closer than is that of *Manuleeae* to most other tribes of *Scrophulariaceae*, and the reduction of *Selaginaceae* to tribal rank must therefore be upheld.

In the last 140 years the classification of *Gesneriaceae* has come full circle. The family first took on its present form when G. Don (1838) united *Gesneriaceae* (New World genera having seeds with endosperm) and *Cyrtandraceae* (Old World genera with seeds lacking endosperm), keeping them as separate tribes with three and four subtribes respectively. Some 40 years later Bentham (1876) shifted the primary emphasis from endosperm to the position of the ovary, and used characters of fruit and disc for his subtribes. As a result, some tribes and subtribes for the first time contained a mixture of Old World and New World genera. Fritsch

(1894—1895) followed BENTHAM closely, but with a little more elaboration.

These were the two systems available for the family when I came to it in the 1930s, through the late Sir ARTHUR HILL's interest in the unequal cotyledons of the seedlings. Almost subconsciously, my mind registered that all the plants with interesting seedlings were from the Old World: the New World genera had equal cotyledons and were something different. Returning to the family in the 1950s, I automatically accepted the Old World genera as a taxonomic group. The justification emerged gradually: all the Old World plants examined became more or less anisocotylous after germination. Neither HILL, nor FRITSCH before him, had realized this. HILL was looking for major differences between the two cotyledons in size and persistence, such as that shown in *Streptocarpus*, *Chirita* or *Monophyllaea*. In some genera (*Boea* or *Cyrtandra* for example) the size difference is small and persistence slight. However, the tendency is pervasive throughout the Old World genera, except the Australasian *Coronanthereae*. In the New World anisocotyly is seen only in the few Central American species of the Old World genus *Rhynchoglossum*. The subfamilies *Cyrtandroideae* and *Gesnerioideae* therefore had to be remodelled so that these different patterns of growth were separated. This was in effect a reversion to the position of 1838, when the crucial character of seedling morphology was quite unknown. It should be emphasized that there are other, less constant, features that help to differentiate the two subfamilies, and also that the use of the seedling character is not a barrier to easy identification. An artificial key to the genera can easily be made without mention of cotyledons.

There was one immediate gain from this re-alignment. Several Old World genera had for long been hovering uncertainly on the borders of *Gesneriaceae*. They all had some anomalous features, though they mostly showed the unilocular ovary alleged to separate *Gesneriaceae* from *Scrophulariaceae*. For most of these genera, study of the seedlings has now been possible, but they are still unknown in the Formosan *Titanotrichum*. Five (*Brookea*, *Charadrophila*, *Cyrtandromoea*, *Jerdonia*, and *Rehmannia*) prove to have normal isocotylous seedlings, and they thus fail this test for inclusion in *Cyrtandroideae*. Now it is quite possible that we shall find some good species of *Cyrtandroideae* which have equal cotyledons: there are many hundreds not yet examined and a taxonomist never expects one single character to hold throughout a large group. But in the genera just mentioned other features make their inclusion in *Cyrtandroideae* doubtful: the equal cotyledons confirm these doubts, and these plants must be excluded. Of course, they might belong to the isocotylous subfamily *Gesnerioideae*, but, without going into details, they are all best accomodated in our present concept of *Scrophulariaceae*.

Thus the value of the anisocotylous character is confirmed. It is not of sporadic occurrence: it is a trend throughout one subfamily. Usually the large cotyledon shows only slight post-germinal growth, more rarely it attains the size of an ordinary foliage leaf, while in at least two groups, in the quite distinct tribes *Didymocarpeae* and *Klugieae*, it reaches maximum expression as the single, and sometimes very large, foliage leaf of the adult plant.

Something of a parallel in remodelling with the help of a vegetative character may be seen in *Zingiberaceae*. In this family a tall leaf-frond is often developed and on it the individual leaves are distichously arranged. The plane of distichy varies, however: it may be parallel or transverse to the direction of the rhizome (WEISSE 1932, 1933). In the old classification of K. SCHUMANN (1904), followed by LOESENER (1930), *Zingiber* itself is associated with *Alpinia* and allied genera: HOLTTUM (1950) maintained its greater resemblance to *Hedychium*, and transferred it to that tribe. The general change of affinity was undoubtedly correct, though separate tribal status for *Zingiber* can be justified (BURTT & OLATUNJI 1972). HOLT-TUM apparently knew nothing about the leaf character, but in fact in the *Hedychieae* and *Zingibereae* the plane of distichy of the leaves is parallel to the long axis of the rhizome, while in *Alpineae* it is transverse to it. In the old system both states of this character were to be found in a single tribe: only after HOLTTUM's revision is it possible to see that there is no switch from one pattern to another in closely related genera: the difference of pattern is in different tribes. For this we can put forward a hypothesis. *Alpineae* are essentially plants of the tropical non-seasonal rain-forest: they have leaf-fronds visible all the year round and their rhizomes are very fibrous and behave rather as organs of spread than as food reservoirs. *Hedychieae* and *Zingibereae*, in contrast, are more often plants of monsoon climates: they have a rest period between the dying down of the old fronds and growing up of the new, and their rhizomes are more fleshy (often with associated root-tubers) and less far-spreading. However in *Hedychieae* there are plants, such as some species of *Kaempferia* and *Boesenbergia*, which have no well-developed leaf-frond and no very marked horizontal rhizome. To these plants the criterion of plane of distichy seems inapplicable. One hypothesis would be that the frondose tropical *Alpineae* gave rise in seasonal climates to non-frondose plants, similar to these *Hedychieae*, and that these in turn gave rise to plants with a leaf-frond. There is no reason why the plane of distichy in the newly evolved frond system should be the same as that in the old. The hypothesis could be investigated further by careful studies in *Kaempferia* and allied genera. If it is disproved that will be an indication that the two groups became separated very early in the evolution of the family. Whatever the outcome, revision of the system was clearly vital to a clear statement of the problem.

These two examples show the value of attention to patterns of growth in defining groups above generic level. They also show that classification between family and genus must provide a system within which the evolutionary problems can be clearly recognized and formulated. Taxonomy at this level is of the utmost importance and this is one of its major functions. It is here, where speculation does not have to be quite so awe-inspiring, nor quite so unprovable, as in the construction of a system for the whole of the angiosperms, that we stand to gain most knowledge of evolution.

In *Gesneriaceae* I have suggested (Burtt 1970) that continued growth of one cotyledon was from the start an advantage, permitting increased photosynthesis before, or during, the organization of the plumular bud. The evolution of the unifoliate species of *Streptocarpus*, or the genus *Monophyllaea*, shows that this feature had considerable potential. The situation in *Zingiberaceae* is different. It may well be that neither plane of distichy has an advantage in itself: it is the formation of a leaf-frond that is advantageous: there happen to be two ways of doing it and both are found. Here the character is a morphological end-point: not one with potential for further elaboration.

The two subfamilies of *Gesneriaceae* have followed very different evolutionary paths, with some convergence but very little parallelism. Within *Cyrtandroideae* (except in *Klugieae*, to which I shall return) there is a marked tendency towards elongation of the ovary and fruit, associated with the restriction of the ovules to the tip of the lamelliform placenta. A variety of fruit forms has developed: some are straight with bilateral dehiscence, some straight with unilateral dehiscence, some twisted in development and dehiscing when the spiral sutures gape on drying; some are indehiscent and even these may be considerably elongated. None of these long-fruited forms is paralleled in *Gesnerioideae*, where fruit evolution has run from a short capsule to a berry and to an inferior ovary.

Corolla-form remains fairly plastic throughout the family. As an adaptation to bird-pollination a red arcuate corolla, associated with similar patterns of pollen-presentation, is developed in *Aeschynanthus* (*Cyrtandroideae-Trichosporeae*), in *Columnea* (*Gesnerioideae-Episcieae*) and *Rechsteineria* (*Gesnerioideae-Gloxinieae*). But the tribes to which these belong are firmly committed to different forms of ovary. This decides the method of ovular protection, whose importance was emphasized by Grant (1950). In *Aeschynanthus* the elongate ovary becomes even more slender and elongate, offering no catch-point to a probing bill. In *Columnea* the roundish ovary is protected by a constriction near the base of the corolla-tube; this acts as a capillary, presenting nectar to the bird well above the level of the ovary. In *Rechsteineria* the ovary is inferior. Thus the pattern of ovular protection depends on the starting point, the evolutionary springboard, from which it developed.

Within *Gesneriaceae-Cyrtandroideae* the tribe *Klugieae* (incl. *Loxonieae*) provides a good opportunity to examine the criteria of tribal status. It is a small tribe of 6 well-marked genera, but there is a remarkable range of morphological character. It stands apart from the rest of the subfamily in the short ovoid ovary which is sharply contracted into the style. The ovary may be completely unilocular with Y-shaped parietal placentae similar to those of *Gesnerioideae* (rather than T-shaped as in most *Cyrtandroideae*), or it may be bilocular with an axile placenta. The inflorescence may be an open pair-flowered cyme, or condensed to a helicoid pseudo-raceme. Fertile stamens may be four or two only, in which case either the anterior pair (in *Rhynchoglossum*) or the posterior pair (in *Epithema*) may persist. Other features are equally varied and *Klugieae* has a wider geographical range than any other tribe: from West Africa to the Solomon Islands, recurring in Central America. It has all the appearance of a relict tribe comprising the surviving members of a numerous and highly diversified group. This tribe is under intensive study by WEBER (1975, 1976).

Meanwhile work on the other tribes, *Cyrtandreae*, *Didymocarpeae*, and *Trichosporeae* has made it more and more evident that they are very closely interrelated. *Boeica* (*Didymocarpeae*) has remarkable similarities to *Rhynchotechum* (*Cyrtandreae*): there is the same short-tubed subregular corolla with four stamens, and the same striking range of indumentum from soft appressed silky hairs to harsh spreading bristles. The linear capsule of *Boeica* contrasts with the round berry of *Rhynchotechum* to give the tribal difference. The link between *Didymocarpeae* (seeds without tails) and *Trichosporeae* (seeds with tails) is surely represented by *Loxostigma* (BURTT 1975). It has one terrestrial species with tail-less seeds and some epiphytic ones with tailed seeds.

Thus a better system than one with four equivalent tribes might mark off *Klugieae* more sharply from the rest. There are two ways of achieving this: either *Klugieae* can be raised to subfamily, or the other three tribes united into one tribe with three subtribes. How are we to decide? And what are the consequences of either step?

The system for *Gesneriaceae* with two subfamilies, *Gesnerioideae* and *Cyrtandroideae*, has the advantage of representing what may have been the fundamental cleavage in the family, caused by the development of seedling anisocotyly. *Klugieae*, though sharing some characters with the *Gesnerioideae*, are strongly anisocotylous and must therefore be regarded as an offshoot of the cyrtandroid stock. Thus to rank it as a subfamily equivalent to the other two might be considered unsatisfactory and it would break the unity of the anisocotylous group. On the other hand, there is in other families an increasing tendency amongst taxonomists to avoid a simple dichotomy at subfamily level. This, of course, becomes particularly obvious in a system such as THORNE's (1968), where a broad family concept results in numerous subfamilies—no less than 21 in *Liliaceae*. Even without that extreme example, there is a notable turning away from simple dichotomy; it has been abandoned, for example in *Commelinaceae* (Brenan 1966), *Meliaceae* (Pennington & Styles 1975; see also Leroy 1976), *Proteaceae* (Johnson & Briggs 1975). There is thus a widespread readiness to compress the evolutionary explosion, that must have been spread over a long period, into that blank space on the paper between family and subfamily headings. That may be justified when we have no clue as to how it took place. In *Gesneriaceae* the seedling morphology supplies such a clue.

The alternative treatment of *Klugieae* is to rank it as a tribe co-ordinate with a tribe *Cyrtandreae*, which would include the narrower *Cyrtandreae*, *Didymocarpeae*, and *Trichosporeae* as subtribes. In 1962 I abandoned subtribes under tribe *Didymocarpeae* (the only tribe large enough to justify their use) because the existing ones were artificial; but there are some 50 genera in the tribe, so that such groupings might well be desirable, when we know how to make them. We lose that opportunity if *Didymocarpeae* is itself reduced to subtribal rank.

To reconstruct the evolution of *Cyrtandroideae* in the simplest way that fits the available evidence, we need to imagine four major branchings. First, that giving rise to the two subfamilies, *Gesnerioideae* and *Cyrtandroideae*, from the basic family stock; second, a branch off *Cyrtandroideae* giving *Klugieae*; third, a branch giving *Didymocarpeae*; fourth, a branch off *Didymocarpeae* giving rise to *Trichosporeae*. Even if subtribes are not reserved for lower groupings, there are only three ranks available for four levels of branching. Thus, as is well known, even at this very simple level a system of classification cannot represent the evolutionary situation.

Perhaps the best course at present is to make no change at all: to retain two subfamilies and four co-ordinate tribes in *Cyrtandroideae*. If desirable for a special discussion *Klugieae* could be designated group A; the other three tribes group B. I do not object to such informal groupings, but I am absolutely opposed to the addition of any new ranks which would carry formal names. Subfamily, tribe and subtribe are quite enough.

To return to the possible subdivision of *Didymocarpeae* and my abandonment of subtribes there. Fritsch had a tribe *Streptocarpeae*, defined solely by the possession of a twisted fruit and this was retained as a unit at subtribal rank by Ivanina (1967) in her carpological study of the family. Such a fruit has probably developed at least 4 times in *Didymocarpeae*. I recently found a letter, on a sheet belonging to the Calcutta herbarium, from C. B. Clarke, the monographer of *Cyrtandreae*, to Sir George King: it is dated 1882. Clarke wrote: "in *Cyrtandraceae* we make rather too much of twisted capsule-valves: I have a section of *Didymocarpus* which according to nature should be in *Boea*, but the capsule valves are perfectly straight". Clarke was quite right. Now, with more species to study, and with a far greater knowledge of other characters, I have been able to remodel some of these genera "according to nature", and three end up with either straight or twisted fruits: the character is not a generic one. This, of course, is the death-knell to its independent use in defining any higher category.

The decision to abandon subtribes in *Didymocarpeae* was not taken because no natural groupings could be made, but because of the number of genera of uncertain affinity that were left. Most of the genera could be grouped into about six subtribes, but there would be about 8 isolated genera which could only form monogeneric subtribes. This I would regard as unsatisfactory, for a system becomes useless when monotypic groups outnumber the remainder.

How far then can classification between order and genus usefully go? We need a measure of the subdivision of a group relative to the number of genera it contains. Here I introduce the term "terminal suprageneric taxon". It is self-explanatory; it refers to the lowest named group above the generic level, of whatever rank it may be. The ratio of genera to terminal suprageneric taxa, and the number of the latter that are monogeneric, sum up a system of classification in a very practical way.

For example, for the tribe *Campanulaceae-Lobelieae*, WIMMER (1957) gives us a system of 20 genera grouped in 14 terminal suprageneric taxa, 11 of them monogeneric. BRENAN (1966) has re-classified the 45 genera of *Commelinaceae* into 15 equivalent groups, 6 of them monogeneric. In *Proteaceae* JOHNSON & BRIGGS (1975) have 75 genera in 33 terminal suprageneric taxa and 14 of these are monogeneric. This work on *Proteaceae* is clearly of the highest quality, that is why it is an important example. In *Gesneriaceae* BENTHAM had 71 genera in 9 terminal taxa; FRITSCH had 84 in 23, 5 monogeneric. New World genera are at present being re-studied by workers in America and I am uncertain about groupings, but I estimate for the whole family 130 genera in 11 tribes: for *Cyrtandroideae* 70 in 4: none monogeneric.

Gesneriaceae is a highly advanced, and presumably recent, family; *Proteaceae* is more primitive and doubtless much older. They are very different kinds of family. Can they be usefully compared in this way?

There are, of course, no universal definitions of subfamily, tribe or subtribe: nevertheless we can strive to make these ranks of equivalent value in different groups. The assessment of the resulting classification in terms of the ratio of genera to terminal suprageneric taxa and the number of these that were monogeneric would then give a measure of comparison of the evolutionary state of two families. There would be, however, no guarantee that the application of equivalent concepts in, say, *Magnoliaceae*, *Orchidaceae*, and *Compositae* would give useful or workable classifications in all those families. If the purpose of classification within a family is to subdivide it into smaller groups that are useful in the process of identification, useful in defining areas for more detailed study, and useful as a whole in summarizing interrelations of the genera, then the standards for the suprageneric groupings must come largely from within.

The use of equal standards (if that were possible) throughout the flowering plants, would result in a special classification useful for one particular comparison: it could serve that purpose without any nomenclature being applied to the taxa between family and genus: one simply needs totals at each level. For the more general needs of taxonomy served by a classification of named categories, the standards most appropriate to the internal structure of the family must be used.

A system between family and genus goes into far too much detail if it reduces the number of names by only a third from the generic level to the next one above it. It is quite wrong to think that all the recognizable degrees of difference must be expressed in the classification. Taxonomy has been called the science of diversity. Taxonomists must think in terms of diversity, and not try to reduce every group to maximum uniformity. It may be necessary to make investigations in that way; but just as we have to dissect the individual parts yet think in terms of the whole organism, so we must analyse into components of maximum similarity but then synthesize into groups of related diversity.

There is one further point about the isolated genera that come to form monogeneric groups. Is there any real reason why these groups should have formal latin names? Would it not be equally satisfactory to use the generic name and simply preface it by the rank at which it was being considered? For example, the recent reclassification of *Meliaceae* (PENNINGTON & STYLES 1975) would read at subfamily level: *Melioideae*, subfam. *Quivisianthe*, subfam. *Capuronianthus*, and *Swietenioideae*. This device would avoid the creation of additional latin names, avoid the farce of trying to decide which characters of a single genus should diagnose a subfamily, and give the immediate information that that subfamily contains only one genus. It would be a device to reduce nomenclatural innovations, and nomenclatural argument about rank [for instance LEROY (1976) regards *Capuronianthus* as a tribe, not a subfamily]. It would not make monogeneric groups more acceptable.

A relatively small number of genera are important in a system of classification not merely because they raise the problem of a monogeneric group, but because they are more or less intermediate between two larger groups.

Jerdonia is a monotypic genus of the hills of S. W. India. WIGHT (1848) placed it only tentatively in *Gesneriaceae*, saying he did not know what else to do with it—but no one since has echoed his doubts. Yet it is clearly out of place in *Gesneriaceae*. It does have a unilocular ovary, but there are 4 parietal placentae, quite unlike the bifid lamellate placentae of *Gesneriaceae*. Furthermore the seed is too large and the endosperm is well-developed and alveolate. It has the characteristic pair-flowered cyme, but this is found in several genera of *Scrophulariaceae*: *Calceolaria*, *Tetranema*, and *Penstemon* at least. *Jerdonia* has an isocotylous seedling. It is one thing to say that *Jerdonia* is better placed in *Scrophulariaceae*; another to decide on its affinity there.

The two subfamilies of *Gesneriaceae* must have had a common ancestral group. It is, as usual, unknown. The character-states that could have given rise to the different expressions now found in these two subfamilies are:

Inflorescence a pair-flowered cyme
Corolla zygomorphic
Aestivation ascending imbricate
Fertile stamens four
Ovary superior, unilocular
Seed with endosperm (alveolate?)
Seedling isocotylous
Stomata anisocytic (usually anomocytic in *Scrophulariaceae*)
Iridoid glycosides absent (often present in *Scrophulariaceae*)

All those features are found in *Jerdonia*. Of course *Jerdonia* has some special features of its own. Those four separate and very simple placentae are not like those of either family; it has pollen in tetrads; the haploid chromosome number is 14, whereas 8 seems probably basic in *Gesneriaceae*. Nevertheless the agreement is remarkable. I think we are justified in regarding *Jerdonia* as the genus of *Scrophulariaceae* which is closest to a missing group from which *Gesneriaceae* may have arisen. There is no single clear-cut distinction, but the two families are maintainable, in fact must be maintained, as distinct entities. Unilocular ovaries occur in *Scrophulariaceae* (quite apart from *Jerdonia*) and a typical bilocular ovary with axile placenta is found in *Gesneriaceae* in *Monophyllaea*, a genus whose enormously enlarged cotyledon makes it an unmistakable member of *Cyrtandroideae*. Speculation that *Jerdonia* is close to the linking group gives no justification for merging the families.

One of the best examples of a small genus that links two families is *Triplostegia*. This has the open cymose inflorescence and smell of *Valerianaceae*; but whereas *Valerianaceae* has an ovary with one fertile cell and two sterile ones, *Triplostegia* has a simple unilocular ovary as found in *Dipsacaceae*. Furthermore, in *Triplostegia* there is an involucel surrounding each flower as in *Dipsacaceae*. This is not just a verbal agreement due to applying the same term to rather different structures. The involucel of *Triplostegia* is just like that of *Scabiosa* (the involucel of *Morina*, which is usually included in *Dipsacaceae*, is rather different in detail). *Valerianaceae* was not separated from *Dipsacaceae* until 1802. Perhaps if *Triplostegia* had then been known the family distinction would never have been made. Now that it has been long accepted, what is the best treatment of *Triplostegia*?

First, I absolutely reject the idea of making it a separate family. Secondly, as a working taxonomist who is constantly called upon for information, references etc., I am very conscious of what I call the historical constraints on any major changes in classification. Were I compiling a completely new system, knowledge of *Triplostegia* might well cause me to unite *Valerianaceae* with *Dipsacaceae*: but that would now be the wrong way to accommodate a genus of two species. The importance of *Triplostegia* is that technically it belongs to *Dipsacaceae*, but there is a strong resemblance to *Valerianaceae*. In *Dipsacaceae* it looks superficially out of place: it is provocative, provides an obvious link to *Valerianaceae*, and can be identified by the aid of numerous existing family keys. Placed in *Valerianaceae* it attracts little attention, for it must be dissected for its special characters to be seen; but all keys to families would have to be altered to accommodate it. I have no hesitation in referring *Triplostegia* to *Dipsacaceae*. As already mentioned it has the typical involucel (and it may be added calyx) of *Scabiosa*. It does not require more than subtribal rank in tribe *Dipsaceae*.

Triplostegia is important from another point of view. It has been suggested that the flower-head of *Dipsacaceae* is a compound structure: a head of heads. This idea derives from the interpretation of the involucel as the involucre of a primary capitulum now reduced to a single flower. *Triplostegia* has a true involucel, but the flowers are solitary in an open cyme. Condensation of this inflorescence would lead to the cymose capitulum of *Dipsacaceae*. There is no question of an inflorescence of capitula being reduced to a secondary head.

The relationship of *Scrophulariaceae* and *Selaginaceae* merits brief attention. The family rank of *Selaginaceae* was upheld by BENTHAM & HOOKER (1876), but it was reduced to a tribe of *Scrophulariaceae* by WETTSTEIN (1891). Subsequently this position has been maintained by most continental authors, but HUTCHINSON (1926) and following him DYER (1975) have again given it family rank.

WETTSTEIN placed his tribe *Selagineae* next to the tribe *Manuleeae*, with which it shares several features including synthecous anthers, and the frequent adnation of bract to pedicel. JUNELL (1961) carried out a critical comparative study of ovary structure, and made an interesting discovery. While *Selaginaceae* has been diagnosed as having a single ovule in a loculus, there are a few species of *Selago* that have two ovules in each: and these ovules are placed medianly, one turned upwards one downwards: neither occupy the apical position of the single ovule of typical *Selago*. For these plants JUNELL made the new genus *Tetraselago*.

Among examples of *Scrophulariaceae* tribe *Manuleeae*, JUNELL examined the monotypic genus *Glumicalyx*. This was important in his final argument, for he found it had fewer ovules than the other genera of *Manuleeae* and that the upper ones were turned upwards, the lower downwards. This seemed to him a step in the direction of *Tetraselago*.

We can now go further (HILLIARD & BURTT 1977). *Glumicalyx* is a genus of six species: three of these are to be added by transfer from the neighbouring genus *Zaluzianskya*, one is new, but the last is to be transferred from *Walafrida*, a genus of *Selaginaceae*, to which it bears no little resemblance but differs sharply in having about 12 ovules in each loculus. In *Glumicalyx* there is thus a range from about 80 down to 12 ovules in each loculus. The upper end of the scale retains contact with the other genera of *Manuleeae*, while the lower looks towards the condition in *Tetraselago*.

While *Glumicalyx* shows an approach to *Tetraselago*, another possible link has been discovered. This is a plant which has to be described as a monotypic new genus: its interesting feature is that each ovary loculus contains up to 6 ovules which are all located on the upper part of the axile placenta, and only the uppermost 1–3 develop into seeds: thus further sterilization of ovules in this ovary might lead to the single apical ovule of *Selago*.

The evidence shows a close relationship between *Manuleeae* and *Selaginaceae*. Without attempting a detailed study it can be said with some confidence that the difference between *Manuleeae* and *Selaginaceae* is no greater than that between *Manuleeae* and its neighbouring tribes in *Scrophulariaceae*, and is much less than that between, say, *Manuleeae* and *Rhinantheae*. Therefore *Selaginaceae* cannot be retained as a separate family. The tribal rank accorded it by WETTSTEIN is the highest that can be justified. The situation here contrasts with the comparisons made between *Scrophulariaceae* and *Gesneriaceae* or between *Dipsacaceae* and *Valerianaceae*. There a single small genus provided a possible link between two well-separated groups. Here the importance of *Glumicalyx* and *Tetraselago* is in showing that a small gap between the two groups in one single character is indeed very small. It is the general close affinity between *Selaginaceae* and *Manuleeae* that justifies ranking them as collateral tribes.

When preparing this paper I jotted down a list of ways in which it might be brought to a conclusion. There was an explanation of the analytic rather than synthetic approach (although analysis and synthesis are but the left and right hand of the taxonomist, whose mind works with both); there was a comparison of the position at family level and below with that among orders and superorders, and of the role of informal systems at both levels. But when these, and other ideas, were committed to paper, each proved to be not an ending but the beginning of something new.

The best conclusion is to reiterate the fundamental importance of sound classification (based on whole plants not on selected characters) within each and every family. It is by working to get the best possible system between family and genus, or in groups of families known to be closely related, that we shall learn about the patterns of evolutionary change. Without a better knowledge at this level we can scarcely hope for a satisfactory superstructure.

I am grateful to Dr. P. KOOIMAN, Delft, for testing *Jerdonia* for iridoid glycosides.

References

BENTHAM, G., and HOOKER, J. D., 1876: Genera Plantarum, **2**. London: Reeve, and Williams & Norgate.

BRENAN, J. P. M., 1966: The classification of *Commelinaceae*. Journ. Linn. Soc. Bot. **59**, 349—370.

BURTT, B. L., 1963: Studies in the *Gesneriaceae* of the Old World: XXIV: Tentative keys to the tribes and genera. Notes Roy. Bot. Gard. Edinb. **24**, 205—220.

— 1970: op. cit.: XXXI: Some aspects of functional evolution. Notes Roy. Bot. Gard. Edinb. **30**, 1—10.

— 1975: op. cit.: XL: The genus *Loxostigma*. Notes Roy. Bot. Gard. Edinb. **34**, 101—105.

— and OLATUNJI, O. A., 1972: The limits of the tribe *Zingibereae*. Notes Roy. Bot. Gard. Edinb. **31**, 167—169.

DON, G., 1838: A general history of dichlamydeous plants, **4**, 643—665. London: Rivington et al.

DYER, R. A., 1975: Genera of South African flowering plants **1**, 558. Pretoria: Dept. of Tech. Agric. Services.

FRITSCH, K., 1894—1895: *Gesneriaceae*. In: Die natürlichen Pflanzenfamilien (ENGLER und PRANTL), IV, **3 b**, 133—185. Leipzig: Engelmann.

GRANT, V., 1950: The protection of the ovules in flowering plants. Evolution **4**, 179—201.

HILLIARD, O. M., and BURTT, B. L., 1977: Notes on some plants of southern Africa, chiefly from Natal: VI. Notes Roy. Bot. Gard. Edinb. **35**, 155—177.

HOLLTUM, R. E., 1950: The *Zingiberaceae* of the Malay Peninsula. Gard. Bull. Singapore **13**, 1—249.

HUTCHINSON, J., 1926: The families of flowering plants **1**, 308. London: Macmillan.

IVANINA, L. I., 1967: The family *Gesneriaceae* (the carpological review). Leningrad: The Academy of Sciences of the USSR (In Russian).

JOHNSON, L. A. S., and BRIGGS, B., 1975: On the *Proteaceae*—the evolution and classification of a southern family. Bot. Journ. Linn. Soc. **70**, 83—182.

JUNELL, S., 1961: Ovarian morphology and taxonomical position. of *Selagineae*. Svensk Bot. Tidskr. **55**, 168—192.

LEROY, J. F., 1976: Essais de taxonomie syncrétique: I: Etude sur les *Meliaceae* de Madagascar. Adansonia sér. 2, **16**, 167—203.

LOESENER, T., 1930: *Zingiberaceae*. In: Die natürlichen Pflanzenfamilien (ENGLER und PRANTL), 2. Aufl., **15 A**, 541—640. Leipzig: Engelmann.

PENNINGTON, T. D., and STYLES, B. T., 1975: A generic monograph of the *Meliaceae*. Blumea **22**, 419—540.

SCHUMANN, K., 1904: *Zingiberaceae*. In: Das Pflanzenreich (ENGLER), Heft **20**. Leipzig: Engelmann.

THORNE, R. F., 1968: Synopsis of a putatively phylogenetic classification of the flowering plants. Aliso **6**, 57—66.

WEBER, A., 1975: Beiträge zur Morphologie und Systematik der *Klugieae* und *Loxonieae* (*Gesneriaceae*): I: Die Spross- und Inflorescenzorganisation von *Monophyllaea* R. BR. Bot. Jahrb. **95**, 174—207.

— 1976: Op. cit. II: Morphologie, Anatomie und Ontogenese der Blüte von *Monophyllaea* R. BR. Bot. Jahrb. **95**, 435—454.

WEISSE, A., 1932: Zur Kenntnis der Blattstellungsverhältnisse bei den Zingiberaceen. Ber. dtsch. bot. Ges. **50 A**, 327—366.

— 1933: Die Art der Distichie an den Achsensprossen von *Zingiber*. Ber. dtsch. bot. Ges. **51**, 13—20.

WETTSTEIN, R. VON, 1891—1893: *Scrophulariaceae*. In: Die natürlichen Pflanzenfamilien (ENGLER und PRANTL), **IV, 3 b**, 39—107. Leipzig: Engelmann.

WIGHT, R., 1848: Icones plantarum Indiae orientalis **4 (2)**, 10, t. 1352. Madras: The author.

WIMMER, F., 1957: *Campanulaceae-Lobelioideae*. In: Das Pflanzenreich (ENGLER), Heft **107**. Berlin: Akademie Verlag.

Address of the author: B. L. BURTT, Royal Botanic Garden, Inverleith Row, Edinburgh EH3 5LR, Scotland, U.K.

Plant Syst. Evol., Suppl. 1, 111—121 (1977)
© by Springer-Verlag 1977

Institut für Holzbiologie und Holzschutz
der Bundesforschungsanstalt für Forst- und Holzwirtschaft,
Hamburg, Federal Republic of Germany

The Anatomy of Secondary Xylem and the Classification of Ancient Dicotyledons

By

H. Gottwald, Hamburg

Abstract: The stem wood of about 700 species belonging to 32 families of the order *Magnoliales* s.l., plus further taxa exhibiting primitive wood anatomical features, was investigated. On this basis, six structural groups can be established each of which show a marked gradation from primitive to advanced stages. Wood structure of *Magnoliales* sensu TAKHTAJAN is only partially primitive, partially moderately derived, while the most primitively structured heteroxylous taxa belong to the "Dillenial-Hamamelidal" and "Theal" group, respectively. Accordingly, there is no compelling evidence to support phylogenetic schemes in which the *Magnoliales* is placed as the only common base for all recent dicotyledons. For this reason, and since there is no structural mechanism which justifies the derivation of the taxa, belonging to the "Dillenial-Hamamelidal" and "Theal" groups which are very primitive in their wood anatomy, from the types embracing *Magnoliales* sensu TAKHTAJAN, a new phylogenetic model assuming an early separation of angiosperms into at least two branches is discussed. Furthermore, possible affinities between primitive and advanced taxa within the different structural groups are shown.

Introduction

The ancient dicotyledons have a special attraction for wood anatomists as, apart from the extraordinary position of this group, most of these taxa are woody plants with a well developed secondary xylem and, furthermore, all the vesselless taxa belong to them. Therefore, these taxa, often regarded as ancestral, offer a suitable structural comparability for tracing the limits and the position of the order *Magnoliales* by means of wood anatomical features.

The secondary xylem of this ancient group was already investigated by McLAUGHLIN (1933) and by LEMESLE (1953) but with very limited material and with other purposes in mind. The present investigation is based only on material of the secondary xylem of adult stem-wood samples belonging to the wood collection of the Bundesforschungsanstalt für Forst- und Holzwirtschaft, which contains almost 45 000 microscope slides covering about 16 000 samples.

Materials and Methods

First, as ancient dicotyledons must be understood all the families which comprise the order *Magnoliales* sensu TAKHTAJAN (1973). Second, all additional taxa which are considered by MELCHIOR (1964) as belonging to *Magnoliales* will be discussed. Third, this group of magnolian allies will be compared with 10 more families which likewise may be called ancient due to their primitively structured secondary xylem. All in all the wood of more than 700 species belonging to 165 genera of 32 families was investigated.

To demonstrate the anatomical details, a two-dimensional model is used in which the six vertical columns represent the various structural principles of the families investigated. These six structural principles, or "structural groups" are classified primarily according to the distribution of the axial parenchyma as for instance as apotracheal diffuse or paratracheal. Further features used are the various distributions of the vessels and specific features of high systematic value such as tannin-tubes in the rays. The headings chosen for these six structural groups e.g. "Theal" or "Laural" are based on the most typical family of each group. These structural principles are only supposed to indicate the plan of architecture or "topographic pattern" of the various taxa and insofar have no direct link with the evolution of anatomical features (Figs. 1a, b). The second dimension (horizontal) is an attempt to show the structural evolution of the various families which in turn is classified into four grades subdividing all structural groups, beginning with one understorey for the vesselless taxa (A). The following three evolutionary grades (B, C, D) are based on the evolution of the vessels from a purely multiple-perforated and scalariform-pitted type ascending to a simple-perforated and alternate-pitted vessel; further, from a pronounced heterogeneous ray developing to a homogeneous one and from a typical fibre tracheid with large bordered pits to a small and scantily pitted cell of a more fibre-like type (Fig. 1a). In making the classification all microscopic features were taken into account with exception of the longitudinal cell dimensions. Such measurements were neglected since reasonable results can be obtained only with large numbers of samples from many trees of the same species and these were not available.

Wood Anatomical Descriptions

Magnoliales sensu TAKHTAJAN. Of the family *Magnoliaceae* 105 species representing all twelve genera, according to the last revision of DANDY (cf. PRAGLOWSKI & DANDY 1974), were investigated including the lesser known genera *Alcimandra*, *Pachylarnax*, and *Tsoongiodendron*, discovered only in 1963.

The investigation clearly demonstrates that all twelve genera are composed, without exception, according to the same principle of structure which is characterized by radially grouped vessels and terminal bands of axial parenchyma. This plan of architecture is sufficiently stable and specific to make it a very simple task to identify any taxa of *Magnoliaceae* as belonging to this family by means of wood anatomical features. This structural type establishes one of the six structural principles denominated "Magnolial". In contrast with this homogeneity of structure, the *Magnoliaceae* embrace a wide range of structural evolution marked by a pronounced progression within the family. This evolutionary drive is mainly visible in the vessels. Here the scalariform perforations change to simple ones and the scalariform pits develop to an opposite arrangement. Similar dynamics to the one linked to the alterations in the vessels can be observed in the rays. Accordingly, the wood structure of the genus *Talauma* is the most primitive and that of the genus *Liriodendron* together with the subgenus *Yulania* of *Magnolia* are the most advanced in the family *Magnoliaceae* due to the non-scalariform pitting of the vessels and the homogeneous rays.

This implies that the wood structure of *Magnoliaceae* is fairly primitive to fairly advanced with a pronounced mosaic-like progression reaching over three evolutionary steps (B, C, D).

The family of *Winteraceae*, the sole homoxylous taxon in *Magnoliales* sensu TAKHTAJAN will be considered, for anatomical reasons, together with the vesselless taxa *Amborellaceae*, *Tetracentraceae*, *Trochodendraceae*, and *Sarcandra* of *Chloranthaceae*. On account of their vesselless state the five taxa must be placed in the lowest of the four steps of evolution. As obvious as this decision is, as uncertain must remain their assignment to any of the six structural groups. Only the genus *Sarcandra* can be correlated fairly well with the "Laural" principle. For the other four homoxylous taxa agreement with the "Magnolial" principle is more probable according to the distribution of the banded axial parenchyma, than with the "Dillenial-Hamamelidal" or "Theal" principles. Furthermore it must be mentioned that *Tetracentron* has vestured vessel pits, a common feature of only the *Lauraceae* and *Myristicaceae* within *Laurales* sensu TAKHTAJAN.

Any possible relationship between these four homoxylous families can only be hypothetical. There are weak indications for an affinity between *Trochodendraceae* and *Tetracentraceae* which share the occurrence of scalariform pitting in tracheids, thus representing the most primitive state of all dicotyledons. On the other hand the families of *Winteraceae* and *Amborellaceae* form a loose structural group with a few more advanced features. Both these statements, however, must remain very vague as the structural heterogeneity within *Winteraceae* does not allow its allocation to a fixed position.

From the pantropical family of *Annonaceae* the secondary xylem of 40 genera was examined. This family forms a very homogeneous structural group in spite of the large number of genera and species. The family is to be placed in a specific "Annonal" structural principle near to the "Magnolial" principle. The progression within the family is only slight in terms of the variation of wood anatomical features, e.g. the pit-diameter decreases from 10 μm in the genus *Uvaria* to the extremely low diameter of less than 2 μm in *Polyceratocarpus*.

The evolutionary position in respect of the anatomical features of all genera of *Annonaceae* is more advanced than that of the *Magnoliaceae*: the vessels always show simple perforations, the vessel-pitting is alternate, the rays are mostly homogeneous and the fibres have only minute pits. In other words: in the *Annonaceae* the progression of the various cell types is well balanced in contrast to those of the *Magnoliaceae* which warrants its placement in the evolutionary grade D.

Of the small and mostly monotypic families the *Himantandraceae* also possesses decidedly advanced features and is to be placed on a level between the *Annonaceae* and the *Magnoliaceae*. As regards the structural principle the *Himantandraceae* tends more toward the "Magnolial" than to the "Annonal" group, as had been suggested by McLAUGHLIN (1933).

The families *Canellaceae*, *Degeneriaceae*, and *Eupomatiaceae* have lesser pronounced structural principles. The *Degeneriaceae* still show a relationship to the "Magnolial" principle which is also indicated by BARANOVA (1972) with regard to the leaf epidermis, and has an evolutionary state in the middle of *Magnoliaceae* in grade C.

The family of *Eupomatiaceae* could be arranged within the "Laural" as well as the "Dillenial-Hamamelidal" principle in a fairly advanced position between grade C and D.

In the family *Myristicaceae* representatives of all 15 genera were investigated. The secondary xylem exhibits a pronounced structural uniformity and requires a separate structural principle between the "Magnolial" and "Laural" groups which is in full accordance with GARRAT (1933). The progression in *Myristicaceae* is indicated by multiple vessel perforations in *Knema* developing to constantly simple ones in *Brochoneura*; the evolutionary state of *Myristicaceae* can be compared with the more advanced part of *Magnoliaceae*. A remarkable feature is the constant occurrence of silica bodies in some genera which in the *Magnoliales* s. l. occur only in the genus *Talauma* of *Magnoliaceae* and in some genera of *Lauraceae*.

One can draw the conclusion that the seven heteroxylous families of *Magnoliales* sensu TAKHTAJAN are spread out over four structural principles of a more or less pronounced structural relationship. In contrast, the four homoxylous families cannot be allocated reliably to any structural principle unlike the homoxylous genus *Sarcandra*. Furthermore, the genus *Talauma* shows the lowest structural evolutionary level of any vessel-bearing taxon of *Magnoliaceae* sensu TAKHTAJAN without, however, reaching the lowest possible position for heteroxylous taxa. Moreover, *Magnoliaceae* is the family with the strongest intra-familial progression and the most pronounced structural mosaic-evolution. Finally it may be said that all features of the secondary xylem correlate completely with the taxonomic concepts of DANDY (1974) and TAKHTAJAN (1973).

Magnoliales sensu MELCHIOR. Of the additional families of *Magnoliales* as circumscribed in "ENGLER's Syllabus", the *Lauraceae* will be considered first: its special anatomical features make it necessary to form a separate "Laural" structural group. In comparison to the previously mentioned taxa, *Lauraceae* possess the greatest structural diversification of all investigated groups due to additional features as, for instance, well-developed paratracheal parenchyma, oil-cells, septations in the fibres and storied rays. A remarkable characteristic of the *Lauraceae* is the occurrence of tyloses in the fibres—besides those in the vessels—which were found for the first time during this investigation and can be interpreted as indicating a slight relation to the *Magnoliaceae*, the only family of all flowering plants where this unique feature has previously been observed (GOTTWALD 1972). The progression within *Lauraceae* is perceptible but mostly restricted to the vessels as is also true of the *Myristicaceae* whose structural advancement is comparable in grade to that of *Lauraceae* (C—D).

The families *Hernandiaceae* and *Calycanthaceae* are characterized by the same "Laural" structural principle and can be placed next to *Lauraceae* according to their evolutionary state.

The monotypic family *Gomortegaceae* is also similar to the *Lauraceae* but due to the pronounced apotracheal parenchyma it seems more reasonable to place *Gomortega* in the "Dillenial-Hamamelidal" group. The advancement of *Gomortega* is comparable to the lesser-developed part of *Magnoliaceae*.

With regard to the family *Monimiaceae* it was not possible to investigate more than about one third of the genera. Here, the *Monimiaceae* will be interpreted in its widest sense and the results of previous investigations by GARRAT (1933) and MONEY (1950) are included. In this sense the family extends from the "Theal" principle through the "Dillenial-Hamamelidal" to the "Laural" principle. Within the *Monimiaceae* a pronounced structural progression, comparable to that of the *Magnoliaceae*, can be noticed, ranging from primitive structures to fairly advanced ones in the genus *Peumus*.

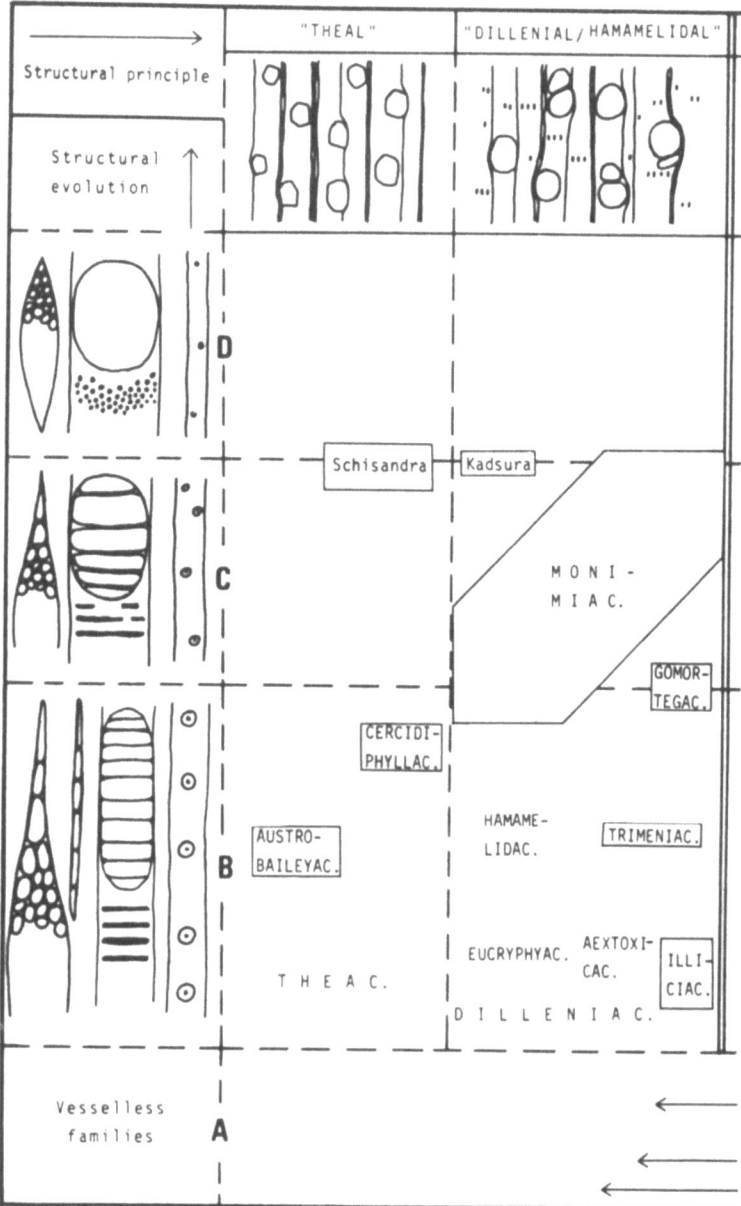

Fig. 1 a

Fig. 1a–1b. Two-dimensional model of ancient taxa including all families of the *Magnoliales* (sensu TAKHTAJAN in thick boxes, sensu MELCHIOR in thin boxes, others without boxes)

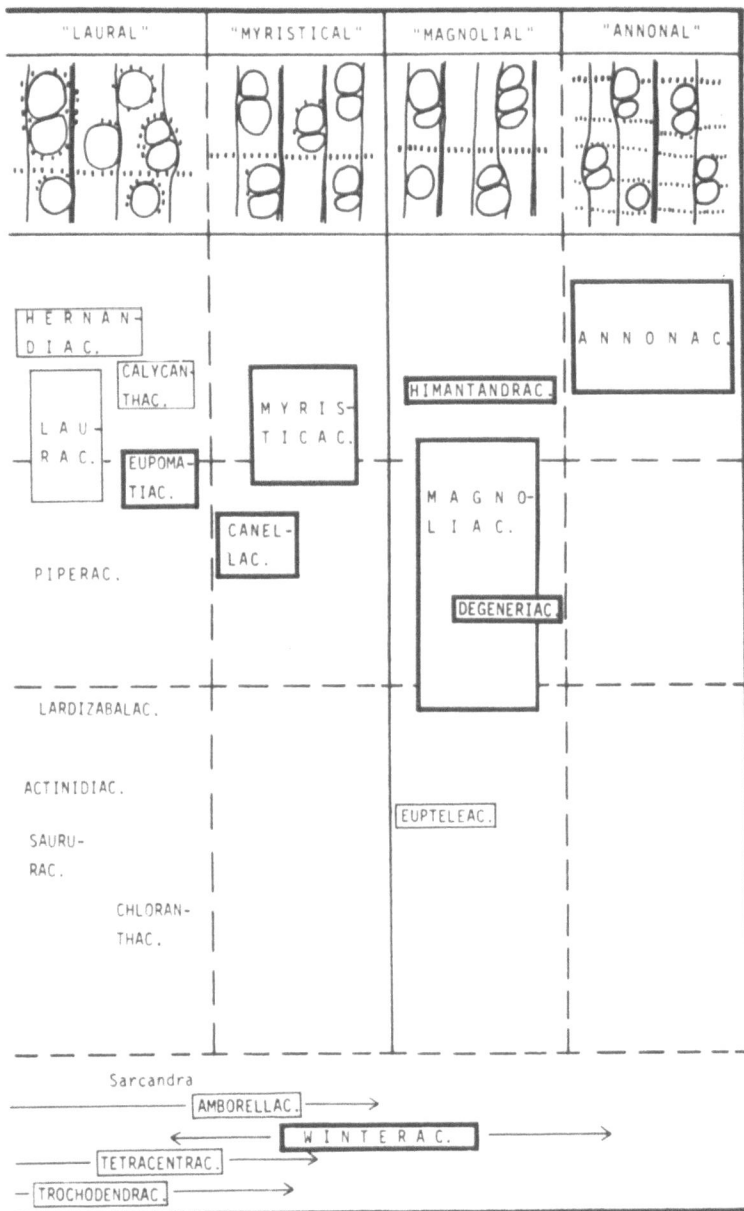

Fig. 1 b

Of the remaining families of *Magnoliales* sensu MELCHIOR only *Trimeniaceae* and *Illiciaceae* can be correlated reliably with a structural principle. They are to be placed in the "Dillenial-Hamamelidal" group in a low or very low position.

The horizontal position of the family *Eupteleaceae* remains uncertain because structural arguments exist for its attribution to the "Dillenial-Hamamelidal" group as well as to the "Magnolial" group, where it can be placed in a relatively isolated position as ENDRESS (1968) has already mentioned. According to its low degree of advancement it must be placed in a primitive position.

The same problem of an undecided structural arrangement exists for the families of *Austrobaileyaceae* and *Cercidiphyllaceae* which, according to their apotracheal parenchyma, have been placed in the "Theal" group close to the "Dillenial-Hamamelidal" group. However, an unclearly developed terminal parenchyma also makes it possible to place *Cercidiphyllum* in the "Magnolial" group as already indicated by McLAUGHLIN (1933).

The placement of *Schisandraceae* within this model is difficult. The genera *Kadsura* and *Schisandra* possess different specialized structures due to their scandent habit which make a separate position for each of them necessary. *Schisandra*, in which no terminal parenchyma can be traced, must be put in the "Theal" group, and *Kadsura*, due to very scanty apotracheal parenchyma in the "Dillenial-Hamamelidal" group in a fairly low position. Further, a pronounced progression is found in the genus *Schisandra* (Figs. 1 a, b).

A survey of these families which MELCHIOR, in contrast to TAKHTAJAN, has incorporated in *Magnoliales* shows a distinct shift to the "Theal" and the "Dillenial-Hamamelidal" structural principles to which no family of the *Magnoliales* sensu TAKHTAJAN could be allocated with certainty. It is also remarkable that the structural evolution within these heteroxylous families included by MELCHIOR in *Magnoliales* is often of a lower level than that of *Magnoliales* sensu TAKHTAJAN.

Families Belonging to Orders Other Than the *Magnoliales*. In a third series, a selection of ten further families belonging to nine orders sensu TAKHTAJAN was investigated. These taxa can also be termed as ancient because of the primitive structural features of their secondary xylem.

Of these additional 10 families none belongs to the "Magnolial", "Myristical" or "Annonal" structural principle. The families of *Chloranthaceae, Actinidiaceae, Saururaceae, Lardizabalaceae,* and *Piperaceae* are to be placed in the "Laural" group with an increasing degree of advancement.

The greatest concentration occurred in the "Dillenial-Hamamelidal" group where the families *Dilleniaceae, Aextoxicaceae,* and *Eucryphiaceae* demonstrate the lowest level of structural evolution of vessel-bearing species, while the *Hamamelidaceae* must be arranged in the same structural group in a slightly higher evolutionary position.

The last family to be considered, the *Theaceae*, has to be placed, due to its specific plan of architecture, into a separate structural group. The very primitive wood anatomical features described by KENG (1962) warrant a low position in the "Theal" structural principle near the *Dilleniaceae* (Figs. 1 a, b).

Results

1. All eight families of *Magnoliales* sensu TAKHTAJAN are spread over four structural groups with a pronounced concentration in the "Magnolial" group.

2. Of these four structural groups the "Magnolial" principle and the "Annonal" principle on one hand, as well as the "Myristical" and the "Laural" principle on the other, can be seen as two pairs of closely related types. Both can be combined again in a wider structural block in which, however, the *Canellaceae* and *Eupteleaceae* hold an isolated position.

3. The vesselless families cannot be clearly linked to one of the six structural principles although some features allow an association with the above mentioned "Magnolial-Annonal-Myristical-Laural" block.

4. Five of the six structural groups have a worldwide distribution; only the "Magnolial" one, consisting of four families, does not occur in Africa including Madagascar.

5. About two thirds of the investigated heteroxyl families—outside the *Magnoliales* sensu TAKHTAJAN—are concentrated in the "Dillenial-Hamamelidal" and the "Theal" structural groups, while only one third belongs to the "Laural" group. That means that all families which MELCHIOR considers as *Magnoliales*, besides those of TAKHTAJAN, are represented in the "Laural-Myristical-Magnolial-Annonal" block.

6. The "Dillenial-Hamamelidal" and the "Theal" structural groups form a second block of differing basic anatomical structure.

7. The evolutionary state of the seven heteroxyl families of *Magnoliales* sensu TAKHTAJAN ranges, in comparison to other ancient families, from an only fairly primitive to a relatively highly advanced level.

8. The most primitive structures of vessel-bearing taxa can be found in the "Dillenial-Hamamelidal" and the "Theal" structural groups, outside the *Magnoliales* sensu TAKHTAJAN. Hence, according to the features of the secondary xylem, the order *Magnoliales* sensu TAKHTAJAN is not at all in accord with his statement of being "the most primitive order of flowering plants".

Conclusions

It seems evident that, within the span of evolution that can be deduced from living plants, no family of the "Theal-Dillenial-Hamamelidal" block has proceeded from the "Magnolial-Annonal-Myristical-Laural" block or vice versa. The *Monimiaceae* can be seen as a possible bridge or a "branching-off family" originally connecting the two main structural blocks at a very early undetermined state of phylogeny.

From the structural point of view these conclusions can lead to the following three interpretations:

Concept I

A complex of various families with the most primitive structures, belonging to the "Dillenial-Hamamelidal" and the "Theal" block, can be regarded as the base for all living dicotyledons. Apart from this base a separate "dead end" has to be considered in which all vesselless families are combined in the lowest possible position.

However, this model of structural evolution would assume that the various families of the other block with the "Magnolial-Annonal-Myristical-Laural" structures (including all heteroxylous taxa of *Magnoliales* sensu TAKHTAJAN) have been derived from the "Dillenial-Hamamelidal" and the "Theal" block. This, however, does not seem possible as such a fundamental structural metamorphosis is not conceivable without a pronounced development of intermediate evolutionary characters for which we have no evidence.

Concept II

Another model would place the taxa with the lowest state of evolution in an appropriately low position with a base consisting of the four exclusively vesselless families from which the other ancient taxa are derived. This also does not seem practicable, since it has not been possible up to now to find evidence for the necessary structural links between the vesselless and the other living heteroxylous families.

Concept III

According to the evidence derived from the anatomical structures of the secondary xylem, a model for the recent dicotyledons is proposed consisting of two main branches representing the two structural blocks and with a third separate dead-end branch comprising the vesselless families. In other words, the anatomical evidence does not support any concept in which all being angiosperms are derived from a single living order or its forerunners (e.g. the *Magnoliales* of TAKHTAJAN). Furthermore, no structural alteration is imaginable which would generate a very primitive structural group from a less primitive one,—if both groups are also distinguished by fundamentally different structural principles. Due to this last mentioned fact a derivation of the *Dillenia-ceae, Theaceae* and others from *Magnoliales* cannot be explained by a "heterobathmic mechanism".

In terms of further evolutionary development the following lines based on anatomical evidence may be suggested: from the "Annonal" group towards *Ebenaceae*; from the "Magnolial" group towards *Rutaceae*; from the "Dillenial-Hamamelidal" group towards *Rosaceae* (p.p.) and from the "Laural" group towards *Leguminosae* (p.p.) and *Boraginaceae*.

Finally it should be strongly emphasized that all the statements made here are based only on the features of the secondary xylem. However, wood anatomy cannot on its own be used to devise and establish phylogenetic models and classificatory systems. Therefore, the wood anatomist must make use of available systems where he either may find confirmation of his own results or else may put forward additional evidence for further discussion and improvements.

References

BARANOVA, M., 1972: Systematic anatomy of the leaf epidermis in the *Magnoliaceae* and some related families. Taxon **21**, 447—469.

ENDRESS, P., 1968: Gesichtspunkte zur systematischen Stellung der Eupte-leaceen. Ber. Schweiz. Bot. Ges. **79**, 229—278.

ENGLER, A., 1964: Syllabus der Pflanzenfamilien (MELCHIOR, H., Hrsg.), 12. Aufl. Berlin: Borntraeger.

GARRATT, G. A., 1933: Bearing of wood anatomy on the relationships of the *Myristicaceae*. Tropical Woods **36**, 20—44.

— 1933: Systematic anatomy of the woods of the *Monimiacaee*. Tropical Woods **39**, 18—44.

GOTTWALD, H., 1972: Tyloses in fibre tracheids. Wood Sci. Technol. **6**, 121—127.

KENG, H., 1962: Comparative morphological studies in *Theaceae*. Univ. Calif. Publ. Bot. **33**, 269—384.

LEMESLE, R., 1953: Les caractères histologiques du bois secondaire des *Magnoliales*. Phytomorphology **3**, 430—445.

McLAUGHLIN, R. P., 1933: Systematic anatomy of the woods of the *Magno-liales*. Tropical Woods **34**, 3—39.

MONEY, L., BAILEY, I. W., and SWAMY, B. C. L., 1950: The morphology and relationships of the *Monimiaceae*. J. Arnold Arb. **31**, 372—404.

PRAGLOWSKI, J., and DANDY, E., 1974: *Magnoliaceae*. World Pollen and Spore Flora **3**, 1—45.

TAKHTAJAN, A., 1973: Evolution und Ausbreitung der Blütenpflanzen. Stuttgart: Gustav Fischer.

Address of the author: HELMUT P. J. GOTTWALD, Wiss. Direktor, Institut für Holzbiologie und Holzschutz der Bundesforschungsanstalt für Forst- und Holzwirtschaft, Leuschnerstr. 91, D-2000 Hamburg 80, Federal Republic of Germany.

Plant Syst. Evol., Suppl. 1, 123—140 (1977)

Department of Botany, University of Canterbury, New Zealand

Ovular Morphology and the Classification of Dicotyledons

By

W. R. Philipson, Christchurch

Abstract: Four types of ovule are defined by integument number and nucellar thickness. Their distribution throughout the dicotyledons is summarized and their importance in defining major subclasses is advocated. In particular, those families with ovules having a single integument are considered to form a complex of orders and families which includes most sympetalous and several polypetalous groups (including *Cornales*, *Araliales*, *Escalloniaceae* and *Pittosporaceae*). This complex (provisionally referred to as the *Unitegminae*) is believed to comprise the modern representative of a distinct evolutionary line within the dicotyledons.

Introduction

Features of ovular morphology have long been used to define taxa at all levels. The Danish botanist WARMING (1878) first clearly stated the distinction between the crassinucellate-bitegmentary ovule and the tenuinucellate-unitegmentary ovule. He recognized the taxonomic importance of these ovular types and named them type dichlamydé and type monochlamydé, these terms being derived from the subdivisions of the dicotyledons of which they were characteristic.

After some decades WARMING (1914) drew together more complete information about these and other types of ovule and was able to relate these to recognized taxa. OSSIAN DAHLGREN (1927) published his careful studies of the nucellus and reassessed the significance of ovular characters in classification. Neither WARMING nor DAHLGREN set out their views as formal systems. This had been done by VAN TIEGHEM (1901) who gave great prominence to ovular (or related) characters in a system which embraced the whole plant kingdom. His complete reliance on ovular types when drawing up his major subdivisions (Orders) resulted in a system that was too rigid and which contained inconsistent and unnatural groupings. It is likely that VAN TIEGHEM's uncritical proposals lead to the taxonomic value of the ovule coming into disrepute. Nevertheless, as I hope to show, several of his principal proposals are helpful. A major obstacle to their general acceptance has been the occurrence of the "sympetalous type" ovule (as the unitegmic-tenuinucellate ovule has frequently been called) within many polypetalous families. It was a review by MAURITZON (1939) which convinced most comparative morphologists that the sympetalous ovule had been evolved in parallel repeatedly. Consequently it has come to have a less dominant role in systematics.

Recently I have suggested (PHILIPSON 1974) that the difficulties seen by MAURITZON would be largely overcome if a distinction were made between polypetalous families in which the so-called sympetalous ovule is constant as opposed to those in which it (or one or another of its features) occurs only sporadically. If this is done, correlations between ovular types and major family complexes become evident. At the same time, and again in the following year (PHILIPSON 1975), I proposed the view that dicotyledonous families in which the sympetalous ovule is dominant form a long-established and independent line of evolution. It should be emphasized that these two proposals were distinct. The first was in the field of classification—the recognition and definition of taxa—even though this was not formalized by applying names to the groups proposed. The second was an interpretation of that classification. (In claiming that the ovule is of value in taxonomy at the higher level, I am in effect saying that it largely conforms with several other independent characters, that it therefore confirms existing boundaries, that it clarifies some doubtful areas, and moreover that it enlarges previous concepts.) In what follows the types of ovule being considered will be described, their distribution among higher categories summarized, and the bearing of this on the framing of major subdivisions of the dicotyledons will be discussed.

Definition of Ovule Types

Consideration of ovule morphology will be limited to two features, namely, the number of integuments and the presence or absence of additional layers within the epidermis at the apex of the nucellus. The combinations of these two characters allow only four types to be defined and in fact all of these occur, though with widely differing frequencies. It would clearly be an advantage to consider a whole syndrome of ovular characters but sufficiently consistent information over a wide sample is not available. No doubt additional characters would reveal other types of ovule, such as that characteristic of the *Centrospermae* or the orthotropous ovule of several apetalous families. The present application of ovular morphology to taxonomy, therefore, must be regarded only as a beginning.

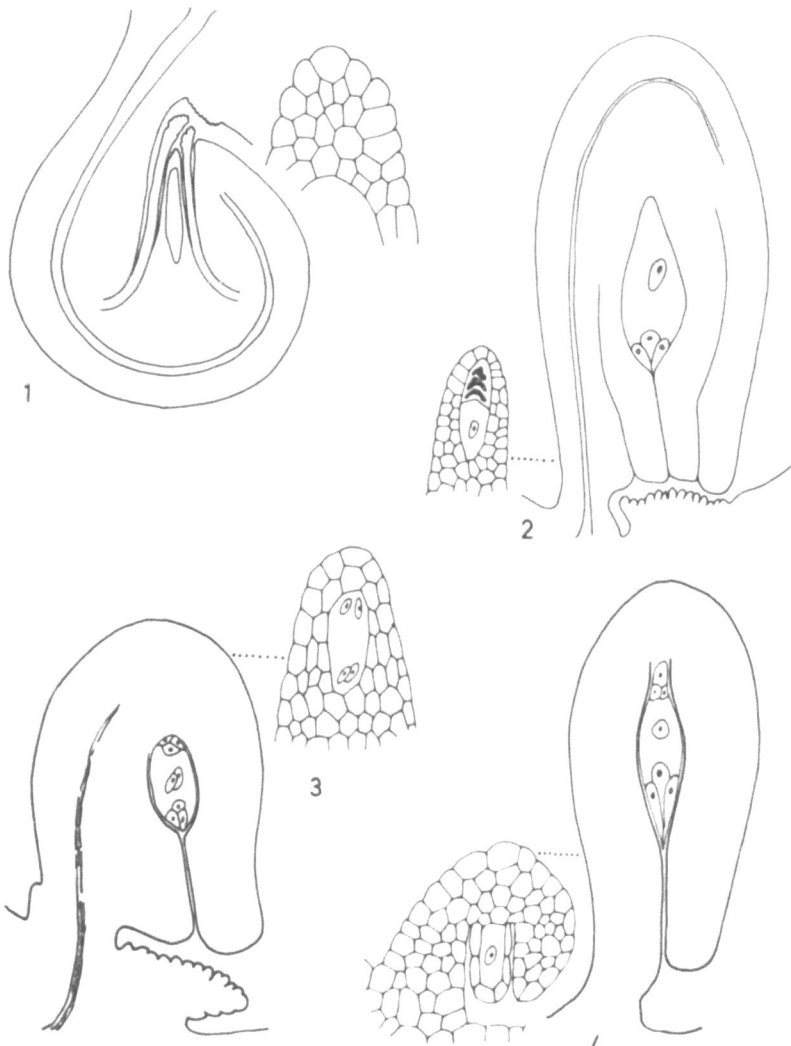

Figs. 1–4. Examples of the four types of ovules. Fig. 1. *Hernandia nymphaei-folia* (PRESL) KUBITZKI, type I; 2 integuments, crassinucellate. Ovule × 40; nucellus × 250. Fig. 2. *Halesia carolina* L., type II; 2 integuments, tenuinucellate. Ovule × 75; nucellus × 250.   Fig. 3. *Griselinia littoralis* RAOUL, type III; 1 integument, crassinucellate. Ovule × 40; nucellus × 250. Fig. 4. *Forstera tenella* HOOK. f., type IV; 1 integument, tenuinucellate. Ovule × 50; nucellus × 250

The four ovule types defined by combinations of the two characters being considered will subsequently be referred to by numbers as follows:

bitegmentary-crassinucellate	Type I
bitegmentary-tenuinucellate	Type II
unitegmentary-crassinucellate	Type III
unitegmentary-tenuinucellate	Type IV

Representative examples of these four types are shown in Figs. 1–4.

It is unlikely that any of these types are strictly homogeneous, with their various parts truly homologous. For example, BHANDARI et al. (1976) suggest that the origin of the integument primordium may prove to be wholly dermal in the sympetalae in general, whereas in other groups with tubular corollas (e.g. *Cucurbitaceae, Convolvulaceae*) it is subdermal and vascularized. Even if much more were known about the morphogenesis of the integuments and the nucellus it is unlikely that many convincing differences could be established between large sub-groups within these types. At least two considerations make such distinctions difficult to prove. In the first place, considerable differences in morphogenesis have been described in ovules whose close relationship cannot be doubted. It is doubtful, therefore, if such differences would ever provide convincing evidence that two forms were unrelated. BOUMAN (1975) has found considerable variation in development within a natural family, the *Cruciferae*. In *Brassica* and *Sinapis* initiation of the outer integument is subdermal; in *Lunaria* it is partly subdermal and partly dermal; while in *Capsella* it is completely dermal. Similarly initial divisions in the formation of the nucellar beak in *Euphorbia* were found by BOR & BOUMAN (1974) to be subdermal, though epidermal cells of the flank later contribute to the beak. In another genus of the *Euphorbiaceae, Codiaeum*, they found the initiation of the beak to be by divisions of both the parietal cells and of cells of the nucellar dermatogen. If such variations occur in related plants, it may prove difficult to demonstrate convincing differences between ovules in relation to these particular characters. Secondly, it must be borne in mind that each ovular type, as defined above, comprises ovules of very different general appearance. As extreme examples, the ovules of *Aralidium* and *Rhododendron* may be compared (Figs. 5, 6). Both are unitegmentary but the integument of *Aralidium* is very thick and is traversed by many vascular bundles (personal observation), whereas that of *Rhododendron* is thin and no vascular strand is present even in the funicle (PALSER et al. 1971). In spite of this great range in form there are good grounds for believing that these two examples represent extremes of a single series. In the

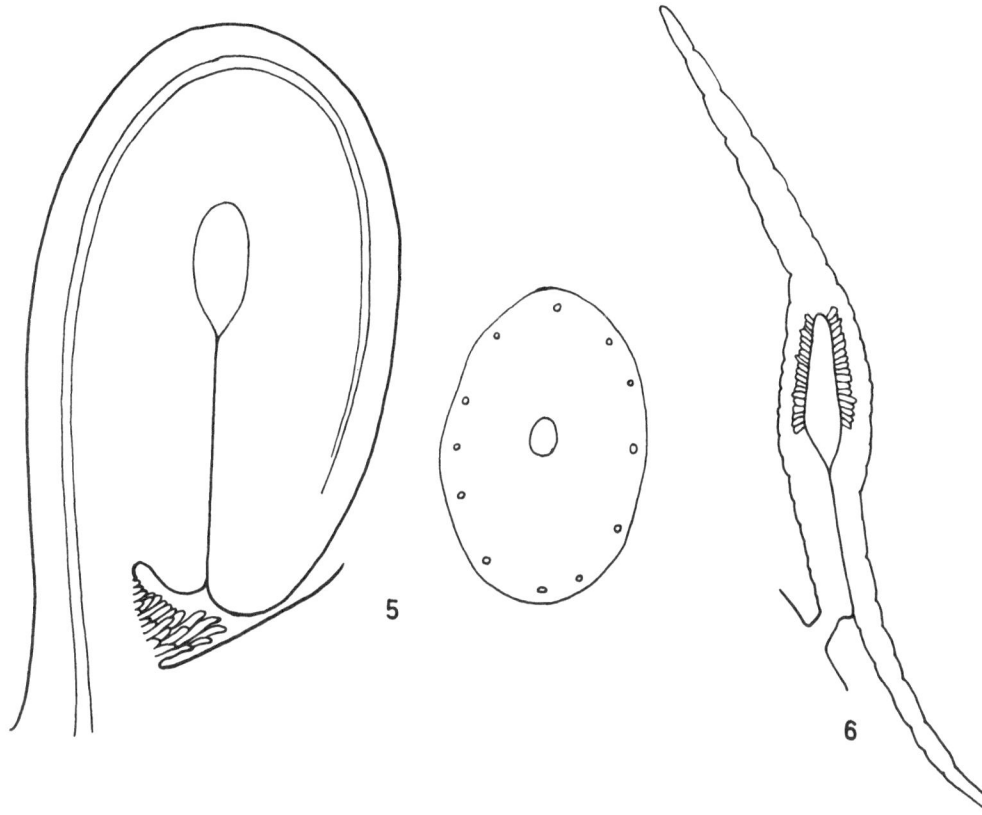

Fig. 5. Ovule of *Aralidium pinnatifidum* MIQ. in sagittal section, ×50, and in transverse section, ×40. Integument massive and with many vascular bundles

Fig. 6. Ovule of *Rhododendron leptanthum* F. v. MULL. ×75. The thin integument is prolonged beyond the micropyle and the chalazal end of the ovule also forms a "tail" in the seed

Araliaceae (in which the integument is thick and the nucellus may be of more than one layer) the nucellus usually degenerates so that the embryo sac comes into direct contact with the integument, the inner layer of which may form a well differentiated endothelium. These are characteristics of the ovule typical of many sympetalous families.

DAVIS (1966) distinguishes between strictly crassinucellate ovules, in which the archesporial cell cuts off a primary parietal cell (which usually divides further) and pseudo-crassinucellate ovules, in which no primary parietal cell is formed, but in which apical cells of the nucellar epidermis divide to form a nucellar cap. In each case the megaspore-mother-cell is capped by more than one layer of cells, though the origin of these cells is different. In the original presentation of this classification (PHILIPSON 1974) this distinction was referred to, but not taken into account in making comparisons. It was felt that insufficient attention had been given to this feature, so that conclusions could not be based on the inadequate information available. DAVIS (1966) recorded pseudo-crassinucellate ovules in few dicotyledonous families, namely *Frankeniaceae*, *Calycanthaceae*, *Podophyllaceae* and *Cobaeaceae* and also in some members of *Ranunculaceae*, *Boraginaceae*, *Ehretiaceae*, *Labiatae* and *Asclepiadaceae* (though this last may have been in error). In all the sympetalous families in this list, the pseudo-crassinucellate condition occurs in one or a few members of otherwise uniformly tenuinucellate relationships (regarding *Cobaeaceae* as a satellite family of the *Polemoniaceae*). The situation in the *Ranunculaceae* (where several genera are pseudo-crassinucellate) is of greater interest, but it scarcely affects the comparisons made here. For these reasons no subdivision of the four ovular types is attempted on the origin of the nucellar tissue.

Distribution of Ovular Types

The distribution among the orders and families of dicotyledons of these two characters, both separately and in association, has been set out by O. DAHLGREN (1927), WUNDERLICH (1959) and PHILIPSON (1974) and was further discussed by R. DAHLGREN (1975b). The last author stresses that the characters are known from too few examples for broad generalizations to be firmly established. Undoubtedly a much larger sample is desirable, but it is felt that the main conclusions reached in this presentation rest on features sufficiently constant to encourage some confidence in their acceptance.

It is not necessary to restate here the detailed information already published by the above authors and also by DAVIS (1966). All that is required is a summary. This can be provided most conveniently by describing the distribution of the combinations of the two characters, that is to say, of the four ovular types (Table 1). The names of taxa are used as in CRONQUIST (1968) unless otherwise stated.

At the risk of over-simplification, the distribution of the ovular types can be described very briefly. Type I is found in nearly all members of CRONQUIST's subclasses *Magnoliidae*, *Hamamelididae* and *Caryophyllidae*, and also in most members of the greater parts of his subclasses *Dilleniidae* and *Rosidae* (Table 1, column 1). However, a substantial number of families of the *Dilleniidae* are characterized by Type II ovules and a few by Type IV (Table 1, columns 2 and 4), and an even larger section of families of the *Rosidae* have ovules of Types III and IV (Table 1, columns 3 and 4). In the remaining subclass, the *Asteridae*, virtually all members have Type IV ovules (Table 1, column 4).

Table 1. The distribution of ovular types among the sub-classes and families
of Dicotyledons (minor exceptions omitted)

(1) **Type** I	(2) **Type** II	(3) **Type** III	(4) **Type** IV
Magnoliidae	*Theaceae*	*Cornaceae*	**Asteridae**
(except *Circeastaceae*	*Guttiferae*	*Davidiaceae*	*Ericales*
Sabiaceae	*Ochnaceae*	*Nyssaceae*	*Diapensiaceae*
Peperomiaceae)	*Marcgraviaceae*	*Alangiaceae*	*Escalloniaceae*
	Lecythidaceae	*Garryaceae*	*Hydrangeaceae*
	Primulaceae	*Araliaceae*	*Philadelphaceae*
Caryophyllididae	*Myrsinaceae*	*Aquifoliaceae*	*Pittosporaceae*
	Theophrastaceae	*Icacinaceae*	*Alseuosmiaceae*
	Coridaceae		*Umbelliferae*
Hamamelidae	*Styracaceae*		*Grubbiaceae*
(except *Myricaceae*	(except *Halesia*)		
Eucommiaceae	*Ebenaceae*		
Juglandaceae	*(etc., etc.)*		
Dilleniididae	*Rafflesiaceae*	*Hippuridaceae*	*Sarraceniaceae*
(except families in	*Byblidaceae*	*Theligonaceae*	*Loasaceae*
upper part of col. 2	*Podostemaceae*	*Bruniaceae*	*Limnanthaceae*
and	*Parnassiaceae*	*Juglandaceae*	*Santalaceae*
Ericales	*Vahliaceae*	*Myricaceae*	*Opiliaceae*
Diapensiales	*Stackhousiaceae*	*Sabiaceae*	*Hydnoraceae*
Sarraceniaceae	*Oxalidaceae*	*Eucommiaceae*	*Circaeastaceae*
Loasaceae)	*Tropaeolaceae*	*Peperomiaceae*	
	Balsaminaceae		
Rosidae			
(except families of			
this sub-class listed			
in other 3 columns)			

The outline of the classification proposed here is that those families
which consistently possess ovules of Type II form a subclass (the
Thealean-Primulalean complex), and that those with ovules of Types III
and IV should be united into another subclass. Since most of what
follows relates to this second subclass, it will be useful to give this
taxon an informal name. Accordingly, the convenient term *Unitegminae* is adopted from the system of van Tieghem (1901).

Before advancing further arguments in favour of this classification,
some additional details of the distribution of ovular types need to be
considered. In each of the last three columns of Table 1, the families
above the dotted line form what appear to be natural and coherent
groups. The families listed below the dotted line represent small, often
highly specialized families, frequently of uncertain relationship, and

for the most part without evident relationships among themselves. Possibly some should be associated with the group above the dotted line, especially as a number are reported to contain iridoid compounds (JENSEN et al. 1975), substances which occur rather consistently within many unitegmic families, as will be discussed later. Even so these aberrant families indicate that morphological modification of the ovule must have occurred repeatedly during the evolution of the dicotyledons. This is shown even more forcibly by the several polypetalous and apetalous families in which aberrant ovular characters occur in only a few of their members. These were referred to by the authors cited above but are omitted here. It is to be expected that their true systematic relationships will become clearer on more intensive study.

Families characterized by Type I ovules will not be considered further. They represent about half of the dicotyledons and clearly include several major taxa. One proposal advanced here is that the families above the dotted line in Table 1, column 2 constitute a major subdivision which spans the boundary between the polypetalous and the sympetalous condition. The affinity of the *Theales* and *Primulales* is acknowledged by several authors and the case for the adoption of this major taxon will not be further advanced here, as the case has already been stated (PHILIPSON 1974). The main part of the proposal is that the groups of families above the dotted line in both columns 3 and 4 of Table 1 together form a single subclass, the *Unitegminae*. The evidence for regarding such a large and diverse group as a single taxon will now be reviewed.

The Unity of the Unitegminae

It is significant that the association of polypetalous families here included in the *Unitegminae* had been recognized previously as a complex of related families without any reference to their ovular morphology. When considering the relationships of the trans-Pacific genus *Griselinia* PHILIPSON (1967) concluded that this genus could not be referred to any existing family, but showed affinity with the *Escalloniaceae*, the *Cornaceae* and the *Araliaceae*. These families were considered to be part of a complex which included the satellite families around the *Cornaceae* as well as families often referred to as the "woody *Saxifragaceae*". Of polypetalous families omitted at that time but now included, the *Pittosporaceae* and *Alseuosmiaceae* are now considered to be close to the complex on other grounds besides ovular morphology. The status of the *Aquifoliaceae* and *Icacinaceae* will be discussed presently as will that of the *Actinidiaceae* and *Grubbiaceae*.

RODRIGUEZ (1971), in discussing the relationships of the *Umbel-*

liferae, considered a rather wider range of families, notably including some sympetalous groups, a topic to which we shall revert presently. Nevertheless, at the base of this flabellate array of families he clearly recognized the same core-complex of *Escalloniaceae—Cornaceae—Araliaceae*, each with their satellite families. RODRIGUEZ also reached these conclusions with only minor reference to ovular morphology. The fact that the ovules of all the members of this complex are unitegmic (and most also tenuinucellate) would be a remarkable coincidence if it were not due to their natural relationship. Only two small families that might be regarded as immediate associates of the complex have Type I ovules. These are *Brexiaceae* and *Grossulariaceae* (in the narrow sense). However, BENSEL & PALSER (1975b) conclude from evidence of floral anatomy that *Ribes* is similar to other members of the *Saxifragoideae*. It may, therefore, have no close relationship with the *Unitegminae*. The association together of *Brexia* and *Ixerba* in a small family may be artificial, but at least *Ixerba* appears correctly placed near the *Escalloniaceae* (BENSEL & PALSER 1975a) and therefore should be accepted as a member of the *Unitegminae* in which the ovule is Type I.

On the basis of the stereostructure of the exine, HIDEUX & FERGUSON (1976) also suggest an evolutionary link between the *Escalloniaceae*, *Cornaceae*, and *Araliaceae*, which together form the core of the *Unitegminae*. They include in their consideration the *Cunoniaceae* and *Saxifragaceae* (sensu stricto), the whole group in their opinion occupying a somewhat central position palynologically.

One of the most telling points in favour of the unity of the *Unitegminae* is the existence of many anomalous genera which cannot be placed with certainty in any of the larger families of the complex but which appear to combine the characters of several of them. TAKHTAJAN (1969) explains the motives behind his close association of *Araliaceae* and *Cornaceae*. The existence of *Mastixia*, *Helwingia* and *Toricellia* nullified his attempts to separate these families. The discussion by RODRIGUEZ (1971) brings out the complexities of the distribution of characters among these genera. RODRIGUEZ also accepted the importance of *Griselinia* as a type in which features of *Araliaceae*, *Cornaceae* and *Escalloniaceae* were blended. He supports the view, already advanced by CRONQUIST (1968) and TAKHTAJAN (1969), that *Corokia* forms a bridge between *Cornaceae* and *Escalloniaceae*. I wish to call attention to another genus which has been variously placed in the *Cornaceae* and *Araliaceae*. The mono-typic Malayan genus *Aralidium* shares a wide range of characters with *Griselinia* and, having no resin canals and a dorsal raphe to the ovule, also approaches the *Cornaceae* more closely than the *Araliaceae*. However, the pits of the fibres of its secondary xylem are unbordered (unpublished observation), a feature characteristic of *Araliales* and contrasting with the bordered pits found in *Griselinia* and other *Cornales*.

The distinction between the fibre-tracheids of the *Cornaceae* and the libriform fibres of the *Araliaceae* is one of several characters which show that the complex of families under consideration is by no means uniform. A general discussion of this heterogeneity will be given later, but the condition in *Aralidium* illustrates that these distinctions can be bridged and do not necessarily prevent the acceptance of the whole complex as a single coherent major taxon. At this level, taxa are not defined by the possession of any one character, or even by a group of characters, but rather by a majority of a whole syndrome of characters. In the particular case of this complex the syndrome is not easily identified because it consists in large part of generalized non-specific characters. RODRIGUEZ (1971) characterized the well-established concept of the *Umbelliflorae* (including *Araliaceae, Umbelliferae* and *Cornaceae* in the widest sense) "by mostly pentamerous or tetramerous flowers with a reduced calyx, free petals and stamens, inferior ovaries mostly with a pulviniform disk or a stylopodium supporting one or more styles, tending to 5, 4, 2, or 1 locules with one pendent, anatropous functional ovule per locule; by drupaceous or schizocarpous fruits (a few baccate), and seed with abundant endosperm and small, straight embryos; by occasional trends to unisexuality and apetaly; and by usually involucrate, cymose, umbellate or capitate inflorescences". To include the *Escalloniaceae* (and its allies) this definition requires only slight modification in respect to ovary position and placentation. When discussing *Griselinia* (PHILIPSON 1967) I summed up the characteristics of these families as "being generally similar in wood anatomy, pollen morphology, and floral and fruit morphology". In now seeking to define this complex by its ovular morphology I am not proposing a single-character classification. Rather, my intention is to use the ovule as the character which serves best to define a group already identified. It serves best because it is precise and also because it coincides more consistently than any other character with the boundaries under consideration.

Even if the coherence of the polypetalous members of the *Unitegminae* were to be accepted, the question of their relationship with the sympetalous families (excluding *Primulales* and *Ebenales*) remains. These families make up the subclass *Asteridae* of CRONQUIST (1968). It is considered that various groups of sympetalous families relate to different polypetalous groups, but that all of these fall within the *Unitegminae*. That is to say, the *Asteridae* are derived by several routes from a common ancestral complex.

Chemical evidence now substantiates the subdivision of the *Asteridae* into two major groups, each with distinct relationships among polypetalous families. HEGNAUER (1969) has published a scheme illustrating the close links between the *Compositae* and the *Araliales* on the one hand, and between the *Caprifoliaceae—Rubiaceae* and the *Cornaceae—Hydrangeaceae* on the other. A very similar scheme is also illustrated by FROHNE & JENSEN (1973: Fig. 98) and JENSEN et al. (1975) consider that the iridoid-containing polypetalous and sympetalous orders to "make up a relatively homogeneous and probably monophyletic group". Another chemical character of this group is the absence of polyines which are frequent in the *Compositae* and the *Araliales*.

Three further links between the unitegmic polypetalous orders and the sympetalous orders will be mentioned. Firstly, the *Empetraceae* and *Clethraceae* are generally associated with the *Ericales*, and a similar affinity is frequently assigned to the *Cyrillaceae*. FAGERLIND (1947) has brought forward evidence, especially embryological, supporting a similar link between the *Grubbiaceae* and the *Ericaceae*. Much earlier LINDLEY (1846) had proposed a relationship between the *Actinidiaceae* and the *Ericaceae*, and a succession of authors until the present day (TAKHTAJAN 1969) have supported this. Secondly, the suggested relationship between the *Aquifoliaceae* and the *Oleaceae* (e.g. WETTSTEIN 1930—1935) was upheld by MAURITZON (1936) mainly on embryological evidence. Thirdly, the two small segregate families *Columelliaceae* and *Alseuosmiaceae* both have sympetalous corollas, but are thought to be most closely related to the polypetalous families *Escalloniaceae* (STERN et al. 1969) and *Pittosporaceae* (CRONQUIST 1968) respectively, and indeed the corolla of the *Pittosporaceae* itself is sometimes incipiently tubular.

Having briefly referred to the ramifications of the complex among sympetalous orders, we may now look for indications of its origins among bitegmic polypetalous alliances. That such relationships exist is suggested by the presence of bitegmic ovules in *Ixerba* and *Quintinia*. The relevance of this type of ovule in *Brexia* and *Ribes* has already been discussed. The transition from ovules of Type I to those of Type IV probably took place in two steps, since a few families of the *Unitegminae* have crassinucellate (perhaps only pseudo-crassinucellate) ovules. It is particularly interesting that the *Araliaceae* stand in this relationship to the *Umbelliferae*, and even within the latter family one example of a pseudo-crassinucellate ovule has been reported (GUPTA 1964). Similarly a gradation from Type I to Type III, with some few examples of Type IV, exists within the *Celastrales*. The inclusion of the *Aquifoliaceae* and the *Icacinaceae* within the *Unitegminae* is debatable, as the acquisition of the single integument may have occurred independently within the *Celastrales*. However, the inclusion of these two families together in the same order appears questionable as iridoid compounds are known in the *Celastrales* only from members of the *Icacinaceae*. This suggests a close tie between this family and the iridoid-containing *Unitegminae*. This still leaves the position of the *Aquifoliaceae* problematical.

The existence of these transitional taxa documents what would in any event be suspected, namely that the unitegmic ovule was derived from the bitegmic ovule. To search for the stock from which the *Unitegminae* arose among extant polypetalous plants is not likely to be productive. A group with simple actinomorphic flowers is probable. One such derivation was suggested by HEGNAUER (1969) when he drew attention to striking chemical resemblances between the *Araliaceae* and *Pittosporaceae* on the one hand and the *Rutaceae* on the other. He had earlier (HEGNAUER 1963) discussed similarities between the alkaloids of the *Rutaceae* and the *Polycarpicae* (*Magnoliales*), and KUBITZKI (1969) developed this suggestion. It might be possible, therefore, to construct on chemical grounds an evolutionary line extending from the Magnolian stock through Rutalean-like plants to polypetalous *Unitegminae* and eventually to the sympetalous families. A diagram of such a scheme has in fact been published by FISH & WATERMAN (1973).

Variability of Some Characters Within the Unitegminae

One of the strongest arguments in favour of the reality of the *Unitegminae* is the almost complete restriction of iridoid compounds to groups with a single integument. Only two reports of iridoid compounds in plants with two integuments are given by PLOUVIER & FAVRE-BONVIN (1972) and JENSEN et al. (1975) in their reviews of the distribution of these substances. Both these exceptional plants (*Liquidambar* and *Daphniphyllum*) are usually placed in the *Hamamelidales*, and both have bitegmic ovules, but *Daphniphyllum* is of very doubtful affinity: it is sometimes placed in the *Euphorbiales* and it should be noted that THORNE (1968) places it next to the *Pittosporaceae*. Although iridoid compounds are so closely associated with unitegmic families, the converse is not necessarily true, for there are many unitegmic taxa from which no iridoid substances have been reported. The distribution of iridoid compounds is not random, for they occur in assemblages of related orders or families. They therefore indicate a subdivision within the *Unitegminae*, defined by the presence or absence of these compounds. It is to be expected that any large section of dicotyledonous orders will be heterogeneous in respect to several characters. Within the *Unitegminae* this is illustrated by various views on the relationship of the *Araliales* to the *Cornales*. EYDE (1967), CRONQUIST (1968) and R. DAHLGREN (1975) diverge from the formerly more generally held opinion that they are closely related. While the separation of these two orders can be based on morphological grounds (e.g. resin ducts, position of raphe), chemical data are more compelling (HEGNAUER 1969, FROHNE & JENSEN 1973, JENSEN et al. 1975). The implications of the distribution of iridoid compounds will first be considered, followed by comparisons with the distribution of a few other characters.

The following summary of the distribution of iridoid compounds relies on the information given by JENSEN et al. (1975) which should be consulted for much detail here omitted (the names of plant groups are used in the sense of CRONQUIST 1968). Iridoid compounds occur in the *Cornales* and in the *Hydrangeaceae*, *Grossulariaceae* and *Icacinaceae* among polypetalous families, and in the following sympetalous alliances: *Gentianales*, *Ericales*, *Scrophulariales*, *Dipsacales*, *Rubiales*, and *Lamiales* (except *Boraginaceae*). Conversely, iridoid compounds are not known to occur among unitegmic polypetalous families in the *Araliaceae*, *Umbelliferae*, *Pittosporaceae* and *Aquifoliaceae*, nor among the following sympetalous families: *Solanaceae*, *Convolulaceae*, *Polemoniaceae*, *Boraginaceae*, *Campanulaceae* and *Compositae*.

Acceptance of this important chemical cleavage within unitegmic groups does not detract from the validity or usefulness of the *Unitegminae*. It is usual for a major group to embrace subdivisions of which the common ancestral stock is unknown and probably extinct. The monocotyledons and dicotyledons within the angiosperms provide a parallel. Nor does the recognition of the larger inclusive group detract from the importance of the evolutionary line marked by the presence of iridoid compounds. Probably we shall never know conclusively whether the acquisition of iridoid substances preceded or followed the reduction in integument number in this evolutionary line. For the resolution of phylogenetic problems events need to be placed in sequence. If the acquisition of iridoid compounds came first, then the group here called *Unitegminae* must be polyphyletic. But I believe the most probable sequence is that the unitegmic ovule arose in a core group from which the whole *Unitegminae* has been derived. Within this core, at an early stage, two divergent lines arose by the acquisition of iridoid compounds. This belief is based on the general similarities between polypetalous members of the complex and also on a preference for the most simple explanation (the principle of parsimony in evolution). In any event, even if the cladistic relations of the members of the *Unitegminae* cannot be clarified, the concept of this group appears a logical major taxon within the dicotyledons. FROHNE & JENSEN (1973: Fig. 98) depict the orders which are included here in the *Unitegminae* as two evolutionary lines linked by a common origin. This stock which gave rise to both the *Araliales* and the *Cornales* must have been unitegmic and thus the whole complex is drawn together.

A contrary opinion has been expressed by JENSEN et al. (1975) who suggest that iridoid compounds developed along an evolutionary line in which the ovules were just about to switch from the bitegmic to the unitegmic condition. The principal factor in support of this view is the presence of iridoid compounds in *Liquidambar* and *Daphniphyllum*, which are bitegmic and probably members of the *Hamamelidales*. If the iridoid compounds are considered to have arisen subsequent to the reduction in integument number, then it would be necessary to postulate an independent origin for these compounds in the *Hamamelidales*, though JENSEN et al. (1975) consider the possibility that these two genera represent relicts of primitive iridoid-bearing groups with bitegmentary ovules. However, a double origin of the iridoids appears more likely than the alternative, which is the independent evolution of the Type IV ovule in the Aralialean and the Cornalean groups—groups which show affinity on a number of other grounds.

It is of some significance that a number of other characters which vary within the *Unitegminae* show no correlation with the distribution of iridoid compounds. Among several such characters the distribution of three will be considered.

Nodal anatomy. Variability of the leaf-trace pattern within families is now known to be much more variable than at the time SINNOTT (1914) first drew attention to the taxonomic value of this character (HOWARD 1974). Nevertheless, a particular nodal type often predominates in a family or large group of families. This is true within the *Unitegminae*, most of which (in both iridoid and non-iridoid containing groups) are basically unilacunar. However, most of the polypetalous members are tri-lacunar (or in the *Araliales*, multi-lacunar). Of the sympetalous groups in which tri-lacunar nodes predominate, one, the *Compositae* is non-iridoid containing, and another, the *Dipsacales*, is iridoid-containing.

Bi- and tri-nucleate pollen grains. Many of the families or orders within the *Unitegminae* include examples of both bi- and tri-nucleate pollen (BREWBAKER 1967), indicating that the step from two or three nuclei has occurred repeatedly within this group. This applies equally to both iridoid-containing and non-iridoid-containing families. Of the principal polypetalous orders of the *Unitegminae* it is only the *Cornales* that have bi-nucleate pollen.

Endosperm types. Most non-iridoid containing families have nuclear endosperm (R. DAHLGREN 1975 b) though there are important exceptions, and both ab-initio cellular and nuclear endosperm occur in the *Solanaceae* and particularly in the *Compositae*. Conversely, most iridoid-containing groups have ab-initio cellular endosperm, but again with notable exceptions (e.g. *Rubiales*, *Gentianales*).

The lack of correlations between the distributions of these various characters serves to remind us that we are dealing with the end products of a long and intricate evolutionary history throughout which the

Angiosperms have displayed great versatility. It is against such a confused background that the ovule and the iridoid substances appear as enduring markers of a main stream of evolution (Fig. 7). Evidently the Type IV ovule, once acquired remained very stable throughout

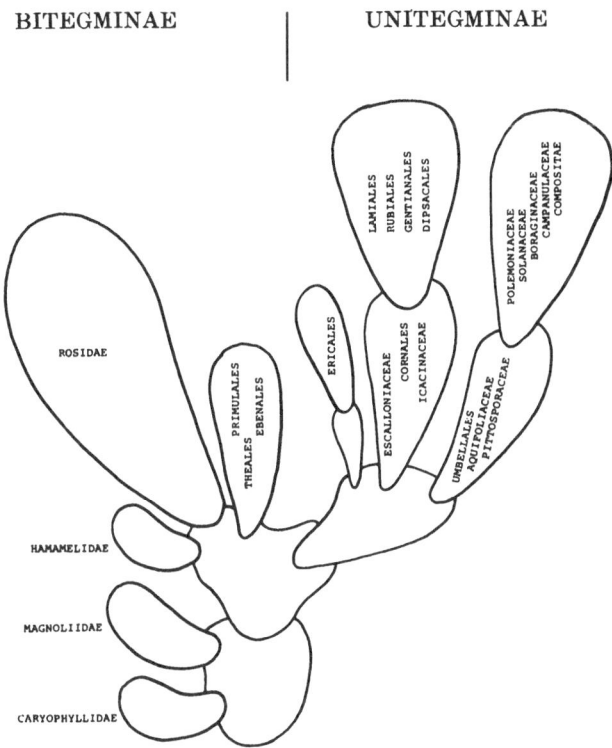

Fig. 7. Schematic representation of some major sub-divisions of the di-cotyledons. Living groups are labelled. The families and orders making up the *Unitegminae* are listed more fully in Table 1. Names of taxa as in CRONQUIST (1968) except that *Boraginaceae* are excluded from the *Lamiales*

massive radiating evolution that produced about half of the existing dicotyledons, including many most successful families. It is to the establishment of this successful dynasty that the sympetalous ovule owes its importance in systematics.

In contrast, the single integument or the thin nucellus evolved repeatedly, as is shown by their sporadic presence in several polypetalous families most of whose members have Type I ovules. The amount of radiating evolution following these events appears to have been slight, and the effect on major classification is, therefore, negligible. The repeated acquisition of these characters is not surprising if, as seems probable, the reduction in integument number has adaptive significance. It is normal within angiosperms for any opportunity to be taken up by several alliances advancing over a broad front. Stebbins (1974) brings to notice an interesting hypothesis deriving the angiospermous ovule from a structure like the cupule of some *Caytoniales* by reduction in the number of ovules per cupule. In the *Caytoniaceae* each ovule was enclosed within its own single integument and several occurred within each cupule. In the *Corystospermataceae* the ovules were reduced to one. The cupule and its contents then become equivalent to a bitegmic ovule. It may then be supposed that one of the integuments is superfluous and consequently would be lost. Moreover, this loss would most likely have occurred not once but repeatedly.

References

Bensel, C. R., and Palser, B. F., 1975a: Floral anatomy in the *Saxifragaceae* sensu lato. I. Introduction, *Parnassioideae* and *Brexioideae*. Amer. J. Bot. **62**, 176—185.

— — 1975b: Floral anatomy in the *Saxifragaceae* sensu lato. II. *Saxifragoideae* and *Iteoideae*. Amer. J. Bot. **62**, 661—675.

Bhandari, N. N., Boumann, F., and Natesh, S., 1976: Ovule ontogeny and seed coat structure of *Scrophularia himalensis* Royle. Bot. Jahrb. Syst. **95**, 535—548.

Bor, J., and Bouman, F., 1974: Development of ovule and integuments in *Euphorbia milii* and *Codiaeum variegetum*. Phytomorphology **24**, 280—296.

Bouman, F., 1975: Integument initiation and testa development in some *Cruciferae*. Bot. J. Linn. Soc. **70**, 213—229.

Brewbaker, J. L., 1967: The distribution an phylogenetic significance of binucleate and trinucleate pollen grains in the Angiosperms. Amer. J. Bot. **54**, 1069—1083.

Cronquist, A., 1968: The Evolution and Classification of Flowering Plants. Boston: Houghton Mifflin.

Dahlgren, K. V. O., 1927: Die Morphologie des Nuzellus mit besonderer Berücksichtigung der deckzellosen Typen. Jb. wiss. Bot. **67**, 347—426.

Dahlgren, R., 1975a: A System of classification of the Angiosperms to be used to demonstrate the distribution of characters. Bot. Notiser. **128**, 119—147.

— 1975b: The distribution of characters within an Angiosperm System I. Some Embryological Characters. Bot. Notiser **128**, 181—197.

Davis, G. L., 1966: Systematic embryology of the Angiosperms. New York-London-Sydney: John Wiley.

Eyde, R. H., 1967: The peculiar gynoecial vasculature of *Cornaceae* and its systematic significance. Phytomorphology **17**, 172—182.

Fagerlind, F., 1947: Die systematische Stellung der Familie *Grubbiaceae*. Svensk bot. Tidsskr. **41**, 315—320.

FISH, F., and WATERMAN, P. G., 1973: Chemosystematics in the *Rutaceae* II. The Chemosystematics of the *Zanthoxylum/Fagara* complex. Taxon **22**, 177—203.

FROHNE, D., and JENSEN, U., 1973: Systematik des Pflanzenreichs unter besonderer Berücksichtigung chemischer Merkmale und pflanzlicher Drogen. Stuttgart: Gustav Fischer.

GUPTA, S. C., 1964: The embryology of *Coriandrum sativum* L. and *Foeniculum vulgare* MILL. Phytomorphology **14**, 530—547.

HEGNAUER, R., 1963: The taxonomic significance of alkaloids. In: Chemical Plant Taxonomy (SWAIN, T., Ed.), 389—427. London-New York: Academic Press.

— 1969: Chemical evidence for the classification of some plant taxa. In: Perspectives in Phytochemistry (HARBOURNE, J. B., and SWAIN, T., eds.), 121—138. London-New York: Academic Press.

— 1971: Chemical patterns and relationships of *Umbelliferae*. In: The biology and chemistry of the *Umbelliferae* (HEYWOOD, V. H., ed.), 267—277. London-New York: Academic Press.

HIDEUX, M. J., and FERGUSON, I. K., 1976: The stereostructure of the exine and its evolutionary significance in *Saxifragaceae* sensu lato. In: The evolutionary significance of the exine (FERGUSON, I. K., and MULLER, J., eds.), 327—377. London: Academic Press.

HOWARD, R. A., 1974: The stem-node-leaf continuum of the *Dicotyledoneae*. J. Arnold Arb. **55**, 125—173.

JENSEN, S. R., NIELSEN, B. J., and DAHLGREN, R., 1975: Iridoid Compounds, their occurrence and systematic importance in the Angiosperms. Bot. Notiser **128**, 148—180.

KUBITZKI, K., 1969: Chemosystematische Betrachtungen zur Großgliederung der Dicotylen. Taxon **18**, 360—368.

LINDLEY, J., 1847: The Vegetable Kingdom, ed. 2. London: Bradbury and Evans.

MAURITZON, J., 1936: Zur Embryologie und systematischen Abgrenzung der Reihen *Terebinthales* und *Celastrales*. Bot. Notiser **1936**, 161—212.

— 1939: Die Bedeutung der embryologischen Forschung für das natürliche System der Pflanzen. Acta Univ. Lund. ser. 2, **35**, 1—70.

PALSER, B. F., PHILIPSON, W. R., and PHILIPSON, M. N., 1971: Embryology of *Rhododendron*. Introduction and ovule, megagametophyte, and early endosperm development in *R. yunnanense*. J. Ind. Bot. Soc. **50 A**, 172—188.

PHILIPSON, W. R., 1967: *Griselinia* FORST. fil.—Anomaly or link. New Zealand J. Bot. **5**, 134—165.

— 1974: Ovular morphology and the major classification of the dicotyledons. Bot. J. Linn. Soc. **68**, 89—108.

— 1975: Evolutionary lines within the dicotyledons. New Zealand J. Bot. **13**, 73—91.

PLOUVIER, V., and FAVRE-BONVIN, J., 1972: Les iridoides et séco-iridoides: répartition, structure, propriétés, biosynthèse. Phytochemistry, **10**, 1967—1722.

RODRIGUEZ, R. L., 1971: The relationship of the *Umbellales*. In: The Biology and Chemistry of the *Umbelliferae* (HEYWOOD, V. H., ed.), 63—91. London: Academic Press.

Sinnott, E. W., 1914: Investigations on the phylogeny of the Angio-
    sperms, I. The anatomy of the node as an aid in the classification of
    Angiosperms. Amer. J. Bot. **1**, 303—322.
Stebbins, G. L., 1974: Flowering Plants, Evolution above the species
    level. Cambridge, Mass.: Belknap Press.
Stern, W. L., Brizicky, G. K., and Eyde, R. H., 1969: Comparative
    anatomy and relationships of *Columelliaceae*. J. Arnold Arb. **50**, 36—75.
— Sweitzer, E. M., and Phipps, R. E., 1970: Comparative anatomy and
    systematics of woody *Saxifragaceae*, *Ribes*. In: New Research in Plant
    Anatomy. Suppl. 1 (Robson, N. K. B., Cutter, D. F., and Gregory,
    M., eds.), Bot. J. Linn. Soc. **63**, 215—237. London: Academic Press.
Takhtajan, A., 1969: Flowering Plants, Origin and Dispersal. Trans.
    Jeffrey, C. Edinburgh: Oliver and Boyd.
Thorne, R. F., 1968: Synopsis of a putatively phylogenetic classification
    of the flowering plants. Aliso **6**, 57—66.
Tieghem, P. van, 1901: L'oeuf des plantes considéré comme base de leur
    classification. Ann. Sci. nat. (Bot.) ser 8, **14**, 213—390.
Warming, E., 1878: De l'ovule. Ann. Sci. nat. (Bot.) ers. 6, **5**, 177—266.
— 1914: Observations sur le valeur systématique de l'ovule. Mindeskr.
    Japetus Steenstrups Føds. **24**, 1—45.
Wettstein, R. von, 1930—1935: Handbuch der systematischen Botanik,
    4th ed. Leipzig-Wien: Deuticke.
Wunderlich, R., 1959: Zur Frage der Phylogenie der Endospermtypen
    bei den Angiospermen. Österreich. Bot. Zeitschr. **106**, 203—293.

Address of the author: Professor W. R. Philipson, Department of
Botany, University of Canterbury, Christchurch, New Zealand.

Plant Syst. Evol., Suppl. 1, 141—153 (1977)
© by Springer-Verlag 1977

Institut für Systematische Botanik der Universität München, Federal
Republic of Germany

Seed Characters in and Affinities Among the *Saxifragineae*

By

J. E. Krach, München

Abstract: Neither the *Rosales* nor the *Saxifragales*, as defined by the
majority of recent authors, can be regarded as being monophyletic. Also
the term "slightly pleiophyletic" does nothing to elucidate the deep gaps
separating the several evolutionary lines usually combined to form the
order (or even suborder) *Saxifragales*.

The clear-cut concepts of an order *Crassulales* leave room for only the
herbaceous families *Crassulaceae* (including *Penthorum* and most of the
Saxifrageae), *Astilbaceae*, and *Peltiphyllaceae*. Some further genera (*Vahlia*,
Eremosyne, and *Cephalotus*) seem to be distantly related to these families.

The *Hydrangeaceae* and the *Escalloniaceae* proper share a remarkable
number of characteristics, which allows their combination as the *Escalloniales*.

The two otherwise very isolated genera *Ribes* and *Phyllonoma* may
be related.

The *Cunoniales* (inc. *Brunellia*, *Bauera*, *Pottingeria* and perhaps the
herbaceous *Francoaceae*) are another closely related group of families.

Only very doubtfully the woody orders *Escalloniales*, *Grossulariales*
and *Cunoniales* possess any closer affinity to the herbaceous families
Saxifragaceae and *Crassulaceae*.

Most of the taxa referred to as saxifragaceous have been so poorly
investigated that it is at present premature to make any far-reaching conclusions regarding their true relationship. Until more knowledge has been
obtained only their evolutionary trends can be suggested without a full understanding of their meaning for the taxonomic treatment of the whole group.

The *Rosiflorae* are still considered by many botanists as a natural
and even as a monophyletic order the broad limits of which correspond
to the "Reihe *Rosales*" of ENGLER (1930), although there is a multi-
plicity of publications for decades which point to the untenability of such
a procedure. In some of the more recent systematic proposals the three
Englerean Unterreihen *Hamamelidineae*, *Rosineae*, and *Saxifragineae*
are treated with otherwise unaltered boundaries as orders.

About the current classification of the genera related to *Hamamelis*, ENDRESS in his contribution to this symposium has said much more than I can. As far as the *Rosineae* in the Englerean sense are concerned I would like to restrict myself to one comment. The lumping together not only of the *Rosaceae* and *Fabaceae*, but also *Neuradaceae, Connaraceae, Chrysobalanaceae*, and *Krameriaceae*, is less and less convincing the more we know not only about the morphology of the flowers, but about the pollen, the anatomy of the wood and the biochemistry of representatives of these groups.

With the following arguments I hope to show also that the *Saxifragineae* of ENGLER (1930), corresponding to the *Saxifragales* of many more recent authors, consist instead of several evolutionary lines which are best not combined into one single order.

If one is in search of the evolutionary lines of a higher taxon it is certainly useful to find a starting-point, common to each of these lines; that is to determine certain basic characteristics from which the development of the recent taxa might have begun. For such a heterogenous taxon as the *Saxifragineae* classified by ENGLER, which envelops *Crassulaceae* and *Cunoniaceae, Pittosporaceae* and *Cephalotaceae* as well, this becomes a rather challenging task. It is best to limit oneself first to subcategories in order to achieve any result at all.

But presumably, merely by being forced to accept narrower horizons one is at first inclined to doubt in the monophyletic structure of the entire group. However, the search for unifying factors in a large group, in itself implies a certain homogeneity within this group. So it seems to me that the process used has itself some prejudicial connotations which cannot entirely be avoided.

A rather broad consensus of basic and essential characteristics within the Englerean *Saxifragineae* can be demonstrated, which applies to more than half of the genera of this taxon:

seeds anatropous, raphe clearly visible in the ripe seed;

two double-layered integuments;

outer epidermis of the seed coat thickened following the *Reseda*-type;

inner epidermis of the seed coat compressed to form a tanniniferous pigment-layer;

seed coat with tannins, but without lignin: cell-walls in the seed coat of the ripe seeds without visible pits;

in the ripe seeds scarcely any nucellar—but at least some endospermous tissue remains;

embryo straight, in the direction of the longitudinal axis of the seed;

cotyledons shorter than hypocotyl, the plumule, if present at all, being only a slight bump;

accumulation of sedoheptulose, no hamamelose;

root tips containing anthocyanin;

largely herbaceous, rarely having secondary xylem and phloem;

tendency to form five-whorled flowers with five parts in each[1].

By this correlation of characteristics the *Crassulaceae*, *Penthorum* and most of ENGLER's *Saxifrageae* are bound together. *Vahlia*, *Eremosyne* and *Cephalotus* might also be assigned to this otherwise very homogeneous group.

For the purposes of this discussion—the division of the *Saxifragales* in the broadest sense into monophyletic taxa—it is relatively unimportant whether one calls this grouping an order or a superfamily. For clarity I will henceforth refer to this group as the *Crassulales*.

Within this order *Crassulales* one can distinguish between several diverging evolutionary pathways. They can be differentiated not only by the presence of characteristics unique to each line but also by the lack of certain characters which are common to the other lines.

The seeds of *Vahlia*, depicted in ENGLER & PRANTL and HUTCHINSON (1964) as anatropic, do not show any traces of a raphe when ripe. This genus is clearly further distinguished by the mode of placentation —two (or three) large placentas hanging from the apex of the loculus— and the tenuinucellate ovules. The inclusion of *Eremosyne* in this group is uncertain. Many of the essential characters allowing a proper systematic placing are still unknown. I have never seen ripe seeds of *Cephalotus*. According to the present descriptions, this genus obviously is no less independent than *Vahlia*.

It is also possible to separate two small families, the *Astilbaceae* and the *Peltiphyllaceae* from the proper core of the order. The independence of these two families seems to have come about through an early departure from the main evolutionary lines.

The great similarity of traits exhibited by each of these families[2] might of course be due to the small number of genera and species in each, and their consequently lower variability.

The members of both these families still live in the (postulated) primary habitat (mesophytic forests) of the ancestors of the whole order *Crassulales*. Thus their development must have been separate from the

---

[1] The number of the parts between 2 and 20 is due to reduction and, also, to secondary multiplication.

[2] A further, more accurate investigation would, in my opinion, settle the question of whether *Peltiphyllum* has one or two integuments. In ripe seeds I have seen four clearly distinguishable cell-layers. In my opinion this is of greater importance than finding (or overlooking) a cuticule between them. It is generally less useful to discuss the number of integuments than to give the number of cell-layers in the integuments, a procedure I consider to yield more information.

evolutionary mainstream of this group, which adapted soon to xeric or montane-alpine biotopes.

A closely related group of genera (united by KLOPFER (1973) into the subfamily *Heucheroideae*) also remained mainly in mesophytic biotopes. They are reminiscent of the *Astilbaceae* and *Peltiphyllaceae* in their biochemistry and morphology (stipules, pinnate laminas).

If one only looks a little deeper than the gross-morphological details, there emerge webs of interrelationships between characters within the taxa concerned that tempt one to conclude that the real *Saxifragaceae* (corresponding to the *Saxifrageae* of ENGLER) and the *Crassulaceae* had a very long common development. I think this hypothesis is better than the two other possibilities (at the usual risk of overestimating the worth of the own findings):

1. a multitude of parallel developments of analogous structures in the *Saxifragaceae* and the *Crassulaceae*,

2. repeated turnoffs from the evolutionary mainstream which lead to the *Crassulaceae* of today by genera we now ascribe to the *Saxifragaceae*.

The contrary—a separate derivation of each of the several subfamilies or tribes of the *Crassulaceae*—is not a possibility at all. According to my knowledge nobody has yet furnished convincing proof that there are subfamilies or tribes within the *Crassulaceae* which are separated from each other by clear and mutually exclusive sets of characters.

Cross-connections of the type which exist between nearly all genera of the *Crassulaceae* (as shown by agreements in single, or sets of, characters) extend to genera of the *Saxifrageae* also. The development of the several evolutionary lines is, in my opinion, best described as a radiation outward from near one point (somewhat like the rays of an umbel).

To combine *Crassulaceae* and *Saxifragaceae* into one single family perhaps carries the process of unification too far. However, I am sure it is closer to the truth (real relationships) than to assign these two taxa to two completely different lines of evolution, and therefore to two separate orders (*Rosales* for the *Crassulaceae*; *Hamamelidales* for the *Saxifragaceae*), as for example GUNDERSEN (1950) did in his modification of the BESSEY-SKOTTSBERG system.

A far more difficult task than the tracing of the evolutionary branchings—in principle already long-known—of the *Crassulales* as accepted here is the identification of a common ancestor of this taxon and any possibly related taxa.

HUTCHINSON's subdivision of the dicots into *Herbaceae* and *Lignosae* is certainly interesting but has gained no general acceptance. According to students of systematics and taxonomy since BAILLON, there remain only a small and unimportant number of herbaceous species related to the

Crassulales[1]. And of this initially small number, some genera have subsequently proved to be not saxifragaceous at all, but to belong instead to several other orders or even subclasses. Therefore it seems necessary to search next-related taxa among the woody plants, particularly because of the rudimentary secondary thickening present here and there in several groups of the *Crassulales*.

Because, sometime after 1872, ENGLER included *Ribes* into his *Saxifrageae*, this genus is generally considered to be the next-of-kin to the *Crassulales*. Only in 1973 did TAKHTAJAN explicitly point to the closer relationship of *Escalloniaceae* to the *Crassulales*. Seed-anatomy can only provide very few indications for such a relationship: thus a unification of *Crassulales* and *Escalloniaceae* into a single order has to be regarded as very problematical, at least, particularly in view of the many divergent characteristics and tendencies.

If one attributes only the correctly placed genera to the *Escalloniaceae*[2] one finds that they can be characterized by the following characteristics:

lamina entire;

no stipules;

formation of an intrastaminal disc;

presence of hamamelose, accumulation of aluminium;

fruit a septicidal capsule (berries are very rare) composed of between five to two, more or less fused carpels;

transition from hypogynous through perigynous to epigynous flowers;

placentas axial or lateral with numerous seeds;

tenuinucellate;

tendency towards forming a single double-layered integument;

tendency to reduction of the embryo-size in ripe seeds;

tendency towards lengthening of the walls of the outer layer of the seed coat, which is formed according to the *Erica*-type.

The seven genera of the *Escalloniaceae* can be easily separated into two tribes: the round-seeded *Cuttsieae* (*Cuttsia, Abrophyllum, Argophyllum*) and the *Escallonieae* (*Quintinia, Forgesia, Escallonia, Valdivia*) with scobiform seeds.

It is possible to expand the family *Escalloniaceae* by adding *Anopterus* in a subfamily of its own (reduction of the number of the carpels to two, fused to a unilocular perigynous ovary, seeds with large wings,

---

[1] Unfortunately the fairy tale of a great similarity between ͵he *Spiraeoideae* and the *Astilbaceae* is still perpetuated, and is used as a basis for establishing a much closer taxonomic relationship between the rather distantly related *Rosales* and *Crassulales* than is in fact justified.

[2] In some current systems this family has been turned into a catch-all for free-floating taxa with no precise taxonomic attachments.

presence of a third layer in the seed coat, extreme reduction of the seed-core, minute embryo).

Theoretically one could add a further subfamily, the *Carpodetoideae* (including *Carpodetus* and *Argyrocalymna*), but these genera more probably belong to the *Ericales* as evidenced by pollen tetrads and hypertrophic cuticule of the nucellus. The round seeds, anatomically somewhat similar to those of the *Cuttsiae*, and the reserve cellulose indicate a special position for these two genera among the *Ericales*, too.

The assembly of characters which typify the *Escalloniaceae*, and which are regarded, not only by myself, as progressive, also fully characterize another family, the *Hydrangeaceae*, from the Englerean *Saxifragineae*. Beneath the cloak of the *Saxifragales* s.l. these two families quite certainly form a genuinely monophyletic taxon, to be referred to here as *Escalloniales*. Within the *Escalloniaceae* the *Escallonieae* have achieved the developmental stratum which is already broadly distributed among the *Hydrangeaceae*. Naturally, the *Escalloniaceae* and the *Hydrangeacae* are characterized by their own phylogenetic trends.

Hydrangeaceae	*Escalloniaceae*
the number of cell-layers in the integument varies according to genus (and investigator); normally only 2 cell-layers remain traceable in the ripe seeds	only one two-layered integument (except *Anopterus*)
petals free	tendency to sympetaly
tendency to reduction of the number of the seeds	many seeds per placenta
endosperm cellular	endosperm nuclear
tendency towards formation of palmately veined leaves	leaves pinnately veined
stamens 5 to many	stamens (4-) 5 (-6)

The uniform and well-definable family *Hydrangeaceae* can, in my opinion, be divided into the two subfamilies centred on *Philadelphus* and *Hydrangea* respectively only if one is willing to restrict oneself to the use of one single characteristic. To do so ignores important evolutionary ties. Thus I cannot support, e.g., HUTCHINSON's splitting of the *Hydrangeaceae* into two families, but I consider the basic place he gave the *Carpentiereae* and their connection by *Carpentiera* to the *Philadelpheae* correct. Anyhow the division into tribes by HUTCHINSON seems to be more natural than that of ENGLER from the view-point of seed-anatomy. On the basis of seed anatomy—and therefore in need of further support—a subdivision of the *Hydrangeaceae* is easily estab-lished into five tribes: *Carpentiereae, Decumarieae, Philadelpheae, Hydrangeae,* and *Dichroeae.*

Kirengeshoma proves not to be a member of this family. I even wonder whether this monotypic japonic genus is a member of the *Saxifragales* in a very broad sense at all. Seed anatomical investigation provides no support for leaving this, often cultivated, garden plant in its attributed place in the system, where it has been an oddity also because of the inconsistency of its appearance with the rest of the group.

In this connection I should like to point to the fact that uncritical acceptance of results and findings in the literature, especially regarding the morphology of *Hydrangeaceae,* may lead to false conclusions. Even the drawings in ENGLER & PRANTL are not always as precise as could be desired, as BUXBAUM (1951) already has explicitly mentioned. These inaccuracies cannot always be attributed to the rather woodcut-like printing methods used in those texts.

Instead of the supposition that there is a closer kinship between *Escalloniaceae* and *Hydrangeaeceae,* one might be inclined to think of them as two evolutionary streams which developed parallel to one another in the southern and northern hemispheres respectively. But even then the development had to begin with woody plants, which bore a large number of crassinucellate, double-integumented seeds in hypogynous capsules fused from 5 carpels with axial placentas. Besides containing a medium-sized embryo, the ripe seeds also had plenty of endospermous —but no nucellar—tissue. Thus they were also likely to be the ances-tors of the *Crassulales.*

Naturally, bearers of those characters are widespread throughout the plant kingdom. Thus the possession, or the possible derivation from possessors of those characters, is not necessarily a sign for an actual relationship to the *Crassulales* or *Escalloniales* either. Among those taxa which have been held to be related with the *Escalloniales,* *Ribes* and the *Cunoniaceae* share a considerable amount of these characters.

The *Pittosporaceae,* whose relationships might be found also within this group, have characteristics intermediate between the primitive ones mentioned above and the progressive ones. The theore-tical possibility that this family had shared part of an evolutionary pathway with the *Escalloniaceae,* which are also inhabitants of the southern hemisphere, cannot be fully denied.

Despite having the basic characters already described, the genus *Ribes* is still rather isolated within the *Rosiflorae,* not only by the morphology of its seeds, but also by a host of other differing characters. The graving suspicion I first got from this, that is that *Ribes* might not belong to this group at all, is based on the fact that many similarities consist of rather primitive characteristics, which are broadly distributed within the *Rosidae-Dilleniidae*, at least among the more primitive taxa.

Meanwhile, however, a number of progressive characteristics that *Ribes* shares with other members of the Englerean *Saxifragaceae* have come to light, making the probability of a parallel development, conceiveable for a single character, in this case rather small. Widespread progressions are: the palmate venation of the laminae, the lack of stipules, the sometimes monosexual flowers, or the development of a berry. Even the shape of the receptacle in *Ribes*, which ENGLER used as a basis to unite this genus with the *Saxifrageae*, might be found among other taxa of the *Saxifragales* in a broad sense with further careful investigation. This hypanthium is otherwise not occurring very widespread, if one excludes the *Onagraceae-Myrtales* which are certainly only very distantly related.

The small subsidiary cells, which also occur in *Escalloniaceae* (which genus?) and *Cunoniaceae* can certainly not be called a primitive character. The indumentum, composed either of long villi with multicellular feet and rounded heads, or—even further developed—composed of hairs with feet recessed in epidermal depressions and with shield- or bowl-shaped heads as occur similarly in *Escalloniaceae* and the as yet poorly known *Pterostemon*, is an equally progressive character. The dwarfish form of the embryo in comparison to the rather large seeds and the massive endosperm, and even the occurrence of rests of nucellar tissue in the ripe seeds, can be ascribed to neoteny.

There is still no consensus as to whether the crystal-layer, so significant for *Ribes*, develops from the external epidermis of the inner integument or from the internal epidermis of the outer integument. The answer to this question may be vital to clarify the relationship of *Ribes* to the *Cunoniaceae*. The development of the inner epidermis of the inner integument into a tannin-layer, however, is a rather widely distributed feature within the *Rosiflorae* (and of course in other subclasses also).

Independent progressions which do not necessarily appear in all subgenera are the development of spines (from epidermal tissue as in *Rosa*!) and of glands (which are sometimes even secretory) on leaves, stems and fruits. *Ribes* also demonstrates notably unique developments in its seed structures: The development of an aril-like tissue from the funiculus occurs nowhere else among the *Rosiflorae* as far as I know, but the name "aril" is unfortunately used for several tissues of heterogeneous origin.

The many-layered outer integument may still be an archaic character. It is equally possible that the development of the highly specialized sarcotesta could be attributed to a secondary proliferation of the cell-layers. The occurrence of still functional chloroplasts even in cells of the ripe seed-coat, can hardly be designated as a new development, but at best as an example of partial neoteny. I am not quite certain, whether the storage of reserve-cellulose in the cell-walls of the endosperm can be similarly explained. This storage of reserve-cellulose occurs in similar structures within the *Montiniaceae* as well, which exhibit comparable structures also in other areas of seed development. A convincing proof of a closer relationship between *Ribes* and the *Montiniaceae* however, is still lacking.

Further, very independent and progressive characteristics are found in the embryogeny which one as yet has not been able to place satisfactorily in any standard category.

I cannot agree with W. H. CAMP (in GUNDERSEN 1950) who states that *Ribes* has been derived from the *Escalloniaceae* as this view is based only on superficial, gross-morphological similarities. Not only are the two families *Escalloniaceae* and *Grossulariaceae* too dissimilar, but the *Escalloniaceae* have also achieved a much higher degree of development in many important aspects. The thought that *Ribes* developed at the time of the uplifting of the Andes as a specialized, low-temperature adapted mountain shrub which then crossed to North America via the Andean bridge and afterwards spread throughout the holarctis is, nonetheless, interesting. The occurrence in South America of only the two subgenera *Parilla* and *Berisia*, which are, according to JANCZEWSKI (1913), highly derived, speaks against CAMP's theory, which also fails to explain the occurrence of one single member of *Parilla* (*P. sardoum*) in another continent[1]. The development of *Escallonia*-like (or, as I would rather have it, *Fuchsia*-like) and, moreover, unisexual flowers in these two subgenera can be easily explained as a convergent development (adaptive evolution with the same pollinators: hummingbirds and Sphingidae).

Besides the far less striking correspondences with the *Escalloniaceae*, *Cunoniaceae*, and, as has now been emphasized, with the *Montiniaceae* too, *Ribes* exhibits an actual relationship concerning seed structure only to one genus within the *Rosiflorae*. This is the small (only 4 species) genus *Phyllonoma*, which is restricted to the Andes between Mexico and Bolivia. The inflorescences of *Phyllonoma*, which are reminiscent of those of *Helwingia*, project from the middle or almost from the top of their leaves.

The seeds of the species (*P. ruscifolia*), which I have investigated, completely lack the crystal-layer so significant for *Ribes*, but this

---

[1] I suppose that the classification of *Ribes sardoum* is also indication of the deeply interwoven relationships of characteristics within the one and single genus *Ribes*. Due to this web, any subdivision of this genus can be based only on the use of a very small number of characteristics, rather than by regarding different character-matrices simultaneously.

evolutionary step—if it is one—is anticipated by the occurrence of "windows" in the crystal-layer even in some of the investigated seeds of *Ribes uva-crispa* (not only the cultivars), and *Ribes diacanthum*. Both these genera share—in addition to characteristics of the flower and the fruit[1]—the following traits:

seed-coat composed of a multi-layered outer and a two- (or three-?) layered inner integument;

outer epidermis of the seeds fully packed with mucilage (myxotesta!);

continual reduction in cell-volume of the outer integument preceding from the outside inwards;

metamorphosis of the cells of the epidermis of the inner integument to a thick tannin-layer which retains its cellular structure;

storage of reserve-cellulose in unpitted cell-walls along with fats and aleurone;

tendency towards development of seeds with small embryos.

Except for the first and fourth, these characteristics have to be viewed as a common progression away from the original characteristics of the ancestral taxon. I am quite certain, that further investigations will reveal further common traits between *Ribes* and *Phyllonoma*. Therefore I find it proper to unite these two genera—and only these two—into the order *Grossulariales*. Both these genera command such a great number of independent characters, however, that they are better each treated as separate monogeneric families.

Potentially, the *Cunoniaceae* possess a seed-coat which has far more layers than that of *Ribes*. This is demonstrated by the seeds of *Schizophragma ilicina*, which are merely four millimeters long, but nonetheless sit inside a stone which is of the size of a peach pit and equally ruminated. Is it conceivable that there is a selective mechanism which in this case could induce such an already well-protected seed to increase the number of layers in its seed-coat?

Thus, for the present, this wealth of layers (approximately 25 cell-layers organized into seven distinct strata) is regarded as a primordial characteristic. Such multilayered seed-coats, however, occur only in species in which the seed itself is not the diaspore. By means of gradual reduction, the seed structure of all the other *Cunoniaceae* in a broad sense (including *Bauera* and *Brunellia* = hereafter referred to as *Cunoniales*) can be derived from the *Schizophragma* model.

ENGLER's classification of the *Cunoniaceae* into tribes is partially based on insufficient knowledge of the individual members. HUTCHINSON

---

[1] *Phyllonoma* had not yet been investigated for other characters. The recent results of MORI and KALLUNKI could no more be included.

completely dispensed with a subdivision, as he often did when uniting a decidedly incongruent conglomerate into a family.

By considering the different degrees of reduction of seed-coat structure (sometimes combined with transformations in single tissues) we can differentiate between the following groups: *Schizophragma*; *Brunellia*; *Aistopetalum*, *Bauera*; *Cunoniaceae* s.s.[1]; *Ceratopetalum*. The *Cunoniaceae* s.s. could be further divided into three groups[2], which partially correspond with the tribes of ENGLER. By this grouping I deliberately avoid defining the single groups as (sub)tribes or (sub)families.

Through a further such reduction one can arrive at the seeds of *Francoa*, although I am not at all certain whether this derivation is entirely hypothetical or not. Yet I cannot deny that it is possible to link this genus to the *Cunoniales*, perhaps as a suborder of its own, but sufficient positive evidence is still lacking.

It is far less difficult to seek out the relatively clear-cut evolutionary trends in the development of the seeds of the *Cunoniales* than to correlate these structures with those of the fruits, flowers and inflorescences. Although the seed of *Schizophragma* within its stone retains primitive characters, I do not believe that the one-seeded drupe of this genus should be regarded in any way as primordial to the *Cunoniales* as a whole. Dehiscent capsules composed of five to two locules containing many seeds are more likely to be primitive. One can regard as progressive the reduction in the number of the seeds partially associated with a reduction in size of the flowers and fruits, and their concentration into a capitulum. Another trend is towards the production of indehiscent fruits, gradually gaining size, which were initially dry (*Aistopetalum*) but then became berries (*Opocunonia*) or drupes (*Schizomeria*).

Based on the previous results it is possible to conclude that from a hypothetical prototype (the so-called prosaxifrageous basis) several main-lines of evolution yielded four currently tangible orders: *Crassulales*; *Escalloniales*; *Grossulariales*; *Cunoniales*. These mainstreams are disparate in the number of their genera and of course in the number of their species. Any attempt to amalgamate these four taxa as suborders[3] into one order called *Saxifragales*[4] is hindered by the suspicion that not these four taxa only evolved from this common ancestor. If one were to imagine an organism from which *Crassulales*, *Escalloniales*, *Grossulariales*, and *Cunoniales* might all have evolved, one can, without

---

[1] Including *Pottingeria*.

[2] I did not get ripe seeds from ENGLER's tribe IV or V.

[3] Despite this I would not object a lowering in rank of each of these categories.

[4] Moreover I do not like the rather vague term *Saxifragales* which, chameleon-like, means something different to every user.

excessive mental gymnastics, imagine that a majority of the taxa of the *Rosidae* could easily have had that organism as an ancestor as well.

From this prototype, so far defined by seed- and flower-characteristics and a few biochemical traits, one can also conceivably derive several certainly distant taxa, if one stipulates the same number of progressions, although not in the same directions. Thus, the doubts whether the four orders treated really had a common ancestor subsequent to the basic branching of the *Rosidae* or even earlier could perhaps be diminished though surely not obviated.

By this I certainly do not mean to imply that all that which has, up to now, comfortably hidden under the blanket name of *Saxifragales*, should remain there. Most of the current systems include under the category of *Saxifragales*, or even *Rosales*, one or another family—not to mention the mostly not-mentioned genera—which have absolutely nothing to do with these taxa. There are other families whose affiliation to these taxa appears to be questionable, because the investigation of other complexes of traits besides the gross-morphology of leaf and flower is still lacking. Thus any broadly supported conclusions are premature. Perhaps in this case it is best to take to heart a command I learnt from H. HUBER: It is far better to admit to an incomplete system of (at least sufficiently) completely known plants than to boast of a complete system of incompletely known plants.

I am very greatly obliged to WILLIAM COLMERS, M.A. for his patient help with the English translation.

References
(For further literature see KRACH 1976)

BUXBAUM, F., 1951: Erneuerung der Systematik der höheren Pflanzen. Wien: Springer.

ENGLER, A., und PRANTL, H., eds., 1930: Nat. Pflanzenfamilien, 2. Aufl., Bd. 18 a: *Saxifragineae*. Leipzig: W. Engelmann.

GIBBS, R. D., 1974: Chemotaxonomy of flowering plants. Vol. I and IV. Montreal and London: McGill-Queen's Univ. Press.

GUNDERSEN, A., 1950: Families of Dicotyledons. Waltham, Mass: Chronica Botanica Co.

HIDEUX, M. J., and FERGUSON, I. K., 1976: The stereostructure of the exine and its evolutionary significance in *Saxifragaceae* sensu lato. In: The evolutionary significance of the exine (FERGUSON, I. K., and MÜLLER, J., eds.). Linnean Society Symposium series no. 1.

HOOGLAND, H. G., 1961: Studies in the *Cunoniaceae* 1. Austr. Journal Botany **8**, 318—341.

HUBER, H., 1963: Die Verwandtschaftsverhältnisse der Rosifloren. Mitt. Bot. Staatssamml. München **5**, 1—48.

HUTCHINSON, J., 1964: The genera of flowering plants. Vol. 1. Oxford.

— 1967: The genera of flowering plants. Vol. 2. Oxford: Clarendon Press.

JANCZEWSKI, E. DE, 1913: Monographie des Grosseillers V. Bull. Internat. Acad. Sc. Cracovie **B 10**, 714—741.

KLOPFER, K., 1973: Florale Morphogenese und Taxonomie der *Saxifragaceae* sensu lato. Feddes Rep. **84**, 475—516.

KRACH, J. E., 1976: Samenanatomie der Rosifloren I. Die Samen der Saxifragaceen. Bot. Jahrb. Syst. **97**, 1—60.

MORI, S. A., and KALLUNKI, J. A., 1977: A revision of the genus *Phyllonoma* (*Grossulariaceae*). Brittonia **29**, 69—84.

TAKHTAJAN, A., 1973: Evolution und Ausbreitung der Blütenpflanzen. Jena: G. Fischer.

Address of the author: Dr. J. ERNST KRACH, Institut für Systematische Botanik, Menzingerstraße 67, D-8000 München 19, Federal Republic of Germany.

Plant Syst. Evol., Suppl. 1, 155—178 (1977)

Lehrstuhl für Zellenlehre der Universität Heidelberg,
Federal Republic of Germany

Transmission Electron Microscopy and Systematics of Flowering Plants

By

H.-Dietmar Behnke, Heidelberg

Abstract: From the large number of TEM investigations of plant cells only two characters have so far proved useful for the classification of higher taxa of flowering plants.

1. Dilated cisternae of endoplasmic reticulum in root cap cells and bundle parenchyma cells are a characteristic feature of members of *Capparales* and will be discussed with respect to their systematic reliability.

2. The micromorphological differences identified within sieve-element plastids represent the only character presently available which is generally applicable to the higher taxa of seed plants: The major distinction between types of sieve-element plastids is based on the presence (P-type) or absence (S-type) of protein accumulations, while their micromorphology and combinations define subtypes of sieve-element plastids. Specific subtypes characterize the *Monocotyledoneae* (P II) and the *Centrospermae* (P III). *Magnoliales/Laurales* (P I) and *Fabales* (P IV) contain distinct subtypes in some of their families and genera, while others have only S-type. *Aristolochiales*, *Vitineae*, *Eucryphiaceae*, *Gunneraceae* contain subtype P I-plastids.

Classification and delimitation of higher taxa in flowering plants can be aided by utilizing the different types of sieve-element plastids, as has been best demonstrated in the *Centrospermae* but is also applicable to monocotyledons and a number of other taxa. An analysis of investigated species and projection into the total number of extant angiosperms results in a distribution of 35% P-type and 65% S-type.

In discussing phylogenetic relationships the plastid data are superimposed on CRONQUIST's, DAHLGREN's, EHRENDORFER's and TAKHTAJAN's angiosperm systems. Finally, arguments for the P-type being the ancestral sieve-element plastid are presented and it is suggested that on multiple occasions the S-type (by loss of protein) has been derived.

Introduction

During the last two decades transmission electron microscopy (TEM) aided by ultramicrotome techniques has proven a powerful tool in plant morphology. Ultrastructural details increased our knowledge of cell organisation and stimulated research in various fields, including, for example, physiology, biochemistry, and genetics. Moreover, these data greatly improved our understanding of the way cells function.   Despite the large number of TEM investigations limited information is relevant for systematics; in addition, most of the data are species-specific and not applicable to higher taxonomic categories.

With regard to seed plants a recent survey on "Electron Microscopy and Plant Systematics" (Cole & Behnke 1975) lists only a few characters which are useful in the classification of families and higher taxa and most of these were determined with the scanning electron microscope (SEM).

Comparative pollen morphology includes by far the most electron microscopic investigations concerned with plant systematics.   The ultrastructure of the pollen wall has been studied since the introduction of TEM into botany, but its systematic implications are interpretable only in combination with results derived from light microscopy (LM), SEM, and TEM studies, and therefore are excluded from the present discussions.   For a comprehensive survey of palynological research concerned with the phylogeny of flowering plants see Walker & Doyle (1975).

Thus, at the present time there are only two TEM-determined characters available, both recently developed, which fit the requirements of our survey: they are applicable to higher categories of flowering plants and are entirely based on TEM methods: 1. Specific dilated cisternae of endosplasmic reticulum occurring in a variety of cell types in at least two families of *Capparales*, and 2. specific plastids present as two main types in the sieve elements of the food-conducting phloem tissue in all taxa of flowering plants (as well as gymnosperms).

Both these characters will be considered in detail and their reliability as systematic markers tested against several current systems for flowering plants.

Material and TEM Methods

TEM studies using the ultrathin sectioning techniques require fresh, living material for the satisfactory preservation of cytoplasmic organelles.

For comparative investigations, material from different sources is best fixed in aldehydes followed by osmic acid, dehydrated by acetone and embedded in epoxy resins, all by standard procedures (e.g. for sieve-element plastids described by Behnke 1975a). Small pieces (about 1–5 mm^3) of the plant material which is penetrated and surrounded by polymerized epoxy resins, after recutting into yet smaller parts (0.1 to 0.2 mm side length) containing the appropriate tissue (e.g. phloem with sieve elements), are sliced into 50 to 100 nm (= mm^{-6}) ultrathin sections. These when viewed with a TEM, in general, give pictures of dissected organelles, not always showing all their specific structures or inclusions. For example, in the investigations of sieve-element plastids (the diameter of which is about 2 μm) 20 to 40 contiguous serial ultrathin sections need to be screened in order to be aware of all their contents. Complete series, however, are difficult to cut in routine preparations and are not needed in most samples, such as those containing either S-type or P II–P IV subtype plastids (see Fig. 25). But if those P I-forms, which contain only tiny protein crystalloids (e.g. in *Laurus* and *Magnolia*), are under investigation, complete series would prove helpful, unless the protein is detected early among the single sections.

Since fresh material is a prerequisite for TEM studies, it is often difficult to obtain the necessary samples for a systematic survey on the distribution of a particular TEM-determined character. Although some plant collections from remote areas such as Australia, New Zealand, and Africa which were sent by Air Mail did survive, in general, our investigations have utilized the limited taxa grown in European Botanical Gardens. The difficulties in obtaining fresh material from Oceania, South America and Africa leave incomplete the investigation of many interesting groups of *Magnoliidae* as well as some of the *Caryophyllidae* (e.g. *Barbeuiaceae, Microtea, Molluginaceae*) and others.

Compared to the problem of obtaining fresh plant material all the other problems associated with the material are insignificant:

Like many other biologists not basically trained in taxonomy I often must rely on the specification of Botanical Gardens, which are not always correct (*Myristica*, in my experience, is a good example: I examined four collections before I obtained one which I now believe is *Myristica fragrans*[1]). There are, of course, errors in collecting species (see, for example, BEHNKE 1976, p. 34 concerning *Phytolacca acinosa*).

TEM-Characters in Higher Taxa of Flowering Plants

1. Dilated Cisternae

Dilated sections of the endoplasmic reticulum with a filamentous content in root cells of *Raphanus sativus* L. were first described and termed "dilated cisternae" (DC) in 1965 by BONNETT & NEWCOMB. Similar structures containing a tubular protein were found in the phloem parenchyma of leaf veins of *Brassica chinensis* L. by FAVALI & GEROLA (1968). IVERSEN (1970 b) initiated the first systematic overlook on the occurrence of DC in *Brassicaceae*; JØRGENSEN et al. (in prep.) detected DC in *Capparis cynophallophora* L.

A compilation of all data concerning DC (cf. Table 1) suggests that these organelles are typical of *Brassicaceae* and *Capparaceae*. Their internal structures (protein filaments or protein tubules) appear to depend on the location of the cells inspected: While in root cap and root epidermal cells the DC content is filamentous, in bundle parenchyma

---

[1] The investigated *M. dactyloides* GAERTN. (Bot. Garten, München) and *M. fragrans* HOUTT. (Garten der Landw. Botanik, Bonn) also differ from the dubious *Myristica species* (which were *Annona*) by the presence of "extruded nucleoli" in their sieve elements.

cells of leaves and stem the DC contain protein tubules. The latter are closely-packed and parallel to the long axis of the cisternae, which at the same time determine their length, while their diameter is about 25 nm (Figs. 1–3).

IVERSEN (1970a) performed a histochemical test for myrosinase and discussed the localization of this glucosinolate-hydrolyzing enzyme in the DC. Since, however, DC were not found in other glucosinolate containing families like *Resedaceae* (IVERSEN 1970b), *Bataceae*, *Gyrostemonaceae*, *Salvadoraceae* (JØRGENSEN et al. in prep.), their involvement in the hydrolysis of glucosinolates is not cogent. At least the cited families can do well without DC. So far, DC represent a micromorphological character for *Brassicaceae* and *Capparaceae*, only.

Table 1. Genera of *Capparales* which contain DC.

Brassicaceae

Allyssoides	
Alyssum	IVERSEN (1970b)
Arabis	
Armoracia	JØRGENSEN et al. (in prep.)
Brassica	FAVALI & GEROLA (1968), IVERSEN (1970b)
Cardamine	
Cardaminopsis	IVERSEN (1970b)
Cochlearia	
Descurainia	
Diplotaxis	CRESTI et al. (1974)
Draba	IVERSEN (1970b)
Eruca	CRESTI et al. (1974)
Iberis	IVERSEN (1970b)
Isatis	BEHNKE & ESCHLBECK (unpubl.)
Lepidium	IVERSEN & FLOOD (1969)
Lunaria	BEHNKE & ESCHLBECK (unpubl.)
Raphanus	BONNETT & NEWCOMB (1965)
Sinapis	IVERSEN (1970b), HAVELANGE & COURTOY (1974)
Sisymbrium	IVERSEN (1970b)
Thlaspi	IVERSEN (1970b), HOEFERT (1975)

Capparaceae

Capparis	JØRGENSEN et al. (in prep.)
Cleome	BEHNKE (unpubl.)

Figs. 1–3. Dilated cisternae (DC) in *Capparales*. Figs. 1 and 2. DC in bundle parenchyma cells of *Raphanus sativus*, *Brassicaceae* (hypocotyledons). Inside the DC are parallel arranged protein tubules (ø about 300 nm; length: same as DC). 1: ×10,000; 2: ×30,000. Fig. 3. DC in phloem-parenchyma cells of *Capparis cynophallophora*, *Capparaceae* (stem). Contents of the DC not dissolved into single protein tubules. ×10,000. ER = endoplasmic reticulum, SE = sieve element; marker = 1 μm

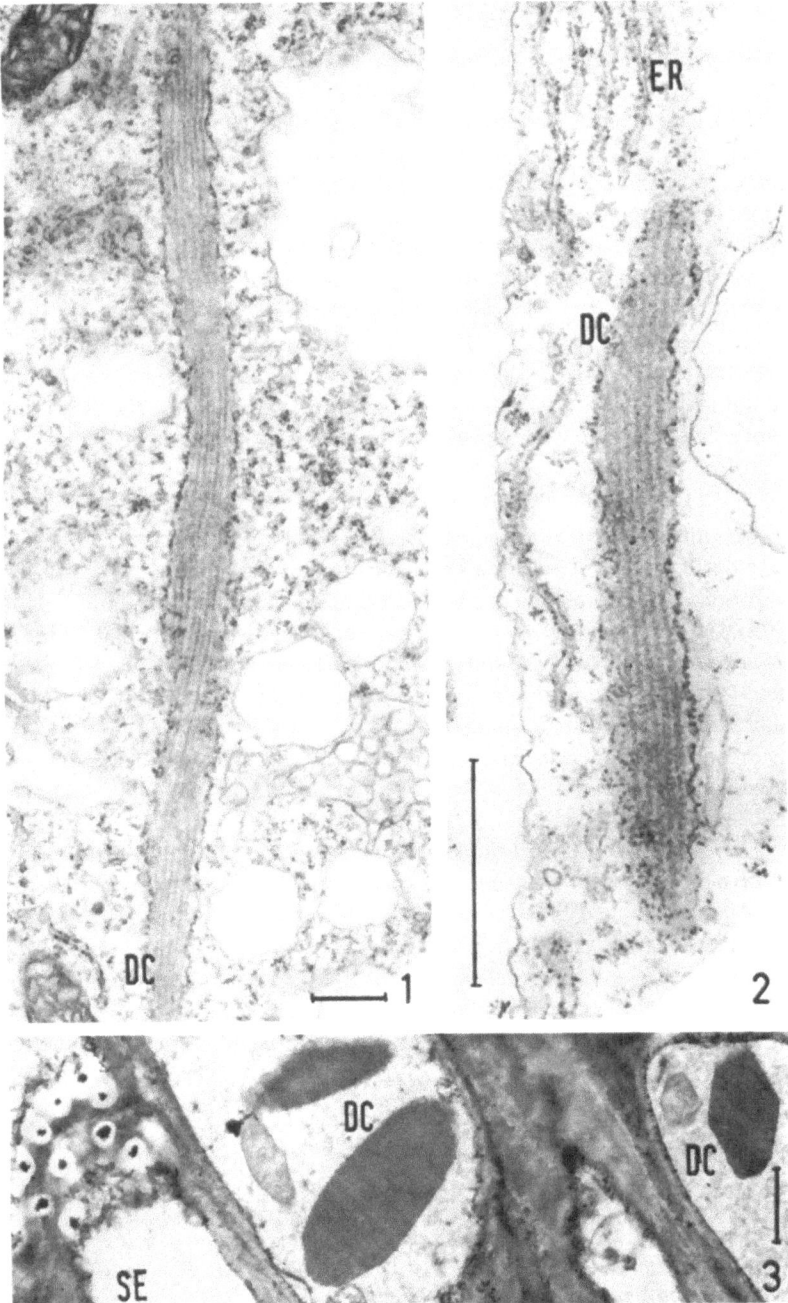

Figs. 1–3

2. Sieve-Element Plastids

Only a decade has passed since sieve-element plastids were first demonstrated to incorporate ultrastructural details which can be used to characterize different plant taxa. The *Dioscoreaceae*, in many morphological and anatomical respects uncommon among monocotyledons, was the first family reported to contain sieve-element plastids whose micromorphology differed from the then known pattern (Behnke 1965, 1967). A subsequent test of 21 families including representatives of nearly all respective orders of *Monocotyledoneae* revealed sieve-element plastids with the same characteristics (Behnke 1969a). Continued research on all dicotyledon subclasses stepwise (follow e.g. Behnke 1969b, 1972, 1975a, 1975b) led to the present knowledge on the distribution of plastid types and subtypes among the flowering plants, based now on 690 species from about 190 families[1].

Types, Subtypes, and Forms of Sieve-Element Plastids

The ultrastructural features of sieve-element plastids which are of systematic value are associated with accumulations of protein and starch.

Starch is present as grains differing in number, size and shape (Figs. 4–6). Single or even all grains of a sieve-element plastid may consist of many darkly stained, small particles (Fig. 6) thus reflecting their glycogene-like chemical composition, while others appear to be solid and evenly formed spherules.

Protein is accumulated as crystalloids and filaments, and the amount, shape and arrangement of both may differ among sieve-element plastids of different taxa.

The protein accumulations are used as the main discriminating features in the classification of types, subtypes, and forms of sieve-element plastids:

Types. Presence of a protein accumulation defines the P-type plastids, whereas absence of any protein accumulation defines the S-type plastids.

In this definition, the starch accumulation is of no primary impor-

---

[1] As of Takhtajan (1973).

Fig. 4. Longitudinal section through stem phloem of *Lindera praecox*, *Lauraceae*: Sieve elements (SE) with many plastids. × 6,000

Figs. 5–7. S-type sieve-element plastids. Fig. 5. *Stephania glabra*, *Menispermaceae*: starch grains (S) of different size. × 20,000. Fig. 6. *Byblis gigantea*, *Byblidaceae*: starch grains (S) composed of globular particles with glycogen-like appearance. × 20,000. Fig. 7. S_0-subtype plastids without any starch grains in *Celtis orientalis*, *Moraceae*. × 20,000.

M = mitochondrium, PC = phloem-parenchyma cell, marker = 1 μm

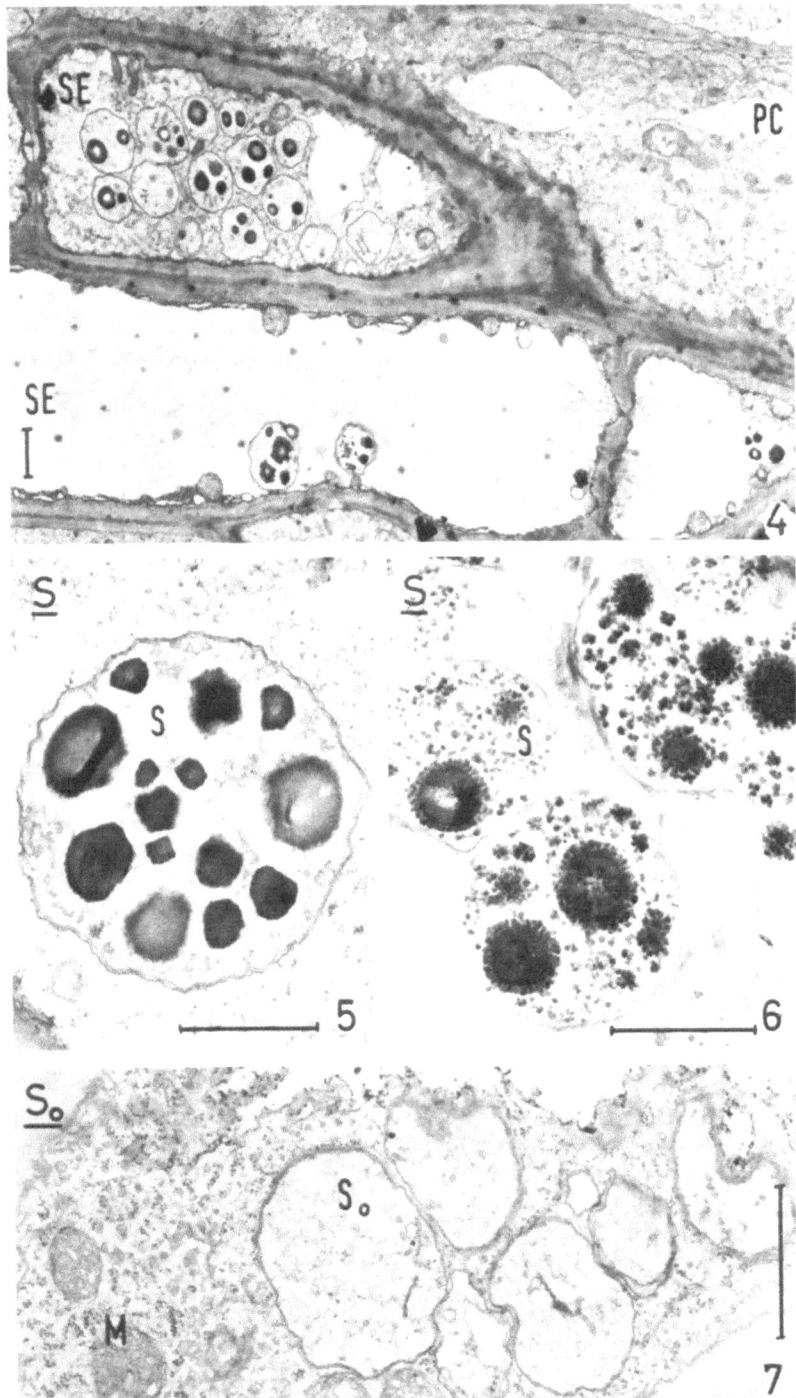

Figs. 4–7

tance for the classification of the plastids since it may be present or absent in both types. If in S-type plastids starch is absent (see e.g. Fig. 7 and Behnke 1973) the plastids are classified as belonging to subtype S_0.

Subtypes. For the division of P-type sieve-element plastids into subtypes some conspicuous protein accumulations and their combina-

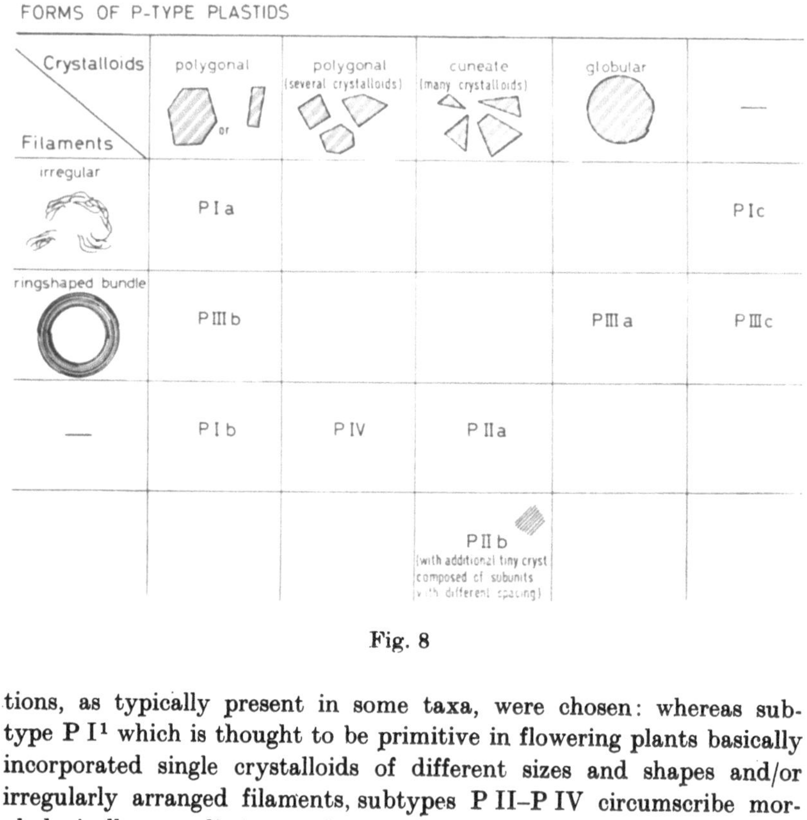

FORMS OF P-TYPE PLASTIDS

Crystalloids / Filaments	polygonal	polygonal (several crystalloids)	cuneate (many crystalloids)	globular	—
irregular	P I a				P I c
ringshaped bundle	P III b			P III a	P III c
—	P I b	P IV	P II a		
			P II b (with additional tiny cryst. composed of subunits with different spacing)		

Fig. 8

tions, as typically present in some taxa, were chosen: whereas subtype P I[1] which is thought to be primitive in flowering plants basically incorporated single crystalloids of different sizes and shapes and/or irregularly arranged filaments, subtypes P II–P IV circumscribe morphologically very distinct and probably more advanced P-type plastids.

Characteristic features are: many cuneate crystalloids in P II (as of all Monocotyledons), one ring-shaped bundle of filaments in P III (as of all *Centrospermae*), many, prevalent polygonal crystalloids in P IV (as of 60% of *Fabales*).

Forms. Most of the subtypes can be subdivided into different forms of P-type sieve-element plastids which are typical of some families or

---

[1] For clarification a consecutive numbering system of all subtypes is introduced; however, this simplified system does not render the previous detailed formulas (Behnke 1975a) invalid.

family groups within the respective higher taxon. Forms are indicated by lower case letters. Out of the number of possible forms resulting from a free combination of crystalloids and filaments, Fig. 8, in a two-dimensional alignment, lists those realized (cf. also Fig. 25). Starch was found to be present at least in some species belonging to forms P I a–c, P II a, P III a, P IV.

The Distribution of P- and S-type Sieve-Element Plastids. Their Subtypes and Forms Among Flowering Plants

P-Type. A brief account for all families of flowering plants which contain P-type sieve-element plastids, also mentioning the appropriate subtypes and forms, is given in Table 2. A description of the subtypes and their main representatives follows here:

P I-Subtype. The P I-subtype is held the most primitive among the P-subtypes of flowering plants. Of all subtypes P I shows the

Table 2. Families of flowering plants which contain P-type sieve-element plastids

1. Dicotyledons

Atherospermataceae	P I a	*Gunneraceae*	P I b
Achatocarpaceae	P III a	*Gyrocarpaceae*	P I b
Agdestidaceae	P III a	*Halophytaceae*	P III a
Aizoaceae	P III a	*Hectorellaceae*	P III a
Amaranthaceae	P III c	*Hernandiaceae*	P I b
Annonaceae	P I a	*Lauraceae* *	P I b
Aristolochiaceae	P I a/P II a	*Leeaceae*	P I b
Basellaceae	P III a	*Magnoliaceae* *	P I b
Brassicaceae *	P I b	*Mimosaceae*	P IV
Cactaceae	P III a	*Molluginaceae*	P III a/b
Caesalpiniaceae	P IV	*Monimiaceae*	P I a/b
Calycanthaceae	P I a	*Nyctaginaceae*	P III a
Canellaceae	P I c	*Petiveriaceae*	P III a
Capparaceae *	P I b	*Phytolaccaceae*	P III a
Caryophyllaceae	P III b	*Portulacaceae*	P III a
Chenopodiaceae	P III c	*Putranjivaceae*	P I b
Didiereaceae	P III a	*Rhabdodendraceae*	P I b
Dysphaniaceae	P III c	*Stegnospermaceae*	P III b
Eucryphiaceae	P I b	*Tetragoniaceae*	P III a
Eupomatiaceae	P I b	*Ulmaceae* *	P I b
Fabaceae *	P IV	*Vitaceae*	P I b
Gisekiaceae	P III a		

2. Monocotyledons   P II,
   all of the families (36, as of Takhtajan 1973) so far investigated, with *Poaceae* belonging to form P II b.

---

* Families which contain P-type and S-type genera.

highest variability within the limits classified earlier: e.g., the sizes and shapes of the single crystalloids within forms P I a and P I b show considerable variation among the taxa to be concerned.

The primary distribution of P I is within *Magnoliales*, *Laurales* and *Aristolochiales*.

Roughly two thirds of the so far investigated families of *Magnoliales* and *Laurales* contain P I sieve-element plastids: *Magnoliaceae* (Fig. 13), *Eupomatiaceae*, *Annonaceae*, *Canellaceae* (Fig. 12), *Monimiaceae* (Fig. 10), *Atherospermataceae*, *Hernandiaceae*, *Gyrocarpaceae*, *Calycanthaceae*, *Lauraceae* (Figs. 14, 15) (see Table 2 for distribution of forms among these families); while the other contain S-type plastids, only: *Myristicaceae*[1], *Winteraceae*, *Austrobaileyaceae*, *Trimeniaceae*, *Chloranthaceae*.

Members of the two families from which the orders *Magnoliales* and *Laurales* derive their names contain both S-type and P I-subtype species and are considered representative for the plastid distribution within these orders. The P-type species of their genera *Magnolia*, *Michelia*, *Laurus*, *Lindera*, *Ocotea*, *Sassafras*, *Umbellularia* all contain form P I b plastids. However, their crystalloids tend to be the tiniest of all of those recorded in P-type plastids. In addition, some crystalloids are so inconspicuous that they are easily overlooked in a routine examination (Figs. 13–15). This faint appearance of protein (as of *Ocotea* and *Sassafras* in Figs. 14, 15) is suggested to be an indication that here, in the *Magnoliales* and *Laurales*, is one of the multiple transitions from P-type to S-type plastids (see also discussion on p. 172).

Aristolochiales (and its family *Aristolochiaceae*) is heterogeneous with respect to P-subtypes: *Aristolochia* (Fig. 9) has P I a plastids whereas *Asarum* forms P II plastids, typical of monocotyledons.

Other higher taxa which contain P I-subtype plastids (form P I b) are: 1. *Eucryphiaceae*, treated either in *Rosales* (CRONQUIST 1968, THORNE 1968), *Saxifragales* (TAKHTAJAN 1973) or *Cunoniales* (DAHLGREN 1975); 2. *Gunneraceae*, included in *Haloragales* (CRONQUIST 1968) or *Hippuridales* (TAKHTAJAN 1973), but raised into the unigeneric *Gunnerales* by DAHLGREN (1975); 3. *Vitaceae* and *Leeaceae*

---

[1] Formerly recorded as P-type (BEHNKE 1971, 1972), see comments under "Material and TEM Methods".

Figs. 9–12. P I-subtype sieve-element plastids. Fig. 9. *Aristolochia argyroneura*, *Aristolochiaceae*: form P I a plastid with polygonal crystalloid (C), irregular filaments (F) and starch grains (S). × 50,000. Fig. 10. *Peumus boldus*, *Monimiaceae*: form P I b plastids with polygonal crystalloids (C) and starch grains (S). × 30,000. Fig. 11. *Cissus antarctica*, *Vitaceae*: form P I b plastids. × 30,000. Fig. 12. *Canella alba*, *Canellaceae*: form P I c plastids with irregular to ring-like arranged filaments (F) and starch grains (S). × 30,000. marker = 1 μm

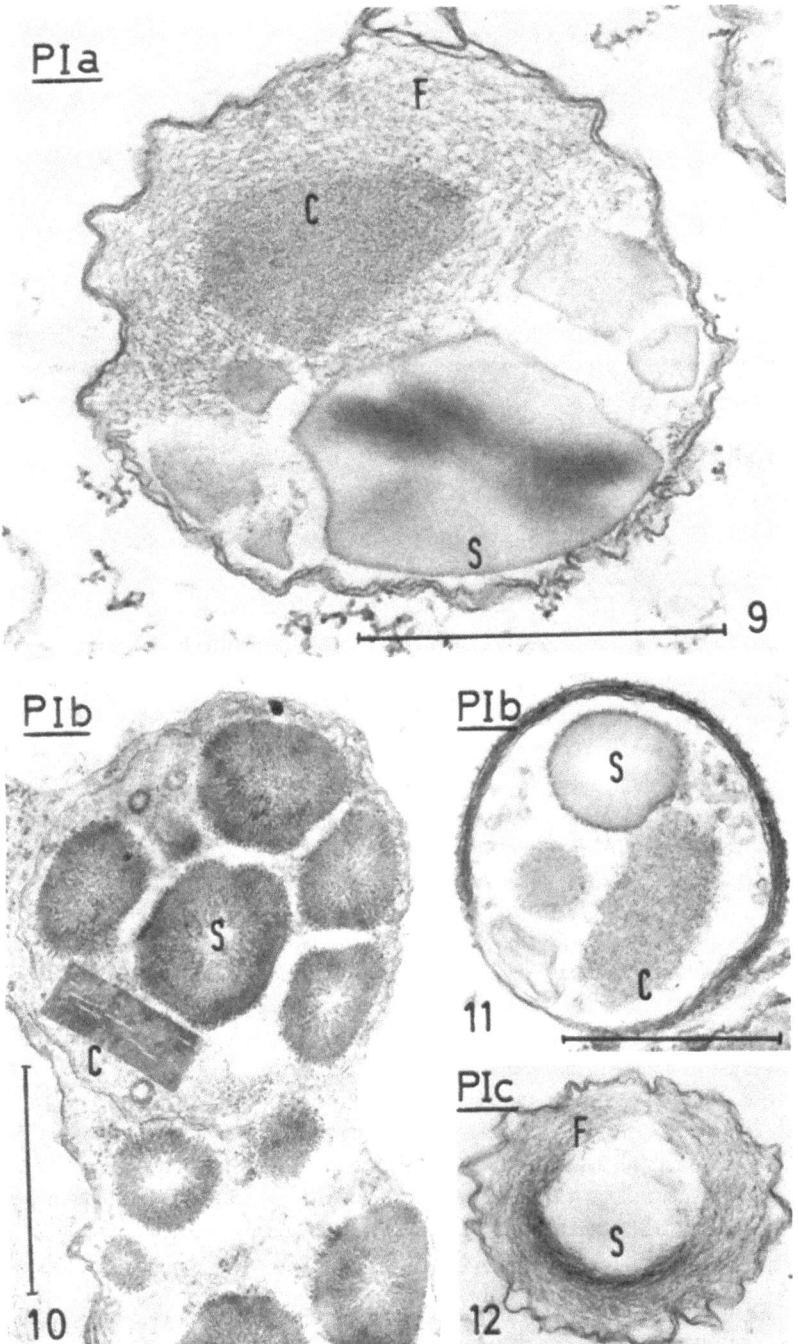

Figs. 9–12

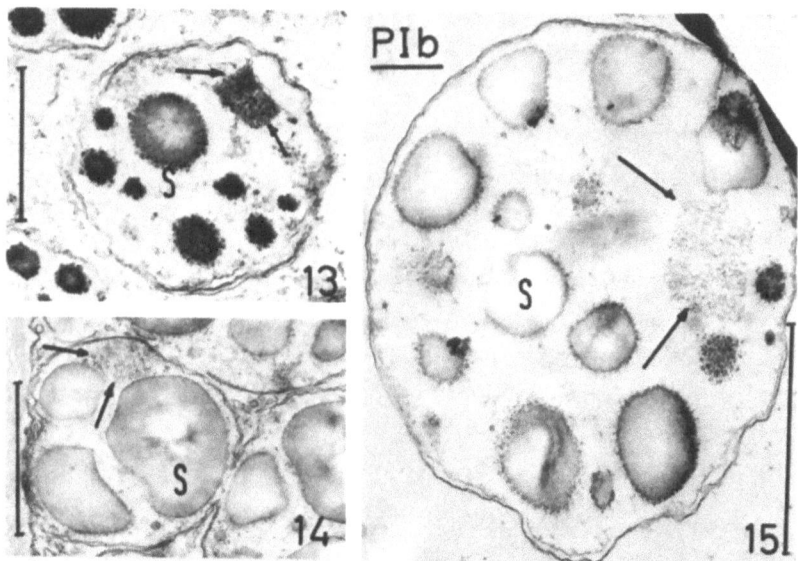

Figs. 13–15. P I-subtype sieve-element plastids with small and inconspicuous crystalloids (arrows): Fig. 13. *Magnolia angustifolia, Magnoliaceae*; Fig. 14. *Sassafras albidum, Lauraceae*; Fig. 15. *Ocotea foetens,
Lauraceae*. Figs. 13 and 14 ×20,000; Fig. 15 ×30,000. marker = 1 μm

usually together with S-type family *Rhamnaceae* aligned within *Rhamnales*, however, placed into *Cornales* by Thorne (1968) and suggested
by Johnston (1973) to be next to *Saxifragales*.

In addition, a few isolated genera scattered throughout dicotyledons
also belong to the group of plants which form P Ib sieve-element
plastids: *Ulmus* (in *Ulmaceae*), *Rhabdodendron* (usually in *Rutaceae*),
Brassica (of *Brassicaceae*), and *Capparis* (of *Capparaceae*).

It is remarkable that this P I-subtype, except for perhaps P IV
(see p. 170) is the only subtype which occurs in genera or families which
are systematically treated near S-type containing taxa.

P II-subtype. Cuneate crystalloids, oriented towards the centre of
the plastids, are the characteristics of the P II-subtype (Fig. 16). The
sieve-element plastids of all of the so far investigated members of *Monocotyledoneae* fall, without exception, within the range of this subtype

Figs. 16–18. P II-subtype sieve-element plastids with cuneate crystalloids
(C): Fig. 16. *Vanilla planifolia, Orchidaceae*; ×40,000. Fig. 17. *Costus
montanus, Costaceae* (with additional starch grains, S); ×15,000. Fig. 18. *Sac-
charum officinarum, Poaceae* (with additional crystalloid composed of substructures with wider spacing than in cunes; see arrow); ×25,000. marker
in Fig. 16 and 17 = 1 μm

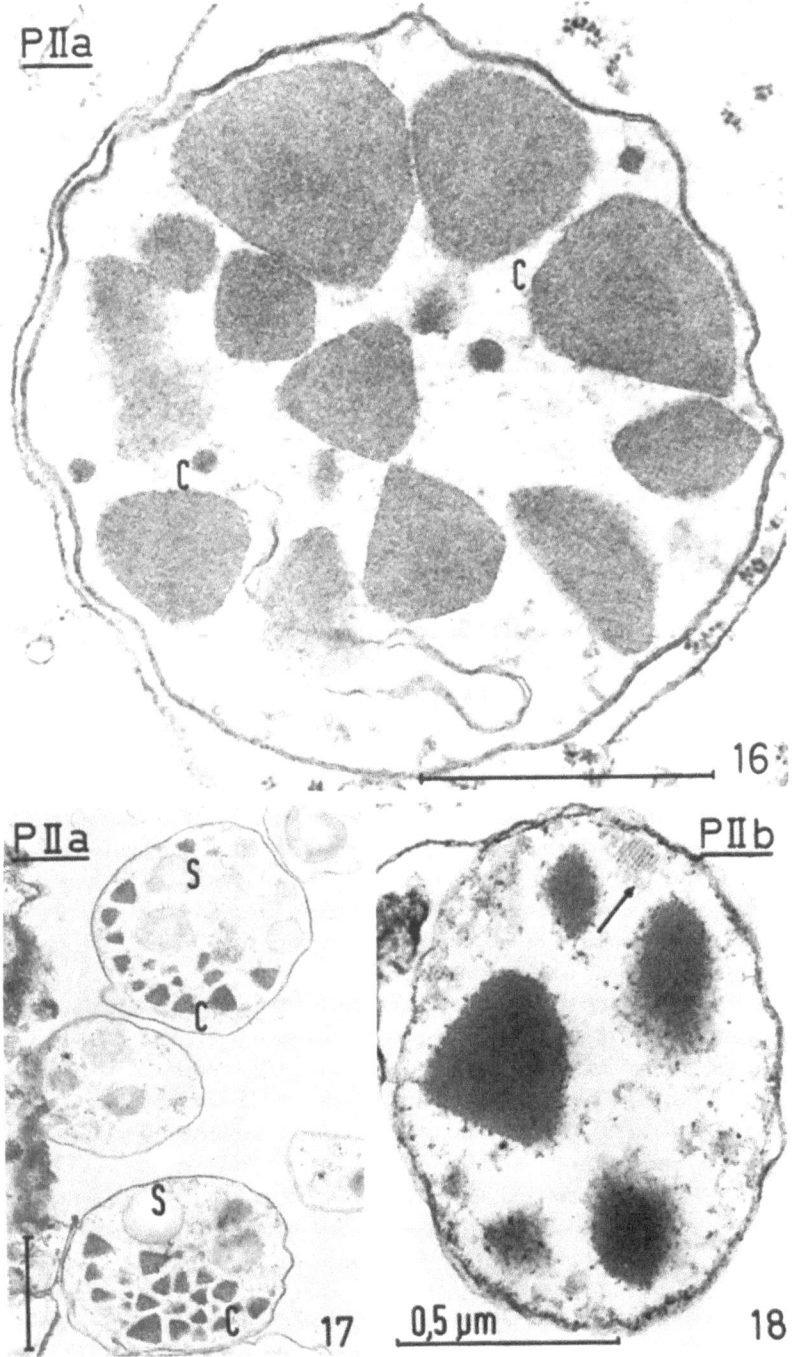

PIIa

C

C

16

PIIa

S

C

S

C

17

PIIb

0,5 µm

18

Figs. 16–18

and there is only one P II-subtype genus which does not belong to the monocotyledons: *Asarum*. Some species, or even entire orders (e.g. *Zingiberales*), have starch accumulations in addition to the cuneate crystalloids of the P II-subtype (Fig. 17).

The family *Poaceae* besides the cuneate crystalloids in their sieve-element plastids contains one or more small crystalloids composed of substructures (Fig. 18) with a wider spacing than in the cunes; these plastids therefore being defined as form P II b.

P III-subtype. The P III-subtype is confined to the *Centrospermae* and in my view can be used for the delimitation of this order (see Mabry & Behnke 1976). A ring-shaped bundle of protein filaments apposed at the periphery of the plastids is the subtype-characteristic (Figs. 19–22). The three presently recognized forms of this subtype differ by the presence or absence and the shape of an additional central crystalloid: form P III a with a globular crystalloid (Figs. 19, 20) in *Achatocarpaceae*, *Agdestidaceae*, *Aizoaceae*, *Basellaceae*, *Cactaceae*, *Didiereaceae*, *Gisekiaceae*, *Halophytaceae*, *Hectorellaceae*, *Molluginaceae* (in part), *Nyctaginaceae*, *Petiveriaceae*, *Phytolaccaceae*, *Portulacaceae*, *Tetragoniaceae*; form P III b with a hexagonal crystalloid (Fig. 21) in *Stegnospermaceae*, *Caryophyllaceae* and *Limeum* (of *Molluginaceae*); form P III c (Fig. 22) without any crystalloid in *Amaranthaceae*, *Chenopodiaceae* and *Dysphaniaceae* (for detailed description see Behnke 1976).

Absence of the P III-subtype has proven to be of considerable help for excluding from the *Centrospermae* the families *Bataceae*, *Gyrostemonaceae*, *Theligonaceae*, *Vivianiaceae* (Behnke 1976) all of which were doubtfully aligned with the *Centrospermae*. With a few exceptions (e.g. *Barbeuiaceae*, *Microtea* of *Phytolaccaceae*, and *Molluginaceae*) the plastid investigation of the *Centrospermae* with about 160 species recorded is regarded as representative.

The division of the *Centrospermae* into three family-groups as done by Behnke (1976), each classified by a distinct form of P III-subtype plastids, almost exactly corresponds to an earlier proposal by Friedrich (1956). In his dendrographic figure, based on flower diagrams, Friedrich (1956: Fig. 5) derives the three suborders *Portulacineae*, *Caryophyllineae*, *Chenopodiineae* from the basic *Phytolaccineae*. Except for the *Gyrostemonaceae* which he keeps in his family list but separates in his diagram and for the *Achatocarpaceae* which are separated in the diagram, but kept within the *Phytolaccineae*, Friedrich's *Phytolaccineae* and *Portulacineae* fit into the form P III a group of Behnke (1976). It is of special interest to note that Friedrich (1956) derives *Caryophyllineae* from *Phytolaccineae* through *Stegnosperma* and part of *Molluginaceae*, where Behnke (1976) found the specific form P III b

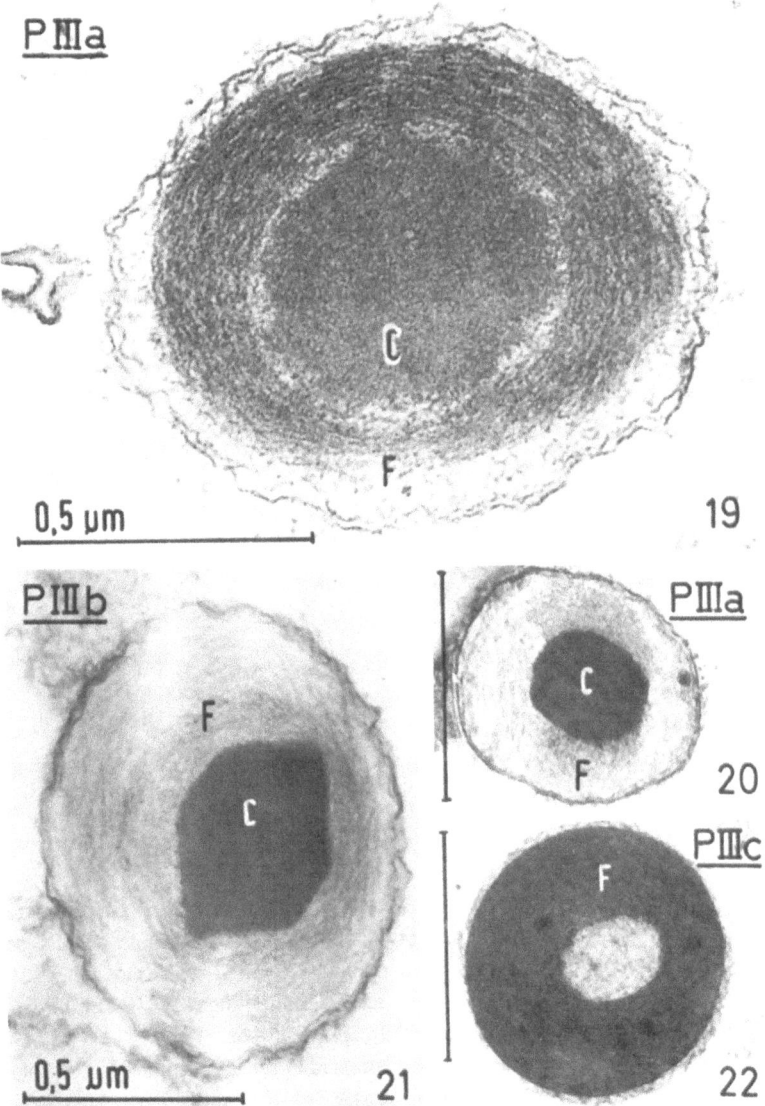

Figs. 19–22.   P III-subtype sieve-element plastids with ring-like bundle of filaments (F):   Fig. 19.   *Decaryia madagascariensis, Didiereaceae*: form P III a with globular crystalloid (C); × 80,000. Fig. 20. *Gibbaeum angulipes, Aizoaceae*: form P III a, but bundle cross-sectioned (F); × 30,000. Fig. 21. *Silene cucubalus, Caryophyllaceae*: form P III b plastid with polygonal crystalloid (C); × 60,000.   Fig. 22. *Alternanthera sessilis, Amaranthaceae*: form P III c plastid without crystalloid; × 30,000. marker in Figs. 20, 22 = 1 μm

plastids. A minor disagreement is in placing *Paronychiaceae* within *Chenopodiineae* (Friedrich 1956) instead of lumping them with *Caryophyllaceae* s. str. because of identical plastid subtypes (Behnke 1976).

   P IV-subtype. Of all the P-subtypes this is the least detected, probably because it is the most restricted. P IV-subtype sieve-element plastids, which contain a few polygonal crystalloids of variable size

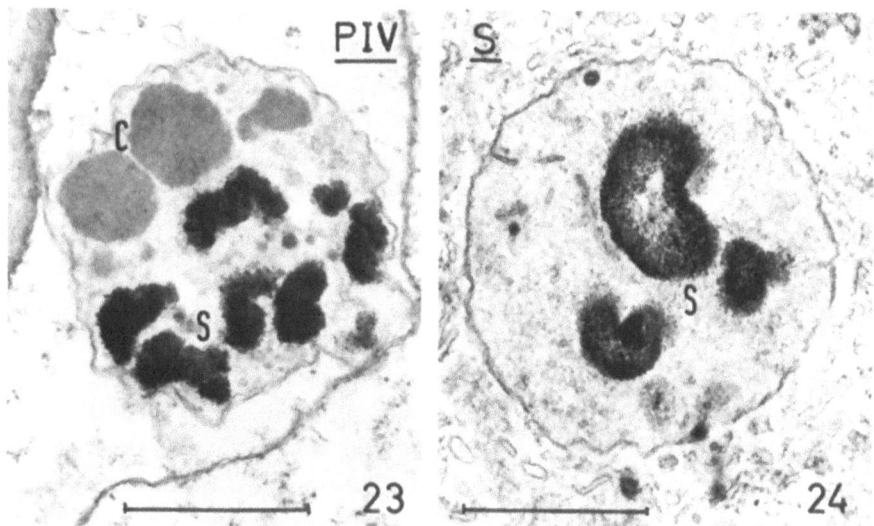

Figs. 23 and 24. Sieve-element plastid of *Fabales*. Fig. 23. P IV-subtype plastid with several polygonal crystalloids (C) and starch grains (S) in *Astragalus cicer*, *Fabaceae*. Fig. 24. S-type plastid of *Hippocrepis comosa*, *Fabaceae*. × 25,000; marker = 1 μm

(Fig. 23), are found in about 60% of the investigated *Fabales*, the others having S-type plastids (Fig. 24). In *Mimosaceae* and *Caesalpiniaceae* only P IV-subtype plastids were found, however, the total number of species (35) studied so far, is insufficient to reach a final decision on the distribution of the S- and P-types.

   S-Type. S-type sieve-element plastids lacking protein deposits but with starch accumulating in grains of different sizes and forms (e.g. Figs. 5, 6, 24) are the most common in flowering plants. A rough calculation based on projections from investigated species indicates that 65% of all flowering plants have S-type plastids.

   Among the subclasses listed by Takhtajan (1973) and their orders investigated the following are almost without exception S-type taxa:
   Ranunculidae: *Illiciales, Nelumbonales, Ranunculales, Papaverales, Sarraceniales.*

Hamamelididae: Trochodendrales, Cercidiphyllales, Eupteleales, Hamamelidales, Eucommiales, Urticales (except for *Ulmus*), *Barbeyales, Casuarinales, Fagales, Betulales, Myricales, Juglandales, Leitneriales* (for detailed list see BEHNKE 1973).

Dilleniidae: Dilleniales, Paeoniales, Capparales (exceptions see under P I-subtype), *Tamaricales, Salicales, Ericales, Euphorbiales.*

Asteridae: Gentianales, Scrophulariales, Asterales.

In the other dicotyledon subclasses the S-type is:

1. Prevailing in **Rosidae**: *Saxifragales* (except for *Eucryphiaceae*), *Rosales, Nepenthales, Myrtales, Rutales, Sapindales, Geraniales, Cornales, Celastrales, Oleales* (but with priority of P-type in *Fabales* and *Rhamnales*, P-type also in *Gunneraceae*).

2. Equal with P-type in **Magnoliidae**: S-type orders: *Piperales, Rafflesiales, Nymphaeales*; some S-type families in: *Magnoliales, Laurales*.

3. In a minority in **Caryophyllidae**: *Polygonales, Plumbaginales*, and *Gyrostemonaceae, Bataceae, Vivianiaceae* of *Caryophyllales* (sensu TAKHTAJAN).

S₀-subtype. Sieve-element plastids which show neither starch nor protein accumulations are recorded as S_0-subtype (see definition on p. 162); these include all investigated species of *Rafflesiaceae, Crassulaceae* and some species of *Ulmaceae, Moraceae, Urticaceae, Rosaceae, Coriariaceae.*

As for the other types and subtypes of sieve-element plastids a reason for the presence or absence of both protein and starch is not evident; different concentrations and qualities of carbohydrates present in the sieve-elements (as listed by KLUGE 1967) also offer no obvious answer.

Interrelationships Between Types and Subtypes of Sieve-Element Plastids

Subtypes and forms of P- and S-type plastids and their suggested interrelationships are presented diagramatically in Fig. 25. Based on the comparatively small number of taxa investigated (about 690 species and 550 genera of 190 families[1]) this compilation might appear to be incomplete. However, the detection within flowering plants of P-type plastids, nevertheless, is probably relatively representative: a comparison with the diagram given in 1975 (Fig. 18 in BEHNKE 1975 b) which included only 500 species from flowering plants easily demonstrates that data for an additional 190 species (with the emphasis on *Rosidae*) did not significantly change the concept.

---

[1] As of TAKHTAJAN (1973).

Therefore, with regard to the origin of plastid types and subtypes in flowering plants we still consider the P-type as primitive and the S-type as derived.

P 1-subtype is by far the most ancestral type of protein-accumulating plastids demonstrating the maximum variability of forms; it is also most often in taxa which are closely associated with others con-

INTERRELATIONSHIPS BETWEEN PROTEIN AND STARCH ACCUMULATING SIEVE-ELEMENT PLASTIDS

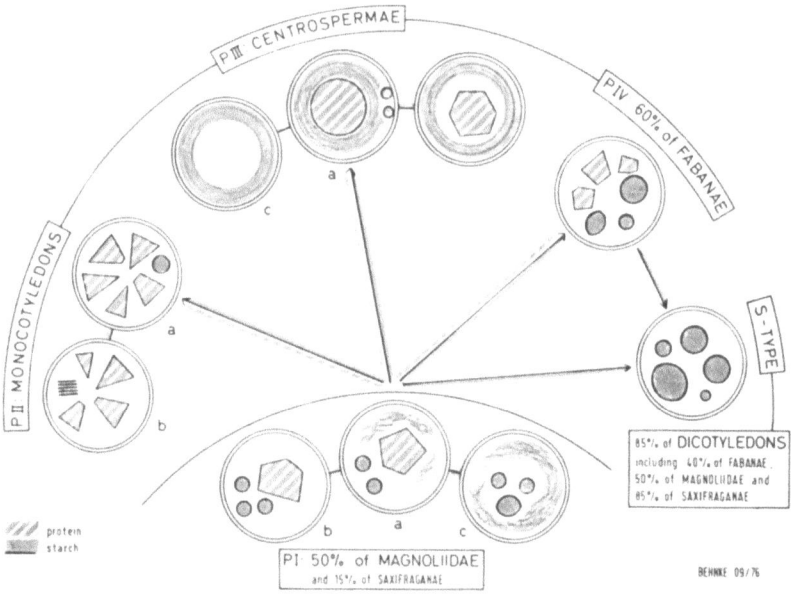

Fig. 25

taining S-type plastids (possibly indicating points of transitions from the P- to S-type, e.g. in *Magnoliaceae* and *Lauraceae*). This subtype is also most similar to the sieve-element plastids of *Pinaceae* (only record of P-type in *Gymnospermae*, Behnke 1974).

Besides P I, subtype P IV also occurs in taxa which are allied with others containing S-type plastids. It is interesting to note that its protein crystalloids could possibly be derived by a splitting and/or multiplication of the single crystalloid in P I.

Both P II- and P III-subtypes are more uniform in their basic pattern, and are therefore especially characteristic for the taxa in which they occur. They are considered to be derived comparatively early from the P I-subtype. There are, in the extant taxa investigated so far, no transitions from P II- and P III-subtypes to the S-type. Indeed, starch has been lost almost completely in these subtypes.

Complete loss of enzymes for the protein accumulations and consequently the formation of the more advanced S-type plastids is suggested to have occurred several times (as indicated in Fig. 25). It is not to be excluded, of course that S-type plastids could also directly be derived from the original angiosperm ancestors.

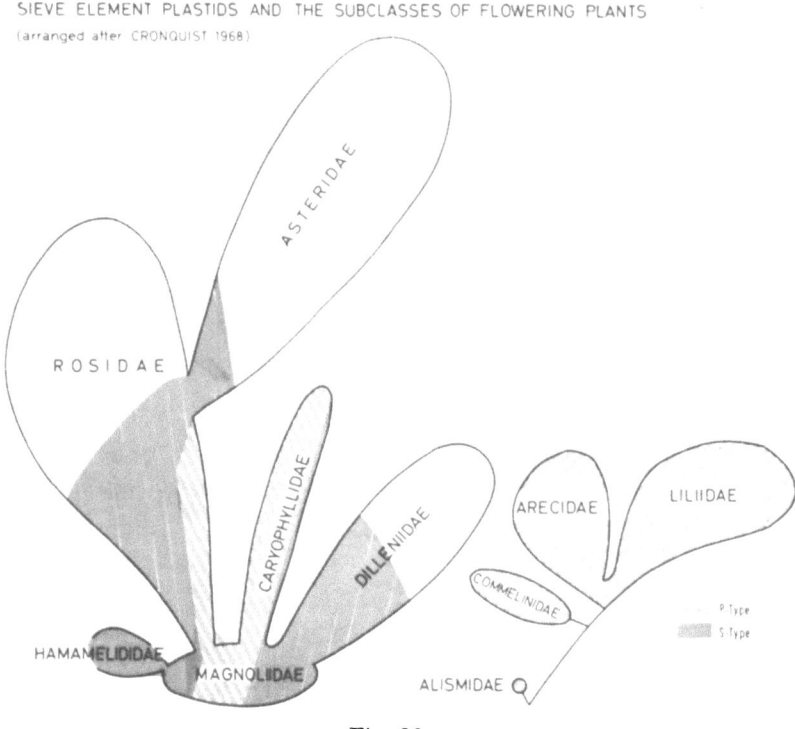

Fig. 26

The Incorporation of the Plastid Data Into Existing Systems of Flowering Plants

CRONQUIST's (1968), EHRENDORFER's (1971) and TAKHTAJAN's (1973) more or less similar systems for flowering plants initially appeared to be adequate for superimposition of the distribution of P-type plastids as a discriminating feature.

At the class level the uniform P-type *Liliopsida* (P II-subtype) are clearly separated from the *Magnoliopsida* with P- and S-type taxa.

As far as subclasses are concerned, there is a restriction of P-type plastids to only three of the six subclasses as e.g. recognized by CRON-QUIST (1968): *Magnoliidae, Caryophyllidae* and *Rosidae* (Fig. 26). The same graph (Fig. 26), however, demonstrates that unlike in *Liliopsida*

SIEVE-ELEMENT PLASTIDS AND THE SUPERORDERS OF DICOTYLEDONS
(arranged after EHRENDORFER 1971)

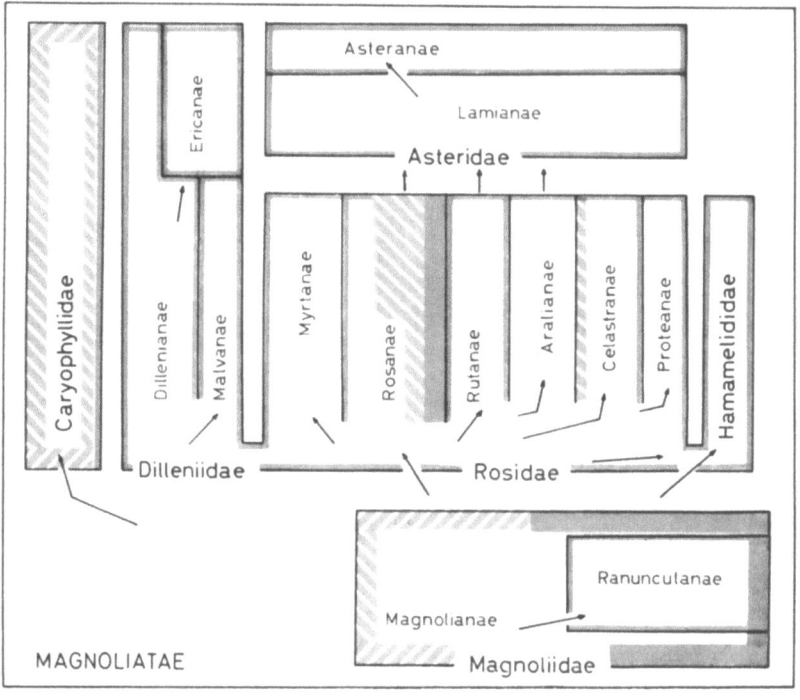

/// = P-TYPE,   ███ = S-TYPE

Fig. 27

these three subclasses also contain S-type plastids (varying from 10% to 85%).

A subdivision of the subclasses into superorders (e.g. demonstrated by EHRENDORFER 1971) confines P-type plastids in *Magnoliidae* to part of the *Magnolianae* but in *Rosidae* still allocates them to two superorders, *Rosanae*, and *Celastranae* (Fig. 27).

Descending to the order level e.g. of TAKHTAJAN's (1973) dendro-gram we are still confronted with the same problem: in addition to the basic *Magnoliales* and *Laurales* there are at least five orders which contain both types of plastids (see Fig. 28): *Caryophyllales*, *Saxifragales*, *Fabales*, *Hippuridales*, *Rhamnales*; probably also *Capparales*.

DAHLGREN's recently developed system of flowering plants (1975) reduces the number of heterogeneous orders by two. His separating

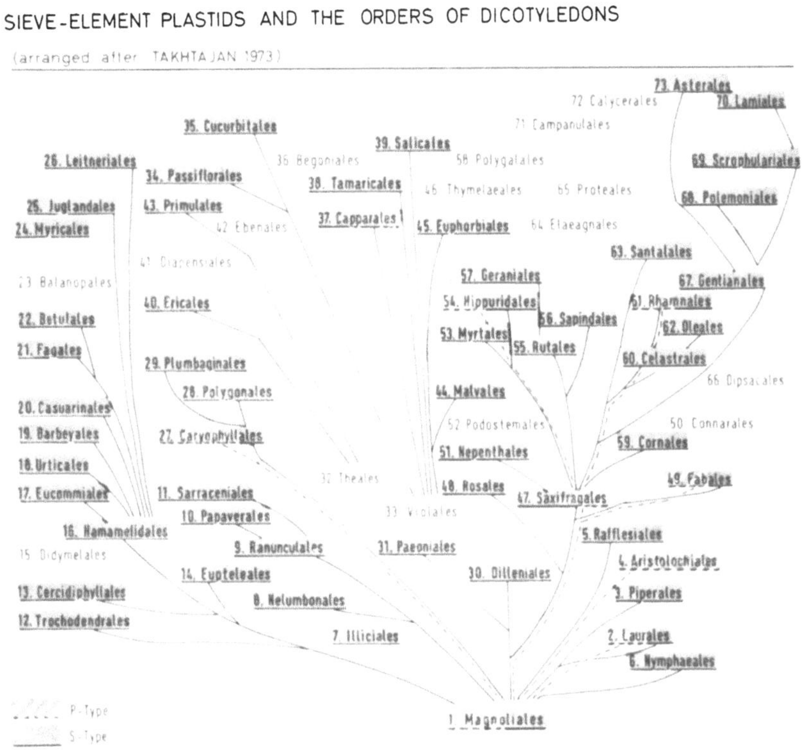

Fig. 28

the P-type family *Gunneraceae* (as order *Gunnerales*) from the S-type families *Hippuridaceae* and *Haloragaceae* and excluding the S-type families *Bataceae* and *Gyrostemonaceae* from *Caryophyllales* closely corresponds to the conclusions derived from sieve-element plastids (see Fig. 1 in BEHNKE & DAHLGREN 1976). The dismemberment of the *Caryophyllidae* (and placing the S-type orders *Polygonales* and *Plumbaginales* next to *Primulales*) also accentuate the specifity of the P III-subtype found in *Caryophyllales*. For a closer connection of those taxa exhibiting the P I-subtype DAHLGREN's (1975) system also does not provide a ready solution.

Considering the difficulties in incorporating sieve-element plastid data into existing systems, a modified alternative is offered (Fig. 29), which at least is in accordance with the distribution of the P- and S-type plastids as presently known.

In this alternative proposal TAKHTAJAN's (1973) division of *Magnoliopsida* into subclasses is maintained since at present S-type plastids cannot be used for classification. Changes in the circumscription of subclasses are undertaken with the *Caryophyllidae* and the

DISTRIBUTION OF SIEVE-ELEMENT PLASTIDS WITHIN FLOWERING PLANTS
(BEHNKE 1976)

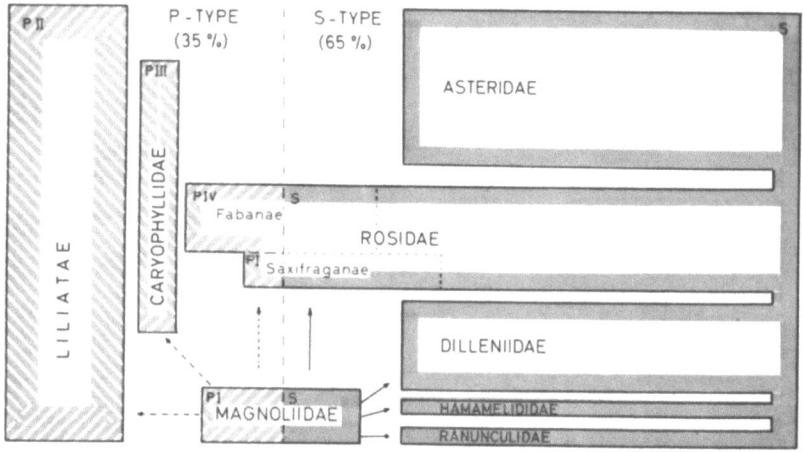

Fig. 29

Rosidae. The *Caryophyllidae* are strictly circumscribed by the P III-subtype containing families of the order *Caryophyllales* see BEHNKE 1976), while the orders *Polygonales* and *Plumbaginales* are included in *Rosidae* (DAHLGREN 1975). In addition, within the *Rosidae* two new superorders are proposed:

A superorder *Saxifraganae* is proposed to include besides all S-type families of *Saxifragales* (sensu TAKHTAJAN) the P I-subtype families *Eucryphiaceae, Gunneraceae, Vitaceae* and *Leeaceae.*

Eucryphiaceae was placed into *Saxifragales* by TAKHTAJAN (1973). WAGENITZ (1975) lists some differential characters which deny a close relationship between *Gunneraceae* and *Haloragaceae* (and *Myrtales* as a whole); DAHLGREN (1975) alignes his *Gunnerales* within *Saxifraganae.* The families *Vitaceae* and *Leeaceae* were proposed by JOHNSTON (1974) to be refered to *Saxifragales.*

Another new superorder *Fabanae* is split from *Rosanae* to include

P IV-subtype containing *Caesalpiniaceae, Mimosaceae* and the *Fabaceae* which contain both P IV-subtype and S-type species.

Except for a few scattered genera these proposals meet the requirements of the distribution of the total of 65% S-type and 35% P-type within flowering plants.

The author is grateful to Prof. T. J. MABRY (Austin, Texas) for many helpful discussions and for linguistic corrections of the English manuscript; to Prof. F. EHRENDORFER (Wien) for comments on the hierarchy and sub-classification of sieve-element plastids; and to Prof. H. MERXMÜLLER (München) for literature comments. Skillful technical assistance by Miss B. SCHMIDT and Mrs. D. LAUPP during the unpublished part of this investigation is acknowledged. Supported by grants from "Deutsche Forschungsgemeinschaft".

References

BEHNKE, H.-D., 1965: Über das Phloem der Dioscoreaceen unter besonderer Berücksichtigung ihrer Phloembecken. II. Mitteilung: Elektronenoptische Untersuchungen zur Feinstruktur des Phloembeckens. Z. Pflanzenphysiol. 53, 214—244.

— 1967: Über den Aufbau der Siebelement-Plastiden einiger Dioscoreaceen. Z. Pflanzenphysiol. 57, 243—254.

— 1969a: Die Siebröhren-Plastiden der Monocotyledonen. Vergleichende Untersuchungen über Feinbau und Verbreitung eines charakteristischen Plastidentyps. Planta (Berl.) 84, 174—184.

— 1969b: Ultrastructure of angiosperm sieve-tube plastids in relation to systematics. Abstr. XIth Intern. Bot. Congress (Seattle), p. 12.

— 1971: Sieve-tube plastids of *Magnoliidae* and *Ranunculidae* in relation to systematics. Taxon 20, 723—730.

— 1972: Sieve-element plastids in relation to angiosperm systematics.—An attempt towards a classification by ultrastructural analysis. Bot. Rev. 38, 155—197.

— 1973: Sieve-tube plastids of *Hamamelididae*. Electron microscopic investigations with special reference to *Urticales*. Taxon 22, 205—210.

— 1974: Sieve-element plastids of *Gymnospermae*: Their ultrastructure in relation to systematics. Plant Syst. Evol. 123, 1—12.

— 1975a: P-type sieve-element plastids: A correlative ultrastructural and ultrahistochemical study on the diversity and uniformity of a new reliable character in seed plant systematics. Protoplasma 83, 91—101.

— 1975b: The bases of angiosperm phylogeny: ultrastructure. Ann. Miss. Bot. Gard. 62, 647—663.

— 1976: Ultrastructure of sieve-element plastids in *Caryophyllales (Centrospermae)*, evidence for the delimitation and classification of the order. Plant Syst. Evol. 126, 31—54.

— and DAHLGREN, R., 1976: The distribution of characters within an angiosperm system. II. Types of sieve-element plastids. Bot. Notiser 129, 287—295.

BONNETT, H. T., and NEWCOMB, E. H., 1965: Polyribosomes and cisternal accumulations in root cells of radish. J. Cell Biol. 27, 423—432.

COLE, G. T., and BEHNKE, H.-D., 1975: Electron microscopy and plant systematics. Taxon **24**, 3—15.

CRESTI, M., PACINI, E., and SIMONCIOLI, C., 1974: Uncommon paracrystalline structure formed in the endoplasmic reticulum of the integumentary cells of *Diplotaxis erucoides* ovules. J. Ultrastruct. Res. **49**, 218—223.

DAHLGREN, R., 1975: A system of classification of the angiosperms to be used to demonstrate the distribution of characters. Bot. Notiser **128**, 119—147.

EHRENDORFER, F., 1971: *Spermatophyta*. In: Lehrbuch der Botanik für Hochschulen, 30th ed., 584—745. Stuttgart: Fischer.

FAVALI, M. A., and GEROLA, F. M., 1968: Tubular and fibrillar components in the phloem of *Brassica chinensis* L. leaves. Giorn. Bot. Ital. **102**, 447—467.

FRIEDRICH, H. C., 1956: Studien über die natürliche Verwandtschaft der *Plumbaginales* und *Centrospermae*. Phyton (Austria) **6**, 220—263.

HAVELANGE, A., and COURTOY, R., 1974: Description et essais de caractérisation cytochimique d'un composant inconnu dans les cellules méristématique de *Sinapis alba* L. (Crucifères). C. R. Acad. Sc. Paris **278**, 1191—1193.

HOEFERT, L. L., 1975: Tubules in dilated cisternae of endoplasmic reticulum of *Thlaspi arvense* (*Cruciferae*). Amer. J. Bot. **62**, 756—760.

IVERSEN, T.-H., 1970a: Cytochemical localization of Myrosinase (β-thioglucosidase) in root tips of *Sinapis alba*. Protoplasma **71**, 451—466.

— 1970b: The morphology, occurrence, and distribution of dilated cisternae of the endoplasmic reticulum in tissues of plants of the *Cruciferae*. Protoplasma **71**, 467—477.

— and FLOOD, P., 1969: Rod-shaped accumulations in cisternae of the endoplasmic reticulum in root cells of *Lepidium sativum* seedlings. Planta **86**, 295—298.

JOHNSTON, M. C., 1974: *Rhamnales*. Encyclopaedia Britannica. Macropaedia Vol. **15**, 793—796.

JØRGENSEN, L. B., BEHNKE, H.-D., and MABRY, T. J., in prep.: Protein-accumulating cells and dilated cisternae of the endoplasmic reticulum in three glucosinolate-containing genera: *Armoracia, Capparis, Drypetes*. Planta.

KLUGE, H., 1967: Untersuchungen über Kohlenhydrate und myo-Inosit in Siebröhrensäften von Holzgewächsen. Dissertation Darmstadt.

MABRY, T. J., and BEHNKE, H.-D. (eds.), 1976: Evolution of Centrospermous Families. Plant Syst. Evol. **126** (1).

TAKHTAJAN, A., 1973: Evolution und Ausbreitung der Blütenpflanzen. Stuttgart: Fischer.

THORNE, R., 1968: Synopsis of a putatively phylogenetic classification of the flowering plants. Aliso **6**, 57—66.

WAGENITZ, G., 1975: Blütenreduktion als ein zentrales Problem der Angiospermen-Systematik. Bot. Jb. Syst. **96**, 448—470.

WALKER, J. W., and DOYLE, J. A., 1975: The bases of angiosperm phylogeny: palynology. Ann. Miss. Bot. Gard. **62**, 664—723.

Address of the author: Prof. Dr. H.-D. BEHNKE, Lehrstuhl für Zellenlehre, Im Neuenheimer Feld 230, D-6900 Heidelberg 1, Federal Republic of Germany.

Plant. Syst. Evol., Suppl. 1. 179—189 (1977)

The New York Botanical Garden, Bronx, New York, U.S.A.

On the Taxonomic Significance of Secondary Metabolites in Angiosperms

By

A. Cronquist, New York

Abstract: I here put forward the speculative interpretation that changes in the major groups of chemical repellents had an important role in the rise of major new groups of dicotyledons. Each set of repellents tends to lose its effectiveness as insects and other predators become resistant to it. A new kind of repellent gives the plants a competitive advantage and permits the evolutionary expansion of a new group. Thus the isoquinoline alkaloids of the *Magnoliidae* gave way to the tannins of the *Hamamelidae*, *Rosidae* and *Dilleniidae*, and these in turn gave way to the iridoid compounds that were most effectively exploited by the *Asteridae*. Within the *Asteridae*, the rise of the relatively recent family *Asteraceae* may relate to a shift to polyacetylenes and sesquiterpene lactones in place of the already less effective iridoids. The pattern is blurred by the continued evolutionary experimentation with new repellents by each major taxonomic group, and by the exploitation of the same new group of repellents by different taxa at about the same time.

Chemical data have until now played only a small role, or virtually none, in the formulation of widely recognized systems of classification of flowering plants. Yet the accumulation of information over the past two or three decades has shown that the chemistry correlates well enough with morphology and other features so that it should be taken into account.

Unfortunately, efforts to incorporate the chemical data into the system of classification have so far met with only limited success. Belief in the overriding, a priori importance of chemistry has led some botanists to taxonomic conclusions completely unacceptable to other botanists who rely more on morphological features.

The benzyl-isoquinoline alkaloids furnish an example of both the usefulness and the problems of the chemical data. It is now well recognized that these alkaloids are widespread in the *Magnoliidae*, but rare elsewhere. They furnish one of the most useful markers of the subclass, and their presence in the *Papaverales* is one of the important features which has led to the general agreement that the *Papaverales* are related to the *Ranunculales* rather than to the *Capparales*. (I here use the name *Magnoliidae* in the broad sense, to include the *Ranunculidae* of TAKH-TAJAN.)

KUBITZKI (1969, 1973) has argued that the distinctive chemistry of the *Magnoliidae* makes this group an unlikely ancestor for other angiosperms. Yet on the basis of comparative morphology the *Magnoliidae* do appear to be phyletically basal within their division; the palaeobotanical data provide stronger support for this point of view year by year, almost day by day. On the basis of the fossil record it is now difficult to challenge the concept that the earliest angiosperms had monosulcate pollen, and it is becoming clear that the monocots began to diverge from the dicots before the origin (or at least before the evolutionary diversification) of triaperturate pollen (DOYLE 1973). The only modern dicots with monosulcate pollen belong to the *Magnoliidae* (WALKER & DOYLE 1975; WALKER 1976). The evidence from leaf architecture is not yet so conclusive, but it is now apparent that early fossil angiosperm leaves find their closest modern counterparts among the *Magnoliidae* (HICKEY & WOLFE 1975). Thus KUBITZKI's conclusion, based largely on chemical evidence, is at odds with the conclusion based on other evidence.

The betalains have been a notorious cause of dispute between chemically oriented and morphologically oriented taxonomists. MABRY & TURNER have adamantly maintained in various publications (e.g., 1964) that the presence of betalains is a sine qua non for admission to the order that has often been called *Centrospermae*. Other botanists (ECKARDT 1976, CRONQUIST 1973) have been equally adamant that it makes no sense to exclude the *Caryophyllaceae* from that order merely because they have anthocyanins instead of betalains. Only very recently MABRY (1976) has concluded that the *Caryophyllaceae* and *Molluginaceae* can be restored, as a suborder, to the same order that includes the other centrospermous families.

Chemically oriented botanists have frequently pointed out the chemical similarity of the *Asteraceae* to the *Apiaceae* (*Umbelliferae*) (e. g., HEGNAUER, 1971). Morphologically oriented botanists have not even deigned to reply in print, because on morphological grounds is seems impossible to associate these two families.

DAHLGREN (1975) has recently assigned major importance to the presence of iridoids and some other compounds in his novel system of classification, but here again the chemically defined groups are morphologically heterogeneous.

So we have a dilemma. The chemical data obviously must be taxonomically important, yet reliance on them to establish evolutionary relationships frequently gives results that are out of harmony with morphologically based systems. We need a new rationale for understanding the significance of chemical data to the general system. I here propose such a new rationale.

It is so difficult to provide a plausible Darwinian interpretation for much of the morphological evolution at the level of families and orders of angiosperms that most phylogenists have concentrated on the establishment of trends and relationships, with little more than lip service to their Darwinian significance. Chemical differences have also often been discussed almost in a vacuum, as if they had no real importance to the plants. Although a Darwinian interpretation of many of the morphological differences still escapes me, I suggest that survival value does provide the key to the evolutionary and taxonomic interpretation of chemical differences among the major groups of angiosperms.

FRAENKEL (1959) and JANZEN (1975) are among those who have emphasized the importance of secondary metabolites in plants as repellents. In accordance with their views, I believe that most of the chemical compounds which have been considered of taxonomic importance at the level of families and above are repellents. They discourage predators or pathogens. The known chemical differences among the major taxa are largely differences in defensive weapons. Even the betalains are reported to be toxic to certain fungi (KIMLER 1975).

Every weapons system has its costs—in the case of plants, metabolic costs. No one plant can afford to maintain a complete, functional array of all possible defensive weapons; the metabolic tax would be insupportable. Adoption of a new weapons system is therefore likely to mean abandonment of an old one. Among the metabolic costs of weaponry I here include not only the direct use of energy to produce the weapons, but also the adjustments and compromises necessary to prevent the plant from being hoist on its own petard. The plant must be able to withstand its own repellents.

It should be no surprise to any student of history that any defensive weapon sooner or later loses its effectiveness. The same may be true of offensive weapons, but we are here concerned with defense. For several hundred years the Great Wall of China served its intended purpose of keeping out the barbarians, but eventually it was breached. No conceivable wall could turn back an airplane. Anti-aircraft guns may discourage trespassing airplanes, but they are useless against ballistic missiles. And so it goes.

We are all acquainted with the decline in effectiveness of DDT over the course of two or three decades, and the appearance of penicillin-resistant strains of pathogenic bacteria over the same time span. Here the massive human disturbance caused rapid evolution of resistance to poisons. Although the rate is slower, the chemical weapons of plants also lose much of their effectiveness as times goes on. The predators and pathogens develop resistance. Compounds that originally served as repellents may even become attractants for specialized predators.

The secondary metabolites of flowers often differ, at least in detail, from those of vegetative organs. Even the different parts of a flower may differ in this way (GIANNASI 1975). The possible biological and taxonomic significance of chemical differences between vegetative and floral parts has scarcely begun to be evaluated, although it seems obvious that flowers have more use for attractants than do vegetative parts. It is the compounds in the vegetative parts that have attracted attention as possible markers of major taxonomic groups. Some of these, such as the betalains, occur in the flowers as well, but others may not. It is clear enough that differences in the flavonoids and other secondary metabolites of flowers are taxonomically significant at the level of genera and species, but it does not now appear that such primarily floral chemical differences are useful at the higher taxonomic levels. More information may of course modify or reverse this view. Кто проживёт увидет.

EHRLICH and RAVEN (1965) recognized an intimate correlation between the evolution of new repellents by plants and the evolution of butterflies capable of feeding (in the larval stage) on plants with such defenses. They further suggested that plants with new repellents "would in a sense have entered a new adaptive zone. Evolutionary radiation of the plants might follow, and eventually what began as a chance mutation or recombination might characterize an entire family or group of related families." The remainder of the present paper largely follows the direction forecast by these authors.

As a speculative interpretation, put forward for consideration by other botanists, I suggest that the evolution of chemical repellents plays an important but complex role in the rise and diversification of new families, orders, and subclasses of angiosperms. When one set of repellents loses some of its effectiveness, the time is ripe for another set. A suitable new set of repellents gives its possessors a competitive advantage and permits their evolutionary expansion. The increased abundance of the new group fosters the evolution of resistance to their repellents by predators. Then another new group of repellents may be exploited, in the continuing struggle between predator and prey. A long period of relative disuse may even permit the revival of an old set of repellents, after the predators have lost resistance.

Evolutionary experimentation with new repellents is constantly going on. Every member of a taxonomic group has the potential opportunity to try something a little different, by way of a single mutation near the end of a biosynthetic chain. Thus the chemical possibilities of a given set of repellents are explored and exploited.

Since every set of repellents gradually loses its effectiveness, opportunities arise for the selective development of new sets. It is not to be expected that the first members of a new set of repellents will be the most useful ones. Therefore the effectiveness of existing repellents must have declined to a level that can be surpassed by one of the less effective members of a potentially more effective new set, or nothing much is likely to happen.

The same set of repellents may be developed and exploited by two or more major taxa at about the same time. Competitive exclusion here takes an oddly delayed form. The evolutionary opportunity is initially open to anything that can produce the right mutations. The opportunity declines in value as the predators become resistant.

Under special local conditions, or with the aid of happenstance fixation of mutations, individual species or even larger groups may turn to repellents already widely used by other taxa. Such re-adopted repellents can not confer any broad-scale advantage on their possessors, and the evolutionary expansion of the taxonomic group that produces them is likely to be blocked by preadapted predators (LEVIN 1976).

The hypothesis provides for all degrees of usefulness of repellents. Some may initially be so much better than their predecessors that they foster tremendous evolutionary expansion of the groups that produce them. The tannins and the iridoids may represent this sort. Others may give their possessors a degree of advantage, but not enough to foster the evolution of a major new group. Such repellents might evolve several times in different major taxa, in each case allowing some expansion of the group before a competitive balance is restored. The triterpenoid saponins might possibly belong to this class. Still others may be very effective, but impose such a heavy metabolic tax that their net value is severely limited. They might be compared to medicines with virtually no margin between the therapeutic and the toxic dose. Repellents of this class might be tried repeatedly, but never lead to the evolution of a major new group. Hydrogen cyanide may belong to this class.

Under this hypothesis, the basic pattern in evolution of repellents is one of successive shifts from one major set to another, in response to progressive increase in resistance by the predators. The pattern is blurred by the constant evolutionary experimentation with new repellents and refurbished old ones. All the major taxa have individual species or genera that produce one or another unusual repellent in addition to or instead of the common ones. When one of these unusual repellents confers a sufficient competitive advantage over a long enough period of time, the small taxonomic group becomes a large one, and the uncommon repellent becomes a common one. Doubtless it is easier for the predators to evolve resistance to some repellents than to others. Therefore the evolutionary expansion of a group of plants with a particular set of repellents will be governed in part by how long it takes the predators to become resistant.

The hypothesis implies that the development of a new set of repellents is neither very hard nor very easy. If it were very hard, then any particular set of repellents would probably evolve only once, and the correlation of chemistry and morphology would be much stronger than it is. If it were very easy, then the old repellents would not persist, but would be abandoned with the speed of mutual fund managers abandoning a falling stock.

I claim no expertise in chemistry, but it seems to me that the genetic difficulties in the development of a new set of repellents are rather modest. Anthocyanins are widespread in the *Magnoliidae* as well as in other angiosperms. Given the machinery to produce anthocyanins, not many extra parts are needed to produce tannins. Mabry (personal communication) has recently concluded that probably only two or three enzymes need be added to the common angiosperm supply to permit the formation of betalains. Likewise it appears that the production of iridoid compounds does not require any great number of enzymes beyond those common to angiosperms in general.

For each of the principal sets of repellents, it appears that most of the necessary equipment is a part of the common heritage of angiosperms, and only one or a few parts need be added to produce a working machine. It is easy to be impressed with the number of links in the biosynthetic chain leading to any metabolite, and equally easy to overlook the fact that most of these links are used in more than one chain. Assembly of a new chain may require the formation of only one or two new links. Yet these new links are not just floating around waiting to be picked up. The proper mutations must occur, and the frequency of these mutations differs from one taxonomic group to another, in accordance with well known principles governing partial genetic control of mutability (Dobzhansky 1941). The number of mutations necessary to permit the plants to withstand their own new set of repellents is unknown, but may be presumed to be more than just one or two.

Thus the formation of new kinds of repellents may be about at the level of difficulty that the hypothesis requires.

Now let us see how these theoretical considerations might apply to the problems at hand. Here again I propose a speculative interpretation, which may or may not prove to be basically correct. It must withstand critical analysis before it can become anything more than a working hypothesis.

I suggest that the *Magnoliidae* are indeed the phyletically basal group of angiosperms. For whatever reasons, they have used benzylisoquinoline alkaloids as their most important defensive weapons. By the time the *Magnoliidae* became common and widespread, in the late Cretaceous, their predators had probably developed a considerable resistance to this set of repellents. The time was therefore ripe to develop and explore a new set. The most immediately available new repellents were the tannins, notably including ellagic acid and the proanthocyanidins. The *Magnoliidae* characteristically produce anthocyanins, and the same basic shikimic acid pathway leads to both anthocyanins and tannins. The *Hamamelidae, Rosidae,* and *Dilleniidae,* which were differentiating from the *Magnoliidae* in the late Cretaceous, all exploited tannins as repellents and discarded the outmoded benzyl-isoquinoline alkaloids.

We may note that STEBBINS (1965), and DOYLE & HICKEY (1976) think that the earliest angiosperms may have been shrubs, and that angiosperm trees came in a little thereafter. If their view is correct, then the change to an arborescent habit may have changed the balance of predators (more insects, fewer vertebrates) and set the stage for insect-oriented rather than vertebrate-oriented repellents. I don't know the relative effectiveness of tannins and benzyl-isoquinoline alkaloids on modern insects and vertebrates, let alone the Cretaceous ones. My suggestion here is merely an example of the sort of approach that might be useful.

In the late Cretaceous, tannins were a refurbished rather than an absolutely new repellent. Some gymnosperms were producing tannins as far back as the late Palaeozoic (NIKLAS & GENSEL 1976), and many conifers continue to do so to this day. Yet the major evolution of insects during much of the Cretaceous period had occurred with angiosperms as the principal hosts, and it can reasonably be assumed that the insects preying on angiosperms in the late Cretaceous were not highly resistant to tannin.

Once the tanniferous groups of angiosperms began to dominate the landscape, the insects and other plant predators had to develop resistance to tannin. In so doing, they probably lost some of their resistance to benzyl-isoquinoline alkaloids. Thus the diminishing abundance of the *Magnoliidae* may have caused a reciprocal increase in the effectiveness of their repellents, which have subsequently been tried out with modest success by a few more advanced groups. Perhaps at some time in the long-distant future the benzyl-isoquinoline alkaloids can again be effectively exploited by a new group of angiosperms.

By about the beginning of the Tertiary period, the insects and other predators may have become fairly resistant to tannins, so that the time was ripe for a new set of repellents. Some members of the *Rosidae, Dilleniidae,* and even a few of the *Hamamelidae* then began to exploit iridoids in place of tannins. In some few of the *Rosidae* the use of iridoids was combined, perhaps fortuitously, with sympetaly, leading to the formation and rapid expansion of the *Asteridae*. The similar combination of sympetaly and iridoid repellents in the *Ericales* (*Dilleniidae*) did not lead to a comparable evolutionary expansion, because the *Ericales* were by then already committed to mycotrophy, limiting their evolutionary opportunities.

Like the tannins before them, the iridoid compounds eventually began to lose some of their effectiveness, and other groups of repellents proliferated. At the end of the Oliogocene, the *Asteraceae* began to exploit polyacetylenes and sesquiterpene lactones together. This very successful combination may have been the most important factor in the explosive rise and diversification of the family.

The principal arsenal of the *Asteraceae* may now already have passed its peak of effectiveness. The evolutionary expansion of the group is so recent that extinction has not yet produced good generic lines, but already the family is experimenting successfully with new defenses. The *Senecioneae*, which are certainly not a primitive tribe, have produced the nearly unique Senecio alkaloids. *Senecio* is the largest genus in the *Asteraceae*, by any reasonable specific standards, and it has become cosmopolitan and ubiquitous. The most recently evolved tribe appears to be the *Lactuceae*, which does not appear until late Miocene, about 15 million years ago (Raven & Axelrod 1974). The *Lactuceae* have largely abandoned polyacetylenes and sesquiterpene lactones, and produce latex instead.

Very recently (Bowers et al. 1976) it has been shown that *Ageratum houstonianum* produces an effective insect anti-juvenile hormone. This substance, a simple chromene, is exciting interest as a possible source of insect-specific pesticides. It remains to be determined how widespread such substances are in the *Asteraceae* or in angiosperms in general. They may or may not have anything to do with the success of the *Asteraceae* in comparison to other families.

Polyacetylenes are exploited by the *Apiales* (especially the *Apiaceae*) and *Campanulaceae*, as well as the *Asteraceae*, but these other groups apparently lack sesquiterpene lactones. It is reasonable to suppose that the absence of the iridoids from the *Campanulaceae* is secondary, because these are present in more primitive families of *Asteridae* and even in the *Goodeniaceae*, which are usually referred to the *Campanulales*. On the other hand, there is no reason to suppose that the precursors of the *Apiales* ever had iridoid compounds. Their most likely ancestors, the *Sapindales*, do not have them. By the time the *Apiales* began to produce polyacetylenes, the opportunity for effective exploitation of iridoids may already have passed.

The *Apiales* have a fairly long fossil history. Fossils considered to represent the *Araliaceae* occur throughout the Tertiary and even into the Upper Cretaceous. It is not at all certain, however, that the earliest members of the group had polyacetylenes. There is a considerable overlap in chemical repellents between the *Araliales* and the presumably ancestral *Sapindales* (HEGNAUER 1971), but the polyacetylenes are not in the arsenal of the *Sapindales*. Thus it is possible that polyacetylenes were not invented in the *Araliales* until long after the origin of the order.

The *Caryophyllales* are chemically noteworthy for the substitution of betalains for anthocyanins in most families. The partial absence of proanthocyanidins and other tannins from the group may be biologically more important than the absence of true anthocyanins. Most of the mechanism for the production of anthocyanins (and proanthocyanidins) is obviously present at least in some members of the order, because some of them produce yellow flavonoid pigments as well as betalains (ALSTON 1967). Although MABRY & TURNER (1964) have proposed that the betalain families are a very ancient group, their known fossil record is less than 60 million years long (MULLER 1970). I suggest that a search for the biological significance of the betalains should concentrate on their repellent (and fungicidal) properties, rather than on their function as flower pigments.

The principles here expounded should be applicable to monocots as well as dicots, but I am not yet ready to propose a comprehensive interpretation. We may note that the intercalary meristem of grass leaves may be a more important factor in the success of the family than any repellent that may be present, but the siliceous deposits in the cells do discourage attacks by chewing insects (FRAENKEL 1959).

I hope that the approach here suggested will permit the harmonious integration of the data on secondary metabolites into the general taxonomic system. The thesis and antithesis of morphological and chemical characters may thus be resolved into a general synthesis.

I thank Dr. DAVID GIANNASI and Dr. KARL NIKLAS for advice and counsel during the preparation of this paper. Responsibility for the contents is, as always, my own.

References

ALSTON, R., 1967: Biochemical systematics. In: Evolutionary biology (DOBZHANSKY, T. S., HECHT, M. K., and STEERE, W. C., eds.), Vol. 1, 197—305. New York: Appleton-Century-Crofts.

BOWERS, W. S., OHTA, T., CLEERE, J. S., and MARSELLA, P. A., 1976: Discovery of insect anti-juvenile hormones in plants. Science **193**, 542—547.

CRONQUIST, A., 1973: Chemical plant taxonomy: A generalist's view of a promising specialty. Nobel Symposium **25**, 29—39.

Dahlgren, R., 1975: A system of classification of the angiosperms to be used to demonstrate the distribution of characters. Bot. Notis. **128**, 119—147.

Dobzhansky, T. S., 1941: Genetics and the origin of species. 2nd. ed. New York: Columbia University Press.

Doyle, J. A., 1973: Fossil evidence on early evolution of the monocotyledons. Quart. Rev. Biol. **48**, 399—413.

— and Hickey, L. J., 1976: Pollen and leaves from the Mid-Cretaceous Potomac Group and their bearing on early angiosperm evolution. In: Origin and early evolution of angiosperms (Beck, C. B., ed.), 139—206. New York: Columbia University Press.

Eckardt, T., 1976: Classical morphological features of centrospermous families. Plant Syst. Evol. **126**, 5—25.

Ehrlich, P. R., and Raven, P. H., 1965: Butterflies and plants: a study in coevolution. Evolution **18**, 586—608.

Fraenkel, G. S., 1959: The *raison d'être* of secondary plant substances. Science **129**, 1466—1470.

Giannasi, D. E., 1975: The flavonoid systematics of the genus *Dahlia* (Compositae). Mem. N.Y. Bot. Gard. **26** (2), 1—125.

Hegnauer, R., 1971: Chemical patterns and relationships of the Umbelliferae. In: The biology and chemistry of the Umbelliferae (Heywood, V. H., ed.), 267—277. Suppl. 1 to Vol. **64**; Bot. J. Linnean Soc.

Hickey, L. J., and Wolfe, J. A., 1975: The bases of angiosperm phylogeny: vegetative morphology. Ann. Missouri Bot. Gard. **62**, 538—589.

Janzen, D. H., 1975: Ecology of plants in the tropics. London: Edward Arnold.

Jensen, S. R., Nielsen, B. J., and Dahlgren, R., 1975: Iridoid compounds, their occurrence and systematic importance in the angiosperms. Bot. Notis. **128**, 148—180.

Kimler, L., 1975: Betanin, the red beet pigment, as an antifungal agent. Bot. Soc. Amer. Abstr. **36**.

Kubitzki, K., 1969: Chemosystematische Betrachtungen zur Großgliederung der Dicotylen. Taxon **18**, 360—368.

— 1973: Probleme der Großsystematik der Blütenpflanzen. Ber. dtsch. bot. Ges. **85**, 259—277 (1972).

Levin, D. A., 1976: The chemical defenses of plants to pathogens and herbivores. Ann. Rev. Ecol. Syst. **7**, 121—159.

Mabry, T. J., 1973: Is the order *Centrospermae* monophyletic? Chemistry in Botanical Classification. Nobel Symposium **25**, 275—285.

— 1976: Pigment dichotomy and DNA-RNA hybridization data for centrospermous families. Plant Syst. Evol. **126**, 79—94.

— and Turner, B. L., 1964: Chemical investigations of the *Batidaceae*. Betaxanthins and their systematic implications. Taxon **13**, 197—200.

Muller, J., 1970: Palynological evidence on early differentiation of angiosperms. Biol. Rev. Cambridge Philos. Soc. **45**, 417—450.

Niklas, K., and Gensel, P. G., 1976: Chemotaxonomy of some paleozoic vascular plants. Part I. Chemical compositions and preliminary cluster analyses. Brittonia **28**, 353—378.

Raven, P. H., and Axelrod, D. I., 1974: Angiosperm biogeography and past continental movements. Ann. Missouri Bot. Gard. **61**, 539—673.

· STEBBINS, G. L., 1965: The probable growth habit of the earliest flowering plants. Ann. Missouri Bot. Gard. **52**, 457—468.

WALKER, J. W., 1976: Comparative pollen morphology and phylogeny of the Ranalean complex. In: Origin and early evolution of angiosperms (BECK, C. B., ed.), 241—299. New York: Columbia University Press.

— and DOYLE, J. A., 1975: The bases of angiosperm phylogeny: palynology. Ann. Missouri Bot. Gard. **62**, 664—723.

Address of the author: Dr. A. CRONQUIST, New York Botanical Garden, Bronx Park, Bronx, NY 10458, U.S.A.

Plant Syst. Evol., Suppl. 1, 191—209 (1977)
© by Springer-Verlag 1977

Laboratorium voor Experimentele Plantensystematiek,
University of Leiden, The Netherlands

Cyanogenic Compounds as Systematic Markers in *Tracheophyta*

By

Robert Hegnauer, Leiden

Abstract: A short chronological review of cyanogenesis (Fig. 1) in vascular plants is given (Table 1). Subsequently the chemistry (Tables 2-4; Fig. 2) and biochemistry (Figs. 3-9) of cyanogenic plant constituents are summarized. Five biogenetical groups (A–E) of cyanogenetic compounds are presently known from vascular plants. The occurrence and function (Fig. 10) of cyanogenesis is sketched and causes of conflicting statements in phytochemical literature are briefly discussed. Finally the presently known distribution of the different pathways, and the resulting constituents, as well as of still unidentified cyanogenic compounds, is given. The paper ends with a short systematic appreciation of cyanogenesis as a systematic marker at higher categories. The differences between *Pteridophyta*, *Gymnospermae* and *Angiospermae* and the similarities between *Liliopsida* and the magnoliid and ranunculid part of *Magnoliopsida* are stressed and the uniformity of *Passiflorales*, including *Flacourtiaceae* as well as the heterogeneity of *Rosaceae* are outlined and discussed from the taxonomic point of view.

Introduction

Since FLÜCKIGER (1883: 722–728, 950–955) reviewed the distribution of hydrocyanic acid in nature, several aspects of the phenomenon cyanogenesis were periodically summarized in literature. In the following discussion of the chemistry and biochemistry of cyanogenesis and cyanogenic constituents and in their evaluation as systematic characters no references are given as a rule. The reader is advised to consult the following recent reviews, when more information or original papers are needed: DILLEMANN (1958), TSCHIERSCH (1967), BUTLER (1969), CONN (1969, 1973), EYJOLFSSON (1970), HEGNAUER (1962–1973a, 1973b), NAHRSTEDT (1973), TAPPER and REAY (1973), GIBBS (1974), SEIGLER (1975).

Cyanogenesis in the Vegetable Kingdom

Cyanogenesis may be defined as the ability of certain plants to release hydrocyanic acid (prussic acid) after injury of cells. Usually cyanophoric plants contain one or several cyanogenic glycosides and

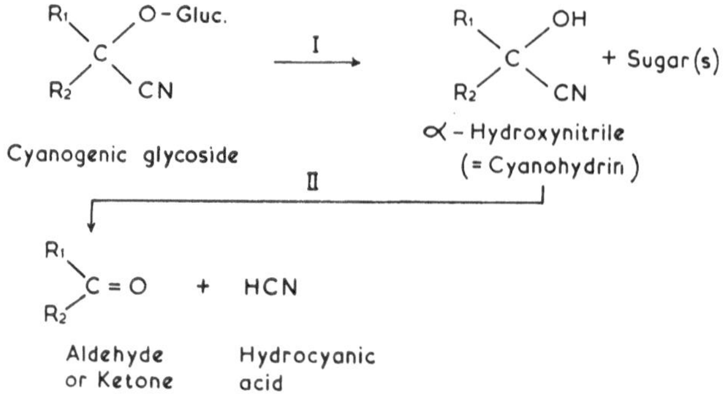

Fig. 1. The process of cyanogenesis operating in most cyanophoric plants. Reaction I: catalyzed by β-glycosidases. Reaction II: the rate of spontaneous release of HCN depends upon temperature and pH; the reaction may also be catalyzed by oxynitrilases

Table 1. Early history of cyanogenesis in the plant kingdom

Year	Species	Plant parts	Family	Authors
1801	*Prunus amygdalus*	Seed	*Rosaceae*	BOHM
1802	*Prunus laurocerasus*	Leaf	*Rosaceae*	SCHAUB
1803	*Prunus persica* and *P. spinosa*	Leaf; flower	*Rosaceae*	SCHRADER
1803	*Prunus armeniaca*	Seed	*Rosaceae*	VAUQUELIN
1812	*Prunus padus*	Bark	*Rosaceae*	BERGEMANN
1812 to 1885	Many *Prunoideae*, *Maloideae* and *Spiraeoideae*	Seeds; leaves; bark; roots	*Rosaceae*	Several authors
1836	*Manihot esculentus*	Tuberous roots	*Euphorbiaceae*	HENRY and BOUTRON-CHARLARD
1840	*Lucuma mammosa*	Seed	*Sapotaceae*	GAYTON
1843?	*Phaseolus lunatus*	Seed	*Leguminosae*	MARCADIEU
1862	*Chardinia orientalis*	Fruit	*Asteraceae*	EICHLER
1862	*Lepidium sativum*	Seeds; seedlings	*Brassicaceae*	SCHULZ
1867	*Ximenia americana*	Unripe fruits	*Olacaceae*	ERNST
1870	*Vicia sativa*	Seeds	*Leguminosae*	RITTHAUSEN-KREUSLER
1871	*Marasmius oreades*	Basidiocarps	*Agaricaceae*	LÖSECKE
1874	*Ipomoea dissecta*	Leaves	*Convolvulaceae*	FLÜCKIGER-HANBURY
1884	*Linum usitatissimum*	Seeds, seedlings, stems	*Linaceae*	JORISSEN
	Aquilegia vulgaris	Flowering plant	*Ranunculaceae*	
	Ribes aureum	Young shoots	*Saxifragaceae*	
	Arum maculatum	Young leaves	*Araceae*	
	Glyceria maxima	Flowering plant	*Poaceae*	

Table 2. Phytochemical history of cyanogenic plant constituents: First two periods (1830–1840: Robiquet and Boutron-Charlard; Henry and Boutron-Charlard; Wöhler and Liebig. 1890–1910: The plant physiology or Treub-Guignard period)

First isolation of pure compound	Name of constituent	Source	Products of enzymatic hydrolysis (A = acetone; B = benzaldehyde)
1830	Amygdalin	Seeds of *Prunus amygdalus* (bitter almonds); *Rosaceae*	HCN + B + 2 glucose (1837); sugar is gentio-biose (1923)
1834/36	Manihotoxin	Tuberous roots of *Manihot esculentus*; *Euphorbiaceae*	Identical with linamarin (1906)
1891	Linamarin	Seedlings of *Linum usitatissimum*; *Linaceae*	HCN + A + 1 glucose
1902	Dhurrin	Young plants of *Sorghum vulgare*; *Poaceae*	HCN + p-OH—B + 1 glucose
1903	Phaseolunatin	Seeds of *Phaseolus lunatus*; *Leguminosae*	Identical with linamarin (1906)
1904	Gynocardin	Seeds of *Gynocardia odorata*; *Flacour-tiaceae*	HCN + dihydroxycy-clopentenone + 1 glucose (1966, 1970)
1905	Sambunigrin	Leaves of *Sambucus nigra*; *Caprifoliaceae*	HCN + B + 1 glucose
1905	Prulaurasin	Leaves of *Prunus laurocerasus*; *Rosaceae*	HCN + B + 1 glucose; is racemized prunasin (1941)
1906	Vicianin	Seeds of *Vicia angustifolia* (*sativa*); *Leguminosae*	HCN + B + 1 arabi-nose + 1 glucose; disaccharide = vicianose (1910)
1907	Prunasin	Twigs of *Prunus padus*; *Rosaceae*	HCN + B + 1 glucose

the corresponding more or less substrate-specific enzymes which catalyze the release of hydrocyanic acid (Fig. 1). Sometimes, however, these enzymes are inactive or totally lacking; such plants may contain large amounts of cyanogenic glycosides without being cyanophoric after damage.

Hydrocyanic acid was first detected in living nature by Bohm in 1801, and amygdalin, the first plant glycoside obtained as a pure and crystalline material, was described in 1830 by the French scientists Robiquet and Boutron-Charlard. Since 1801 seeds, flowers, leaves

Table 3. Phytochemical history of cyanogenic plant constituents: Third period (The toxicological or RIMINGTON and STEYN-FINNEMORE period)

First isolation of pure compound	Name of constituent	Source	Products of enzymatic hydrolysis (OH—B = hydroxy-benzaldehyde)
1935	Acacipetalin	Leaves of South-african *Acacia* taxa; *Leguminosae*	HCN + isobutyric acid (= dimethylketene + H_2O) + 1 glucose; revised structure: HCN + methyl-acrolein + 1 glucose (1975)
1936	*Goodia*-glu-coside	Leaves of *Goodia lotifolia*; *Leguminosae*	HCN + *p*-OH—B + 1 glucose (= *p*-gluco-syloxybenzaldehyde-cyanohydrin)
1936	Phyllanthin (renamed taxiphyllin in 1964)	Leaves of *Phyllanthus gasstroemii*; *Euphorbiaceae*	HCN + *p*-OH—B + 1 glucose
1936	Zierin	Leaves and twigs of *Zieria laevigata*; *Rutaceae*	HCN + *m*-OH—B + 1 glucose
1937	Acalyphin	Leaves of *Acalypha indica*; *Euphorbiaceae*	Structure still unknown
1938	Lotaustralin	Herb of *Lotus austra-lis*; *Leguminosae*	HCN + methylethyl-ketone + 1 glucose
1948	Lucumin	Seeds of *Calocarpum sapota* (= *Lucuma mammosa*); *Sapota-ceae*	HCN + benzaldehyde + disaccharide; sugar is primverose (1971)

and (or) barks of many rosaceous plants were shown to be strongly cyanophoric. Two tropical alimentary crop plants, Cassava and Lima bean, belong to the first non-rosaceous taxa which were demonstrated to be cyanophoric. The early history of research on cyanogenesis is summarized in Table 1.

After JORISSEN had shown in 1884 that several non-rosaceous European plant species release hydrocyanic acid (Table 1), plant physiologists, toxicologists and phytochemists became much interested in cyanogenesis and its causes. Between 1885 and 1910 Dutch (e.g. VAN ROMBURGH, GRESHOFF, TREUB) and French (e.g. GUIGNARD, HÉBERT) scientists detected many new cyanophoric plants. In 1906 GRESHOFF published a list of cyanogenic taxa containing 179 species representing 34 families of Angiosperms and one family of Fungi.

Table 4. Phytochemical history of cyanogenic plant constituents: Fourth period (The period of biogenetical and ecological investigations)

First isolation of pure compound	Name of constituent	Source	Products of enzymatic hydrolysis (OH—B = hydroxy-benzaldehyde)
1966	*Nandina*-glu-coside	Leaves of *Nandina domestica*; *Berberidaceae*	Identical or enantiomeric with *Goodia*-gluco-side
1966	Proteacin	Seeds of *Macadamia integrifolia*; *Proteaceae*	HCN + p-OH—B + 2 glucose; one glucose linked to the phenolic OH
1969	Triglochinin	Herb of *Triglochin maritima*; *Juncaginaceae*	HCN + triglochinic acid (ketene + H_2O) + 1 glucose; seco-dhurrin structure (1972)
1969	Barterin (= Bar-terioside)	Rootbark of *Barteria fistulosa*; *Flacour-tiaceae*	HCN + hydroxycy-clopentenone + 1 glu-cose (1970)
1969	Deidamin (re-named deida-clin in 1970)	Leaves of *Deidamia clematoides*; *Passi-floraceae*	HCN + cyclopente-none + 1 glucose (1970)
1969	Cyanolipid	Seedoil of *Cordia verbenacea*; *Boraginaceae*	HCN + hydroxy-methylacrolein + 2 fatty acids (C_{16}–C_{22})
1970	Cyanolipids	Seedoils of many *Sapindaceae*	HCN + a C_5-aldehyde + one or two fatty acids
1971	Tetraphyllin-A ⎫	Fruits of *Tetrapathaea tetrandra*; *Passi-floraceae*	Identical with deidaclin
1971	Tetraphyllin-B ⎭		Identical with barterin
1972	*Thalictrum*-glucoside	Herb of *Thalictrum aquilegifolia*; *Ranunculaceae*	A monomethylester of triglochinin; methyla-tion probably during isolation! (1976)
1973	Holocalin	Seeds of *Holocalyx balansae*; *Legumi-nosae*	HCN + m-OH—B + 1 glucose
1974	Cardio-spermin	Herb of *Cardio-spermum hirsutum*; *Sapindaceae*	HCN + hydroxyme-thylacrolein + 1 glu-cose
1975	Dihydroaca-cipetalin	Leaves of South-african *Acacia*-taxa; *Leguminosae* and leaves of *Hetero-dendron oleaefolium*; *Sapindaceae*	HCN + isobutyralde-hyde (= dihydroac-rolein) + 1 glucose
1976	*Sorbaria*-CN-glucoside	*Sorbaria arborea*; *Rosaceae*	HCN + hydroxyme-thylacrolein + glu-cose + p-OH-benzoic acid

To-day we know about 25 cyanophoric species of Fungi, about 50 species of cyanogenic ferns, 6 species of cyanogenic Gymnosperms and about 2000 species of cyanogenic Angiosperms.

Chemistry and Biochemistry of Cyanogenic Plant Constituents

Structural studies with cyanogenic plant constituents started in 1837, when Wöhler and Liebig showed that amygdalin is composed

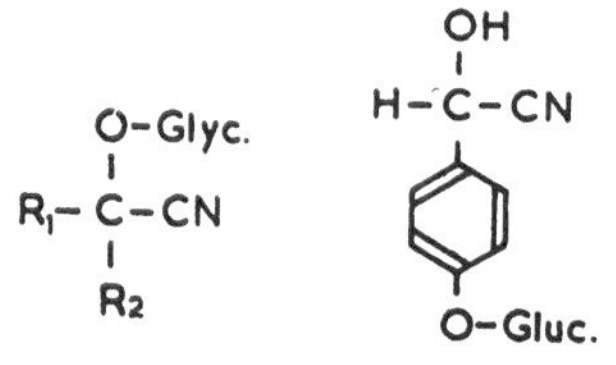

Fig. 2. Presently known chemical types of cyanogenic plant constituents. I: cyanohydrin glycosides (= glycosides of α-hydroxynitriles = α-hydroxycyanides); the usual type of cyanophoric plant constituents (compare Figs. 1, 3, 4, 6, 7, 8, 9). II: unstable cyanogenic compounds (α-hydroxy group free; compare Figs. 5 and 6; sugar, if present, combined with a phenolic hydroxyl). III: glycosides of α-unsaturated α-hydroxycyanides (i.e. glycosides of ketene cyanohydrins; cf. Fig. 7; hydrolysis generates highly unstable ketenes which by addition of water generate stable organic acids). IV: cyanolipids of seed oils (n may be 14, 16, 18 or 20; cf. Fig. 8)

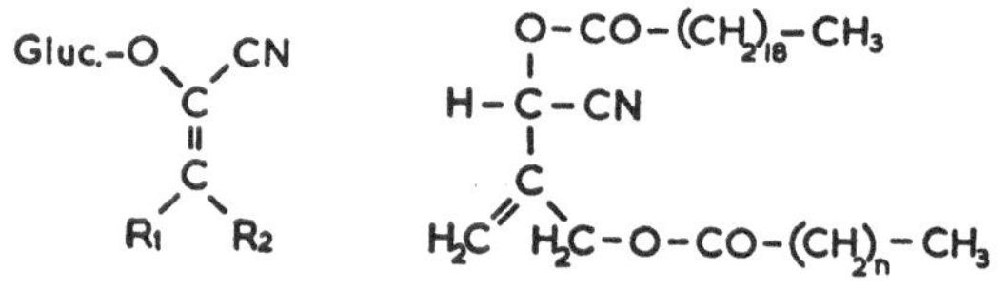

of HCN, benzaldehyde and glucose. These German chemists simultaneously detected a protein-like compound in almonds which catalyzes hydrolysis of amygdalin; this proteinaceous matter was called emulsine. It was in fact the second enzyme to be described in biochemical literature. Since 1837 the chemical investigation of cyanogenic plant constituents advanced slowly and rather irregularly. In fact, the four periods sketched in Tables 2–4 can be discerned.

As a result of the endeavour of many scientists working during the four periods outlined in Tables 2–4 the chemistry, biochemistry and biology of cyanophoric plant constituents became known to some extent.

Chemistry of Cyanogenic Compounds

Formally the four chemical types of cyanophoric compounds represented in Fig. 2 may be distinguished.

For the plant taxonomist, however, biogenetic pathways and their resulting products are more reliable characters than mere structural details and chemical reactivities of the individual plant metabolites, which allow their chemical typification.

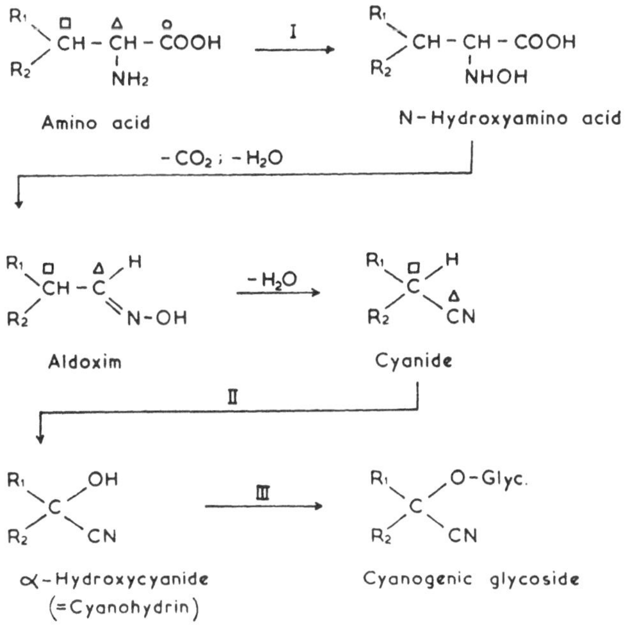

Fig. 3. The biosynthesis of cyanogenic glycosides (chemical type I; cf. Fig. 2). I: this reaction is probably catalyzed by highly substrate-specific enzymes which, however, usually do not discern between valine and iso-leucine; the enzymes catalyzing the reactions have not yet been isolated. II: enzymes catalyzing this reaction have not yet been characterized. III: more or less substrate-specific glucosyltransferases have been isolated from several cyanophoric plants

Biochemistry of Cyanogenic Compounds

ROSENTHALER (1923) suggested that the amino acids phenylalanine, tyrosine and valine might be the precursors of prunasin, dhurrin and linamarin. In the meantime this hypothesis was extended and refined by the experimental work of many biochemists. Today several steps of cyanoglycoside synthesis by plants are rather well known (Fig. 3).

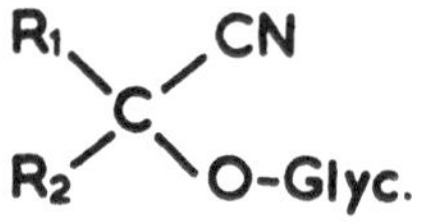

Biosynthetic Group	R_1	R_2	Sugar (= Glyc.)	
Phenylalanine A	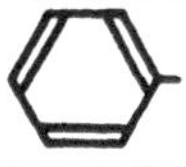	-H	Glucose Gentiobiose Vicinose Primverose	Prunasin und Sambunigrin Amygdalin Vicianin Lucumin
(see fig. 5)	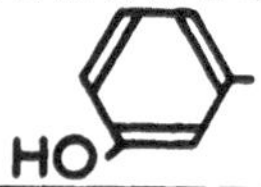	-H		Holocalin and Zierin
Tyrosine B (see fig. 5)	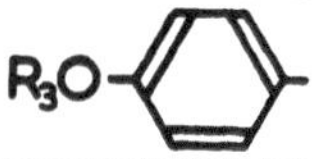	-H	Glucose	R_3 = H: Taxiphyllin and Dhurrin R_3 = Glucosyl: Proteacin
Valine and Isoleucine C	CH_3-	$-CH_3$	Glucose	Linamarin
	C_2H_5-	$-CH_3$	Glucose	Lotaustralin
Leucine D (see fig. 7)	$CH_3-CH(CH_3)-$	-H	Glucose	Dihydroacacipetalin
	$CH_2=C(CH_3)-$	-H	Glucose	Acacipetalin
	$CH_2=C(CH_2OR_4)-$	-H	Glucose	R_4 = H: Cardiospermin R_4 = p-Hydroxybenzoyl: Sorbaria-CN-glucoside
Cyclo-pentenyl-glycine? E (see fig. 8)	$-CH=CH-CH_2-CH_2-$	R_1+R_2	Glucose	Deidaclin
	$-CH=CH-CHOH-CH_2-$	R_1+R_2	Glucose	Barterin
	$-CH=CH-CHOH-CHOH-$	R_1+R_2	Glucose	Gynocardin

Fig. 4. The 5 biosynthetic main groups (A–E) of cyanophoric constituents of vascular plants. N. B.: References to other figures in column 1 should be corrected as follows: group A and B (see Fig. 6); group D (see Fig. 8); group E (see Fig. 9). In column 4 read vicianose instead of vicinose

According to the amino acids involved five main biosynthetic groups of cyanophoric compounds have been demonstrated to occur in *Tracheophyta* (Fig. 4) and an additional group is present in Fungi (Fig. 5).

The greatest number of presently known cyanogenic glycosides belongs to the aromatic groups A and B (Fig. 4). The differences be-

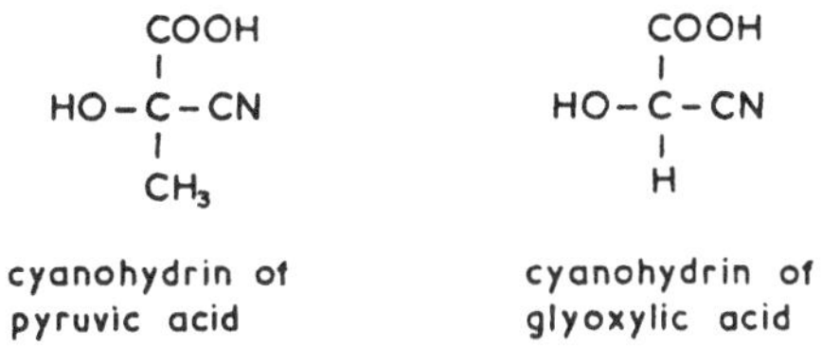

COOH
|
HO – C – CN
|
CH₃

cyanohydrin of
pyruvic acid

COOH
|
HO – C – CN
|
H

cyanohydrin of
glyoxylic acid

Fig. 5.   Cyanogenic constituents of Fungi (TAPPER & MACDONALD 1974)

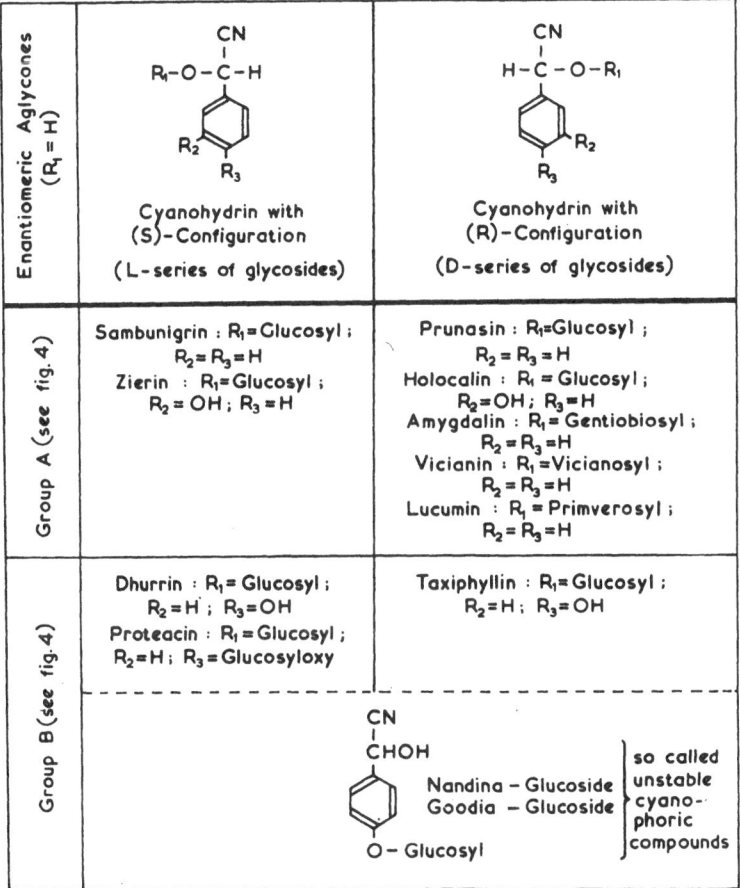

Fig. 6.   The presently known aromatic cyanogenic glycosides (groups A
and B of Fig. 4)

tween individual compounds relate to the configuration of the agly-
cones, the nature of the sugars, and in group B to the positions of the
attachment of glucose (Fig. 6). Another variant of group B is repre-

sented by triglochinin (Fig. 7) which was shown to be an isomerized seco-dhurrin (or seco-taxiphyllin).

Only two glucosides, linamarin and lotaustralin, are known from group C (Fig. 4). They often occur together suggesting the already

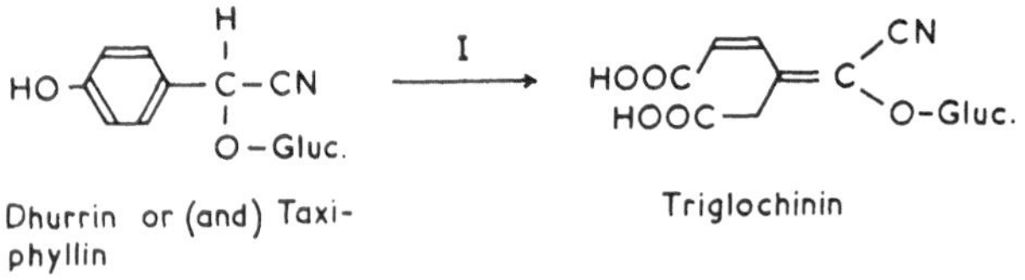

Dhurrin or (and) Taxi-phyllin   Triglochinin

Fig. 7. Triglochinin belongs to the tyrosine-derived cyanogenic glucosides (group B of Fig. 4). I: the sequence of reactions leading to triglochinin involves ring fission; therefore triglochinin is a so-called *seco*-compound

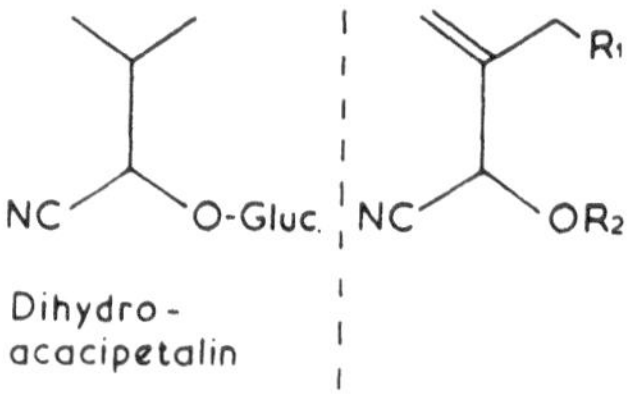

Cyanophoric compounds	R_1	R_2
Acacipetalin	H	Glucosyl
Cardiospermin	OH	Glucosyl
Sorbaria-CN-Glucoside	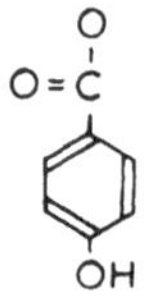 (see figure)	Glucosyl
Cyanolipids	$O-CO-[CH_2]_{18}-CH_3$	$CO-[CH_2]_{18}-CH_3$

Dihydro-acacipetalin

Fig. 8. The recently discovered and rapidly growing group of leucine-derived cyanogenic compounds (group D, Fig. 4). Concerning cyanolipids it should be noted that they may be spontaneously cyanophoric (cyanohydrin hydroxyl not esterified) or non-cyanogenic (no hydroxyl in α-position). *Sorbaria*-CN-glucoside according to NAHRSTEDT (1976)

mentioned (Fig. 3) lack of specificity of the amino acid aminohydroxylase concerned.

Group D (Fig. 4) comprises momentally 4 glucosides and several cyanolipids (Fig. 8). Most probably the number of leucine-derived cyanophoric plant constituents will increase considerably in future.

The strange cyclopentenoid representatives (Fig. 9) of group E of Fig. 4 form a puzzling cluster of cyanophoric glucosides. Theoretically they could be derivatives of the non-proteinogenous amino acid cyclopentenyl glycine. The latter once was detected in seeds of a species of *Hydnocarpus*, a genus known for the presence of gynocardin-type glucosides. At present, however, experimental research about the biosynthesis of these compounds is still lacking.

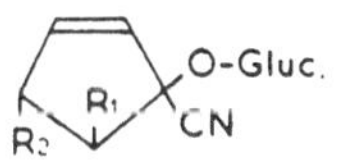

Glucosides	R₁	R₂
Deidaclin (= Tetraphyllin A)	H	H
Barterin (= Tetraphyllin B)	H	OH
Gynocardin	OH	OH

Fig. 9.   The presently known representatives of the gynocardin group
(E in Fig. 4) of cyanogenic glucosides

Occurrence, Function, Distribution and Systematic Meaning of Cyanogenesis and Cyanogenic Compounds in Vascular Plants

On p. 196 the number of presently known cyanophoric species of vascular plants is given as 2056. This is a very rough estimate. There are several reasons why more precise statements are impossible, some of which deserve to be mentioned shortly in this context.

(a) Formerly most phytochemists paid little attention to serious plant identification; a considerable number of taxa is probably erroneously reported as cyanophoric.

(b) Because of the enormous amount of nomenclatural and taxonomic synonymy many taxa are reported twice or several times under different names as cyanophoric in original literature and reviews.

(c) Research workers interested in cyanophoric plants applied different methods to demonstrate cyanogenesis. Some of the procedures used lack specificity for HCN. Consequently the literature about cyanophoric plants contains a number of false positive reports.

(d) The sensibility of analytical methods for the detection of HCN varies widely and depends upon many factors. In our context, a plant should release at least 10–20 mg HCN per kg fresh weight. This implies that one gram of fresh plant material usually reacts positively when the GUIGNARD-MIRANDE- or (and) the FEIGL-ANGER-MIRANDE-test (FIKENSCHER & HEGNAUER 1977) are applied. If more sensitive procedures are used, most or possibly all plants can be shown to release trace amounts of HCN (ROSENTHALER 1923, GEWITZ et al. 1974). From the systematic and ecological point of view, however, only accumulation of cyanophoric constituents represents a character worthwhile of consideration.

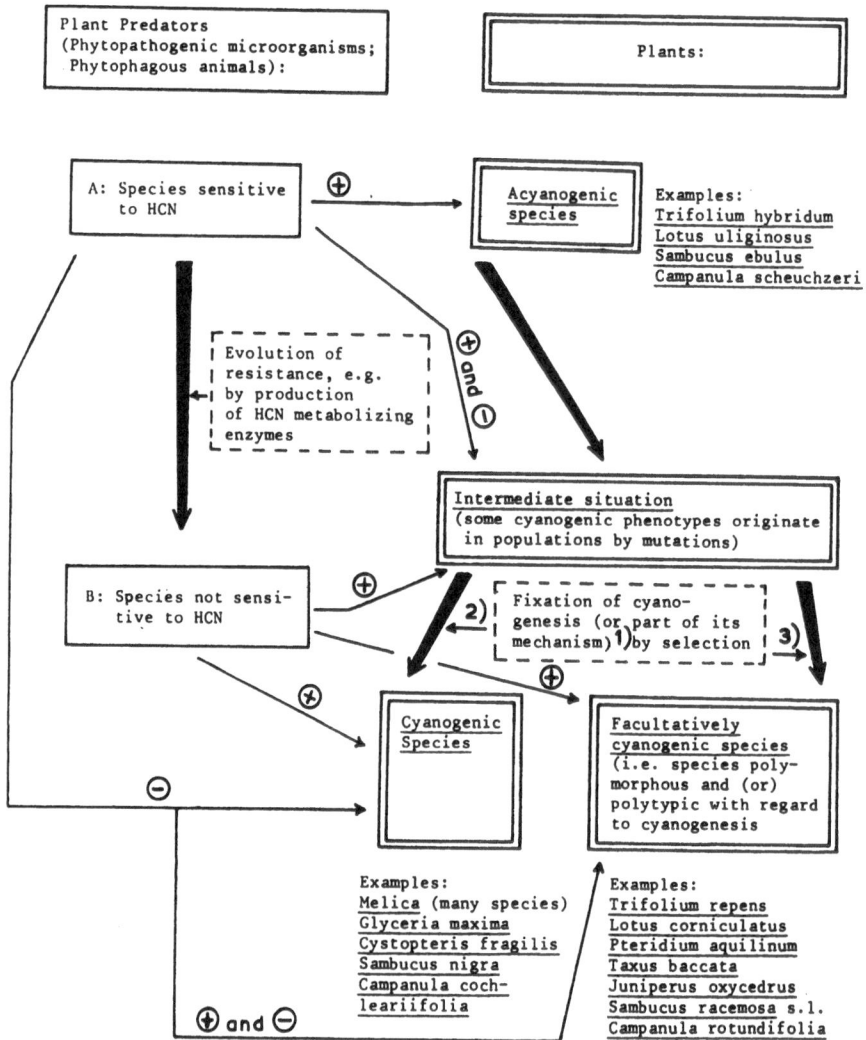

Fig. 10. Possible role of cyanogenesis in vascular plants (modified after JONES 1972, 1973). (+) Plant potentially suitable as host (food). (—) Plant not suitable as host (food). 1. e.g. plants which produce cyanogenic glycosides, but no splitting enzymes. 2. Evolution of very efficient mechanisms preventing autointoxication by HCN (e.g. rapid conversion of HCN to asparagine via β-cyanoalanine). This mechanism is fixed within the taxon because it seems to be efficient in all habitats occupied by the taxon. 3. Evolution of similar detoxification mechanisms which, however, are still subject to selection, because in some biotopes part of the mechanism may be inhibited, and the system, therefore, becomes deleterious to the taxon concerned

(e) Reports in literature are inconsistent for a number of taxa. Besides factors such as mentioned sub (a) to (d) other possibilities should always be considered as eventual explanations for contradictions. Many plants accumulate cyanogenic compounds only in one or a 'few organs; if different plant parts are tested, the results may be different.

Table 5. Variation of cyanogenesis within one species: Field and herbarium observations with *Centaurea scabiosa* L. (sensu lato)

Collection year	Locality [1]	Cyanogenesis-tests performed	
		in the field [2]	in the laboratory [3]
1946	Kt. Tessin, CH	—	1 very strong
1947	Kt. Tessin, CH	—	1 negative
1947	Kt. Graub., CH	—	1 negative
1962	Tolkamer, NL	—	1 negative
1966	Kt. Glarus, CH	—	1 negative
1967	Kt. Glarus, CH	—	1 negative
1967	Kt. Graub., CH	1 negative	1 negative
1968	Kt. Tessin, CH	—	1 negative
1969	Alpes Maritimes, F	—	1 negative
1969	Kt. Glarus, CH	2 negative	1 negative; 1 very strong
1972	Kt. Glarus, CH	3 negative, 1 positive	3 negative; 1 very strong
1972	Kt. Graub., CH	1 positive	1 very strong
1973	Kt. Graub., CH	1 positive	1 very strong
1974	Kt. Glarus, CH	6 negative, 5 positive	5 negative; 6 very strong
1974	Kt. Graub., CH	6 positive, 1 negative	6 very strong; 1 negative

[1] Most collections and tests performed in Switzerland; "Kt. Glarus" (includes lake Walensee); "Kt. Graub." means valley of the Rhine between Chur and Sargans.

[2] Guignard-Mirande-Test; mostly without adding emulsin.

[3] Leaves of herbarium material; some droplets of water and some mg of emulsin added; observation after 4 hours; Guignard-Test.

Moreover, accumulation of cyanogenic plant constituents is often conditioned in a taxon-specific manner by ecological factors and the age of the plant investigated and its different parts. Therefore the same plant may react positively or negatively depending on the timing of an investigation. Finally, genetical variation within species with regard to cyanogenesis is much more frequent than is generally assumed (Fikenscher & Hegnauer 1977). Genetical polymorphism and poly-typism enable population geneticists and ecologists to deal with cyanogenesis and its function (JONES 1972, 1973; Fig. 10). At the same time variability of cyanogenesis within species may be a significant feature in experimental plant systematics and a nuisance when used as a character in α-taxonomy (e.g. Table 5).

Distribution and Systematic Implications

Without any doubt the number of cyanogenic taxa and cyanogenetic compounds known will increase considerably in the future. Nevertheless the facts available at present and summarized in the following suggest that the biogenetical groups A to E represent characters worth to be considered by plant taxonomists.

Pteridophyta. Cyanogenesis has been reported in literature for representatives of the genera *Asplenium, Blechnum (Lomaria), Cystopteris, Dryopteris, Lindsaea, Polypodium, Pteridium, Schizaea* and *Thelypteris*, and observed by the author in *Microgramma lycopodioides* (L.) COPEL[1]. Prunasin was isolated from species of *Cystopteris* and *Pteridium*, and vicianin from several species of *Davallia*. Within the range of *Pteridophyta* only some Ferns seem to produce and accumulate cyanophoric compounds and phenylalanine seems to be the only precursor used for this purpose.

Gymnospermae. Cyanogenesis occurs in needles and young twigs of *Juniperus oxycedrus* L., *Metasequoia glyptostroboides* HU et CHENG and 4 species of *Taxus*. In the taxa mentioned tyrosine is the precursor of the cyanogenic glucoside present.

Angiospermae. In the main group of modern vascular plants cyanophoric constituents occur very erratically and all compounds known nowadays are represented. One of its main groups, *Liliopsida* (formerly *Liliatae*), was already analyzed by HEGNAUER (1973b). Cyanogenic taxa are relatively frequent in *Araceae, Juncaceae, Juncaginaceae* (incl. *Lilaea*), *Poaceae* (= *Gramineae*) and *Scheuchzeriaceae*. Outside these families cyanophoric species are very rare. From *Commelinaceae*, e.g., one species, *Tinantia erecta* SCHEIDEW., was demonstrated to be cyanogenic by several authors. Tyrosine-derived cyanophoric glucosides (dhurrin or its enantiomer taxiphyllin; triglochinin) were observed in all species of *Liliopsida* which were investigated in some detail. Obviously *Liliopsida* use tyrosine only for the synthesis of cyanogenic compounds.

Most of the presently known cyanogenic plant constituents are restricted to the second major group of Angiosperms, the *Magnoliopsida* (formerly *Magnoliatae*). A detailed analysis of the 7 subclasses (TAKH-

---

[1] See Addenda.

TAJAN 1973) of *Magnoliopsida* reveals many gaps in the chemical identification of cyanophoric plant constituents, and some biochemical tendencies which the present author esteems significant to plant taxonomy. The presently known facts are summarized next.

The compilation makes use of the system of TAKHTAJAN (1973) and is based on HEGNAUER (1960 and 1962–1973a and unpublished own observations) and GIBBS (1974; only positive results of GIBBS himself utilized). In the compilation subclasses, superorders, orders, families and, in brackets, the number of genera are given. The following abbreviations are used: n.i. = cyanogenic compounds not yet characterized; p.c. = cyanogenesis should be confirmed; P = phenylalanine-derived compounds (Fig. 6); D = tyrosine-derived glucosides (Fig. 6); T = triglochinin variant of D (Fig. 7); Li = biogenetical group C (Fig. 4); L = leucine-derived glucosides (Fig. 8); CN-Lip = cyanolipids (Fig. 8); G = gynocardin group of glucosides (Fig. 9); ? after group of compounds (e.g. G?) means that the identification should be confirmed. Statements concerning groups of cyanophoric constituents within families do not mean that their presence was established for all species and genera known to be cyanophoric. *Magnoliaceae* (2) T, e.g., refers to the presence of triglochinin in *Liriodendron tulipifera* L.; the cyanophoric species *L. chinensis* SARG. and *Magnolia sprengeri* PAMPAN have not yet been thoroughly investigated.

Magnoliidae-Magnolianae: *Magnoliales*: *Magnoliaceae* (2) T, *Degeneriaceae* (1) n.i., *Annonaceae* (1) n.i., *Canellaceae* (1) n.i., *Winteraceae* (1–2) n.i.; *Laurales*: *Calycanthaceae* (2) n.i., *Lauraceae* (1) p.c.

Ranunculidae-Ranunculanae: *Ranunculales*: *Menispermaceae* (1) n.i., *Ranunculaceae* (6–11) D and T, *Nandinaceae* (1) D, *Berberidaceae* (1) n.i.; *Papaverales*: *Papaveraceae* (4) T (*Eschscholtzia*) and Li? (*Papaver nudicaule*), *Fumariaceae* (1) n.i.

Hamamelidae-Hamamelidanae: *Trochodendrales*: *Trochodendraceae* (1) n.i.; *Hamamelidales*: *Platanaceae* (1) n.i.[1]; *Urticales*: *Ulmaceae* (1) n.i., *Moraceae* (1) n.i.; *Juglandales*: *Juglandaceae* (1?; possibly false positive GUIGNARD-test reaction).

Caryophyllidae-Caryophyllanae: *Caryophyllales*: *Gyrostemonaceae* (1) n.i., *Molluginaceae* (1) n.i., *Cactaceae* (1) n.i., *Amaranthaceae* (3) n.i., *Chenopodiaceae* (5) n.i.

Dilleniidae-Dillenianae: *Theales*: *Clusiaceae* (1) n.i.; *Violales*: *Flacourtiaceae* (15) G; *Passiflorales*: *Passifloraceae* (incl. *Paropsieae*) (6–8) G, *Turneraceae* (3) G?, *Malesherbiaceae* (1) G?; *Cucurbitales*: *Cucurbitaceae* (1) n.i.; *Capparales*: *Capparidaceae* (3) n.i., *Tovariaceae* (1)

---

[1] See Addenda.

n.i., *Brassicaceae* (10) n.i., *Resedaceae* (2) n.i. (in *Capparales* cyanogenic glycosides and [or] hydrocyanic acid are probably by-products of glucosinolate synthesis or decomposition).—***Ericanae:*** *Ericales: Ericaceae* (3) n.i., *Epacridaceae* (2) n.i., *Pyrolaceae* (2?; possibly false positive Guignard-test reactions); *Ebenales: Sapotaceae* (4) P.—***Malvanae:*** *Malvales: Elaeocarpaceae* (2) P?, *Sterculiaceae* (1–2) n.i., *Malvaceae* (1) n.i.; *Euphorbiales: Euphorbiaceae* (11) Li, D and possibly P.

   Rosidae-Rosanae: *Saxifragales: Cunoniaceae* (1) p.c., *Davidsoniaceae* (1) p.c., *Escalloniaceae* (1) n.i., *Itéaceae* (1) D ?, *Grossulariaceae* (1) D?, *Hydrangeaceae* (2) P and possibly D, *Crassulaceae* (7–8) n.i., *Saxifragaceae* (2) n.i.; *Rosales: Rosaceae* (many genera and species of *Spiraeoideae, Prunoideae* and *Maloideae*; from *Rosoideae* only *Kerrieae, Adenostomeae* and *Cercocarpinae* are cyanogenic) P in *Prunus, Maloideae* and a few species of *Spiraeoideae* and L in *Sorbaria* (compare Fig. 8) and possibly in other taxa of *Spiraeoideae* and the few cyanogenic taxa of *Rosoideae* mentioned above; *Fabales: Mimosaceae* (4) P and L and possibly Li, *Caesalpiniaceae* (3) P, *Fabaceae* (approximately 20) P, D and Li; *Nepenthales: Droseraceae* (3?; probably false positive Guignard-test reactions).—***Myrtanae:*** *Myrtales: Myrtaceae* (1) P, *Melastomataceae* (3) P?, *Oliniaceae* (1) P, *Onagraceae* (3) n.i., *Lecythidaceae* (1) n.i.; *Hippuridales: Haloragaceae* (3) n.i.—***Rutanae:*** *Rutales: Rutaceae* (2) P; *Sapindales: Sapindaceae* (14) L and CN-Lip; *Geraniales: Linaceae* (1) Li; *Polygalales: Tremandraceae* (2) n.i.—***Aralianae:*** *Cornales: Apicaeae* (1) D (needs to be confirmed).—***Celastranae:*** *Celastrales: Aquifoliaceae* (1) p.c., *Icacinaceae* (1) P?, *Salvadoraceae* (1?; see remarks sub *Capparales*), *Celastraceae* (1) n.i., *Corynocarpaceae* (1) n.i.; *Rhamnales: Vitaceae* (1) n.i.; *Santalales: Olacaceae* (2) P.—***Proteanae:*** *Proteales: Proteaceae* (10 D).

   Asteridae-Lamianae: *Dipsacales: Caprifoliaceae* (1–2) P; *Gentianales: Gentianaceae* (3) n.i., *Apocynaceae* (1) n.i., *Asclepiadaceae* (3) P?, *Rubiaceae* (5) P?; *Polemoniales: Convolvulaceae* (2) P? and Li?, *Boraginaceae* (3) CN-Lip[1]; *Scrophulariales: Solanaceae* (1) n.i., *Scrophulariaceae* (1–2) P, *Myoporaceae* (1) P, *Acanthaceae* (3) n.i.; *Lamiales: Verbenaceae* (2) n.i., *Lamiaceae* (3) n.i.—***Asteranae:*** *Campanulales: Campanulaceae* (1) n.i., *Goodeniaceae* (1–2) n.i.; *Asterales: Asteraceae* (15–20) P and Li.

   Consultation of the foregoing compilation of cyanogenic vascular plant taxa reveals a number of systematically interesting facts. These are summarized in the following conjectures.

   (a) Each of the three major taxa of vascular plants, *Pteridophyta* (phenylalanine only), *Gymnospermae* (tyrosine only) and *Angiospermae*

---

[1] See Addenda.

(all pathways illustrated by Fig. 4), has its own biochemical characters with regard to production of cyanophoric compounds.

(b) The aromatic glycosides (Fig. 6) occur very erratically in vascular plants, but are present in all their divisions, classes and subclasses. Most probably they represent the phylogenetically oldest cyanophoric plant constituents among the compounds treated in this paper.

(c) Within Angiosperms *Liliopsida* and the first two subclasses of *Magnoliopsida*, which represent together WETTSTEIN's *Polycarpicae*, seem to use exclusively the tyrosine-pathway (Figs. 4, 6, 7) for the synthesis and accumulation of cyanogenic constituents. This feature might be systematically significant.

(d) Within *Magnoliopsida* phenylalanine-derived compounds are the more frequent and the more wide-spread aromatic cyanogenic constituents, if *Polycarpicae* are excluded.

(e) Group C (Fig. 4) seems to parallel group A with regard to erratic occurrence and wide distribution within *Magnoliopsida*. The fact that linamarin and lotaustralin generally occur together suggests that one or more enzymes engaged in their biosynthesis do not discriminate between valine and isoleucine. This might be caused by a structural similarity of the two amino acids, i.e. branching of the chain at C-3.

(f) Leucine-derived compounds (Figs. 4 and 8) occur in leaves of *Rosaceae* (*Sorbaria*: *Spiraeoideae*), *Mimosaceae* and *Sapindaceae* as glucosides, and in seeds of many *Sapindaceae* as cyanolipids. Perhaps all taxa mentioned belong to one and the same evolutionary line of *Magnoliopsida* as suggested long ago by HALLIER f.

(g) The gynocardin group of constituents (Fig. 9) is highly characteristic of *Passiflorales* and of two tribes of *Flacourtiaceae* (*Oncobeae* and *Pangieae*). This character represents an extremely valuable systematic marker within *Dilleniidae*.

(h) Contrary to the statement of HEGNAUER (1960) cyanogenesis— if adequately and carefully used as a character in plant systematics— may prove to be of considerable value to a natural classification of plants, even at the higher systematic levels. Seemingly the cyanophoric constituents show what was called by the present author (HEGNAUER 1971) an asystematic distribution of secondary plant metabolites. Nevertheless it is my conviction that a thorough analysis of such characters can reveal tendencies of biochemical evolution and, therefore, ultimately might furnish arguments which can be used in efforts to improve presently accepted classifications.

(i) The facts known to-day suggest that more than one biosynthetic group (Fig. 4) of cyanophoric compounds occur only in very large genera or families belonging to *Dilleniidae* (*Euphorbiaceae*), *Rosidae*

(*Rosaceae, Acacia, Fabaceae*) and *Asteridae* (*Asteraceae*). In such taxa the chemical character a c c u m u l a t i o n of c y a n o g e n i c c o m p o u n d s is taxonomically significant at infrafamiliar levels. Linamarin, e.g., is a character of many *Loteae, Trifolieae* and *Phaseoleae* in *Fabaceae* and of *Calenduleae* in *Asteraceae*. In *Rosaceae* most cyanophoric taxa with the basic chromosome number x = 9 do accumulate neither amygdalin nor prunasin (e.g., species of *Exochorda, Gillenia, Sorbaria, Kerria, Neviusia, Rhodotypos, Coleogyne, Adenostoma* and *Cercocarpus*). Whether most of them contain compounds similar to the *Sorbaria*-CN-glucoside (NAHRSTEDT 1976) has still to be demonstrated.

(k) The erratic occurrence of cyanogenesis in *Boraginaceae* (cyano-lipids in the seed oil of *Cordia verbenacea*[1]; not yet identified compounds in seedlings of *Borago officinalis*) deserves the attention of phytochemists and taxonomists. The systematic position of this family of *Asteridae* is still uncertain.

I thank Dr. A. NAHRSTEDT, Institut für Pharmazeutische Biologie, Freiburg i. Br., for information and a manuscript copy on *Sorbaria*-CN-glucoside, and Dr. M. ETTLINGER, Chemical Laboratory II, University of Copenhagen, for a stimulating discussion of likely biosynthetic routes to gynocardin-type glycosides. I am obliged to Dr. B. HENNIPMAN, Rijks-herbarium, Leiden, for the identification of the Brazilian fern *Microgramma lycopodioides* (L.) COPEL.

References

BUTLER, G. W., 1969: Metabolism of cyanoglucosides, mustard-oil glyco-sides and selenium-containing compounds in plants. Proc. Roy. Austra-lian Chem. Inst. **36**, 65—70.

CONN, E. E., 1969: Cyanogenic glycosides. Agric. Food Chem. **17**, 519—526.
— 1973: Biosynthesis of cyanogenic glycosides. Biochem. Soc. Symp. **38**, 277—302.

DILLEMANN, G., 1958: Composés cyanogénétiques. In: Handbuch der Pflanzenphysiologie (RUHLAND, W., et al., eds.), Band VIII (redigiert von MOTHES, K.), 1050—1075. Berlin-Göttingen-Heidelberg: Springer.

EYJOLFSSON, R., 1970: Recent advances in the chemistry of cyanogenic glycosides. Fortschr. Chem. Org. Naturstoffe **28**, 74—108.

FIKENSCHER, LUCIE H., and HEGNAUER, R., 1977: Die Verbreitung der Blausäure bei den Cormophyten. 11. Mitteilung. Über die cyanogenen Verbindungen bei einigen *Compositae*, bei den *Oliniaceae* und in der Rutaceen-Gattung *Zieria*. Pharm. Weekblad **112**, 11—20.

FLÜCKIGER, F. A., 1883: Pharmakognosie des Pflanzenreiches, 2. Aufl. Berlin: R. Gaertner's Verlagsbuchhandlung.

GEWITZ, H.-S., et al., 1974: Presence of HCN in *Chlorella vulgaris* and its possible role in controlling the reduction of nitrate. Nature **249**, 79—81.

---

[1] See Addenda.

GIBBS, R. D., 1974: Chemotaxonomy of Flowering Plants, 4 volumes. Montreal: McGill-Queen's University Press.

HEGNAUER, R., 1960: Die systematische Bedeutung des Blausäuremerkmales. Pharm. Zentralhalle **99**, 322—329.

— 1962—1973a: Chemotaxonomie der Pflanzen, Bände 1—6. Basel-Stuttgart: Birkhäuser Verlag (Systematic distribution of cyanogenesis and cyanogenic constituents; consult the heading Cyanogenese in indices).

— 1971: Pflanzenstoffe und Pflanzensystematik. Naturwissenschaften **58**, 585—598.

— 1973b: Die Verbreitung der Blausäure bei den Cormophyten. 10. Mitteilung. Die cyanogenen Verbindungen der *Liliatae* und *Magnoliatae-Magnoliidae*: Zur systematischen Bedeutung des Merkmals der Cyanogenese. Biochemical Systematics **1**, 191—197.

JONES, D. A., 1972: Cyanogenic glucosides and their function. In: Phytochemical Ecology (HARBORNE, J. B., ed.), 103—124. London-New York: Academic Press.

— 1973: Co-evolution and cyanogenesis. In: Taxonomy and Ecology (HEYWOOD, V. H., ed.), 213—242. London-New York: Academic Press.

NAHRSTEDT, A., 1973: Cyanogene Glykoside in höheren Pflanzen. Pharmazie in unserer Zeit **2**, 147—155.

— 1976: Ein neues cyanogenes Glykosid aus *Sorbaria arborea* (*Rosaceae*). Z. Naturforsch. **31 C**, 397—400.

ROSENTHALER, L., 1923: Zur Prüfung der Treubschen Hypothese. Biochem. Z. **134**, 215—224.

SEIGLER, D. S., 1975: Isolation and characterization of naturally occurring cyanogenic compounds. Phytochemistry **14**, 9—29.

SCHÜTTE, H. R., 1973: Biosynthese von cyanogenen Glykosiden und Senföglucosiden. Fortschr. der Botanik **35**, 103—119.

TAKHTAJAN, A., 1973: Evolution und Ausbreitung der Blütenpflanzen. Jena: VEB Gustav Fischer Verlag.

TAPPER, B. A., and MACDONALD, M. A., 1974: Cyanogenic compounds in cultures of a psychrophilic basidiomycete (snow mold). Canad. J. Microbiol. **20**, 563—566.

— and REAY, P. F., 1973: Cyanogenic glucosides and glucosinolates (mustard oil glycosides). In: Chemistry and Biochemistry of Herbage (BUTLER, G. W., and BAILEY, R. W., eds.), Vol. 1, 447—476. London-New York: Academic Press.

TSCHIERSCH, B., 1967: Blausäure und Blausäureglykoside, eine Übersicht. Pharmazie **22**, 76—82.

Addenda

(1) **Cyanogenesis in ferns**: The following genera should be added to the list given on p. 204: *Actinopteris*, *Cheilanthes*, *Pteridanetium*, *Microlepia*, *Pteridrys*, *Stenochlaena*, *Campyloneurum*, and *Phlebodium*, and the number of presently known cyanogenic species approaches 50 (HARPER, N. L., et. al., Phytochemistry **15**, 1764—1767 [1976]; HARPER, N. L., SEIGLER, D. S., Econ. Botany **30**, 395—407 [1976]; Proc. Okla. Acad. Sci. **56**, 95—100 [1976]).

(2) *Platanus* contains triglochinin and dhurrin (Fikenscher, L. H., Ruijgrok, H. W. L., Planta Medica **31**, 290—293 [1977]).

(3) Cyanolipids do not occur in the seed oil of *Cordia verbenacea* (Seigler, D. S., Biochem. Systematics Ecology **4**, 235—236 [1976]).

Address of the author: Prof. Dr. R. Hegnauer, Laboratorium voor Experimentele Plantensystematiek, Schelpenkade 14 a, Leiden, The Netherlands.

Plant Syst. Evol., Suppl. 1, 211—226 (1977)
© by Springer-Verlag 1977

Departamento de Botânica,
Faculdade de Ciências Médicas e Biológicas de Botucatu, Brazil

Some Aspects of Beetle Pollination in the Evolution of Flowering Plants

By

Gerhard Gottsberger, Botucatu

Abstract: Pollination by beetles seems to have strongly influenced the evolution of angiosperm flowers. Beetles are a predominant group of potential visitors and pollinators of flowers since earliest times, so cantharophily is apparent within groups of flowering plants of the most diverse evolutionary levels. In the subclass *Magnoliidae*, cantharophily is a dominant feature of many families, but even in these archaic groups a more open, unspecialized type of beetle pollination and a more specialized one can be distinguished. Specialization here was probably connected with increase in flower size, numerical increase and grouping and flattening of the sexual organs, etc. The strongly protogynous flower attracts beetles through imitative odours, to which the insects are already conditioned in their other activities. Secondary polyandry, such as occurs in the more basic groups of *Rosidae, Dilleniidae* and *Caryophyllidae*, is in many cases related to cantharophily and may find its functional explanation in this mode of pollination. This probably somewhat more recent radiation into beetle-pollination might have caused the stamens to increase in number in order to save some of them from the crude visitors. Cantharophily in secondarily polyandrous groups is also frequently connected with the protogynous condition of flowers and with odours which act directly on the instincts of the visitors. During these new waves of flower-biological radiation other insects besides beetles must have already been in existence. The pollination of primitive *Rosidae, Dilleniidae* and *Caryophyllidae* is therefore much less exclusive compared with the *Magnoliidae*, viz., beetles and other insects often frequent flowers jointly. Since beetles have continued as the predominant insect group until today, cantharophily can be observed also in advanced groups of angiosperms. In this case, cantharophily is no longer a sign of primitiveness as it is in the *Magnoliidae* and to some extent probably also in secondarily polyandrous groups, but a relatively recent adaptation into a still existing ecological niche.

Introduction

In a meeting such as the present one in which the evolution and classification of higher categories is to be illuminated from different points of view, it seems appropriate to speak about the functional aspects of flowers.

14*

It has been shown many times and by many examples that morphological, anatomical, chemical and physiological characteristics of flowers largely correspond to the necessities of pollination and that many of these features must have evolved in response to the agents involved.

From the time that the idea of co-evolution of plants and of pollination agents gained probability, it was considered significant to study these correlations for at least two reasons. First, functional aspects underline structural ones and so may show us the ecological background for flower characteristics. Second, reproductive organs of plants and their function can be seen in a historical context. The different pollination agents (excluding here the relatively timeless wind and water) evolved in different ages. For example, within insects, beetles are especially old, but butterflies and social bees are more recent. The appearance of new groups of potential agents of pollination certainly had very strong effects upon flowering plants. Waves of radiation into new ecological conditions must have occurred during the historical development of angiosperms.

It was certainly tempting to use flower-biological concepts for phylogenetic reasoning. Thus, we often say without further consideration that because beetles are an old group, plants with cantharophily are always derived from phylogenetically old stocks, or, since birds and mammals are more modern animals, therefore we should at least suspect that plants with pollination by birds and mammals should be considered as derived phylogenetically from more modern groups, and so on.

This kind of reasoning was used to a large extent and probably worked in many cases. However, we largely neglected to distinguish between organisms which probably remained more or less in the original condition, and others which perhaps because of a changing environment had to adapt to new conditions. Today, flower-biological concepts are changing to a form that better fits into the whole framework.

I shall now attempt to demonstrate some of these new concepts as I personally see them at the moment. The main example will be cantharophily, which is a feature of many primitive angiosperms, but which also accompanies the differentiation of flowering plants up to its most evolved members. The main part of this paper will be to show you the supposed evolution of flowers as far as cantharophily is concerned. The second subject of this symposium, classification, will not really be considered, although cantharophily would be a good example to use in dealing with this question. Secondary polyandry, a distinctive feature in the more primitive members of some subclasses and used even as a differential-diagnostic character is, in my opinion, also largely connected with beetle pollination and may find its functional explanation through this mode of pollination.

Beetles and Beetle Flowers

The trouble with talking about beetles is that we know so little about them. Our knowledge of the sense-physiology of beetles in general and of flower beetles in particular is very rudimentary and not comparable with what we know about honeybees, bumblebees, butterflies, moths and even flies.

The orthognathic position (perpendicular to body axis) of mouth parts, which is the original condition and common in the majority of beetles, limits the length of the mouth parts (FAEGRI & VAN DER PIJL 1971: 114). Consequently such beetles are able to visit only relatively flat flowers which are somewhat dish-, bowl- or brush-shaped. The famous *Nemognathus* and others, because of their extremely prolonged mouth parts, which are in a prognathic position (parallel to body axis), are also able to explore deep-shaped flowers. But these are rare and extremely specialized cases. On their flower-visits beetles may feed on pollen, on floral organs like petals, tepals, stamens and carpels, or on nectar as far as it is available. Very often they damage the flowers considerably.

How are such beetles attracted by flowers? Data are accumulating which show that most beetle flowers emit characteristic odours, which are usually somewhat similar to those of fresh or rotten fruits, or aminoid, similar to carrion, decayed fish, sperm, or the beetles themselves. Both fruity and aminoid odours of flowers seem to stimulate strongly the sexual instincts of beetles (PORSCH 1950: 290; GOTTSBERGER 1970; etc.); copulation is induced and very often even egg deposition occurs within the flowers.

It is very often said that the general syndrome of beetle pollination and beetle flowers is rather unspecialized due to the lack of specialization of the visitors. Some authors (VAN DER PIJL 1969: 91; FAEGRI & VAN DER PIJL 1971) divide beetle pollinated flowers into two groups. The more primitive one has larger flowers, frequently amorphic or haplomorphic sensu LEPPIK, and the second and more advanced one has small flowers condensed in inflorescences. Protogyny is a common feature in beetle flowers.

Beetle Pollination in Different Subclasses of Angiosperms

I now would like to describe cantharophily as it appears in the different subclasses of the *Magnoliatae*. In order to keep the discussion short, I shall restrict myself to dicotyledons.

Having studied the phylogenetically old *Magnoliidae*, which have so many characters in common with the assumed ancestors of the angiosperms, it has been learned that beetle pollination is a dominant and very often exclusive feature within this group. But unfortunately, studies have not started at the right end and the knowledge now represented in textbooks refers to highly specialized cases of cantharophily. Until recently we were aware of pollination methods only in the *Eupomatiaceae*, *Calycanthaceae*, *Magnoliaceae*, *Annonaceae*, and *Nymphaeaceae*, which in my opinion are all cases of a more or less advanced type of cantharophily.

But what is about pollination in *Winteraceae*? *Winteraceae* are considered as being especially primitive because of their wood, their unifacial stamens and carpels, their placentation, etc. (see Gottsberger 1974). The high number of primitive characters in the *Winteraceae* is perhaps explained partly by the fact that they are high palaeopolyploids with the highest polyploid chromosome numbers yet known in angiosperms. Palaeopolyploids tend to preserve more primitive characters than diploids (or low polyploids) which have diverged more actively (Ehrendorfer et al. 1968: 349).

For a primitive group we might expect also a primitive mode of pollination. The only thoroughly investigated case of pollination in the *Winteraceae* is that of *Drimys brasiliensis* (Gottsberger et al., in progress). This species belongs to the New World section, *Drimys*, which because of its morphology, anatomy, cytology and chemistry is considered more primitive than the Old World section *Tasmannia* (see Kubitzki & Vink 1967: 14; Ehrendorfer et al. 1968: 338), nowadays treated by some authors as a separate genus, *Tasmannia* (see Ehrendorfer 1976: 229). The hermaphrodite flowers of *Drimys brasiliensis* show a beetle pollination syndrome which, however, is different from that of other cantharophilous *Magnoliidae*. The sexual organs are exposed and unprotected during the whole anthesis in the permanently open flower. The flower odour is not fruit-like or aminoid but sweet and recalls a "typical" blossom odour. Such an absolute odour (Faegri & van der Pijl 1971: 87—89) probably has no "meaning" outside the flower, and unless an insect has "learnt" this odour from the flower, it will not start any reaction. As we will see later on, the more specialized flowers in respect to beetle pollination, such as these of the *Annonaceae*, *Magnoliaceae*, *Eupomatiaceae*, *Calycanthaceae*, and *Nymphaeaceae*, emit, as far as it is known, so-called imitative fruit-like or aminoid odours which "imitate" odours to which the insects are already conditioned by instincts and by "experience" in their other activities, and which would normally start an instinctive "meaningful" reaction (Faegri & van der Pijl 1971: 87). Flowers with such imitative odours frequently attract a crowd of beetles. The flowers of *Drimys brasiliensis*, with their sweet absolute odour attract only a few species of beetles, which do not further harm the flowers but eat only the pollen.

Being acquainted with the pollination syndrome in other families of the *Magnoliales*, *Laurales*, and *Nymphaeales*, I want to emphasize the difference between their specialized type of cantharophily and the open, obviously less specialized one manifested in *Drimys brasiliensis*.

A very fine case of specialized cantharophily is to be seen in the *Annonaceae*. In species of *Guatteria*, for example, the open but still developing flowers remain dish-shaped and greenish during several weeks. At anthesis the petals turn yellowish and fold together to form a pollination chamber. It is from this moment on or a little later that the flowers start to emit a strong fruit-like odour which particularly attracts beetles that normally inhabit fruits and feed on them. This kind of odour can be considered an imitative one. The fruit-beetles act and react in the flower as they do on their normal substrate, feeding on the soft odoriferous perianth or stamens, copulating inside the dark pollination chamber and depositing their eggs. The closing of the flower has a somewhat protective effect, as larger and even more destructive beetles have difficulties in entering the flower interior with the sexual organs and thus very often only gnaw the soft perianth. Further, the dark, fragrant floral chamber possibly has similar properties to the dark, fragrant interior of fruits where the beetles normally live, thus attracting these fruit beetles quite precisely by deceit (GOTTSBERGER 1970). Further examples of an advanced mode of pollination in which imitative fruit- or also carrion-like odours act upon the instincts of fruit or carrion beetles can be found in most other *Annonaceae* species (see WESTER 1910; UPHOF 1933; CORNER 1940; ZIMMERMAN 1941; VAN DER PIJL 1953; KRAL 1960; GOTTSBERGER 1970). The closed flowers of *Annonaceae*, contrary to the open, unprotected flowers of *Drimys*, receive frequent visits of mostly gnawing and thus destructive pollinators; it is noticeable that their many stamens and carpels with hard heads are very densely aggregated on the floral axis before disintegration.

Species of *Magnolia* have been recently investigated in respect to their pollination (THIEN 1974). The beetles which enter the flowers to feed on secretions of the gynoecium, stigmatic papillae, petal secretions or pollen, may also cause considerable damage (KNUTH 1904: 303; HEISER 1962: 262). I have already suggested before (GOTTSBERGER 1974: 463) that most *Magnolia* species may function by deceit-attracting fruit-eating or otherwise specialized beetles. This is now partly confirmed (THIEN et al. 1975); not only are some of the pollinators beetles that visit decaying fruits, but the floral odours of *Magnolia* contain many methyl esters which are present also in the odours of many fruits.

A very similar mode of pollination characterizes the Brazilian *Magnoliacea Talauma ovata* (GOTTSBERGER, in progress). The large cream-white tepals open at the apex to form a small entrance through which the beetles crawl into the interior of the flower. Just as in *Magnolia* species, beetles seem attracted by strong fruit-like odours. In an early stage the odour is clearly apple-like, sometimes mixed with the odour of terpentine, sometimes like an indefinable fruit ester, and at a later time of anthesis, because of accumulation of the beetle's excrements, the odour changes to a pungent one, somewhat similar to a mixture of fish and the beetles themselves. The flowers of *Talauma*, like most Magnolias, *Annonaceae* or other cantharophilous flowers of the *Magnoliidae*, are protogynous. The beetles trapped by the strong imitative odours may deposit pollen on stigmata in the earlier stage of flower development and receive new pollen shortly before they leave when the stamens have dehisced. At the later phase, their bodies usually have become sticky from crawling over the non-sweet stigmatic secretions, so that pollen easily adheres to them. After anthesis and opening of the flowers, the heavy damage done by the beetles becomes evident. Tepals are eaten and stamens and carpels are gnawed. For the time being, and until there is proof, one may at least suspect that the beetles involved in the pollination of *Talauma* are also such species as live and breed on fruits. Just as in flowers of *Annonaceae* we discovered beetle larvae in shed floral axes and in undeveloped carpels. The beetles not only cause crossing of flowers in *Talauma* but also selfing, which was proved experimentally. Self-compatibility is another widespread feature of cantharophilous *Magnoliidae* (see HEISER 1962; GOTTSBERGER 1970; THIEN 1974; etc.).

With CARLQUIST (1969: 340) I share the idea that the large number of sexual organs, the flattening of the stamens, their dense aggregation along the floral axis and consequently the large dimensions of *Magnoliaceae* flowers might well be interpreted as an adaptation better to escape from damage by specialized, crude pollinators (see GOTTSBERGER 1974: 463).

Beetle pollination syndromes in *Eupomatia* (*Eupomatiaceae*), *Calycanthus* (*Calycanthaceae*), *Victoria* and other *Nymphaeaceae* are comparable insofar as their flowers all offer special nutritious tissues to the visitors. In *Eupomatia laurina* beetles eat the flat inner staminodes (HAMILTON 1897), in *Calycanthus occidentalis* the beetles feed on special food bodies on the staminodes, stamen tips and innermost petals (GRANT 1950 a), and in *Victoria amazonica*, carpellary appendages in the flower cavity provide food for the beetles (PRANCE & ARIAS 1975). The protogynous flowers of the above-mentioned species emit strong fruit-like odours and usually attract a mass of beetles which are kept for some time in the flower cavity. Perigyny may be a mode of ovule protection in these species (see GRANT 1950 b).

There is certainly still very much to be done in pollination studies on the more tropical *Magnoliidae*, especially in *Magnoliales* and *Laurales*, of which only a few have actually been investigated. Therefore, I do not wish to leave behind this uniform picture of exclusiveness of beetle pollination in primitive angiosperms, and will mention also some exceptions. For example, species of *Lauraceae* are visited chiefly by beetles and flies (KNUTH 1904, CORNER 1940). On the other hand, in flowers of *Monimiaceae* and *Siparunaceae* I have never observed beetles at all. In the species of *Mollinedia* (*Monimiaceae*) with urceolate receptacles, pollination of unisexual flowers is accomplished by *Thysanoptera*. The female insects bore holes into male and female buds and deposit their eggs in the interior of the receptacles. Their larvae start development in buds and become adult when flowers open. Afterwards they fly out, go to other flowers and are responsible for the pollen transfer (GOTTSBERGER, in progress). This strange mode of pollination reminds one of the much more sophisticated mode in *Ficus*, where the urceolate syconia are used as breeding places by wasps.

But let us return to the main point. After having discussed some pollination syndromes of entomophilous members of the *Magnoliales*, *Laurales* and *Nymphaeales* we may summarize and try to interpret our results. *Drimys*, a member of *Winteraceae*, the presumably most primitive present-day angiosperm family, has an open type of beetle pollination without further specialization. I believe that the unifacial stamens and carpels have remained as an archaic feature in its flowers because of its unspecialized cantharophily. The beetles attracted casually by a flowery odour, and not by deceit, are mostly small and do not further harm the sexual organs of the flowers. Because of this less destructive mode of pollination I think that there was and is no selection pressure to enlarge, to flatten or to increase the number of the sporophylls or to aggregate them densely along the floral axis.

On the other hand, flattened structures, and/or a high number of sporophylls, are characteristic for flowers of the other families mentioned above, which all show a specialized cantharophily. Their flowers utilize deceit, attracting fruit-, or carrion- or dung-beetles. The side effects of these more crude visitors seem to be compensated by larger flowers, by the production of food bodies or thick petals, by a semi-inferior ovary, or by connective shields, or by mechanisms like closing which help to exclude the larger beetles. The flowers offer protection, alimentation, breeding places and imitation of the normal substrate of the beetles.

Returning once more to the *Winteraceae* I wish to draw attention to another feature which seems to underline the primitiveness of flower structures in this family. The dioecious species of *Tasmannia*, contrary to *Drimys*, have anemophilous, unshowy and unisexual flowers (PERVUKHINA 1967). The fact of manifestation of anemophily already in *Winteraceae* was used, however, in a sense to show that primitive angiosperms were not quite fixed as to one or the other mode of pollination even in their earliest times. Doubts were brought up as to what extent entomophily is indeed the original and basic condition in angiosperms. I believe that another way of interpretation can be given. If the *Winteraceae* are really such a primitive and unspecialized group as we believe them to be, then, and although the present-day members are now fixed in their archaism, their ancestors probably had plasticity enough to differentiate quite actively. It is a principle in evolution that usually the more basic and unspecialized groups or members of a group are able to diverge actively, whereas the advanced and more specialized ones are often more fixed. The only possible way of flower-biological radiation in the earliest times of angiosperms was probably from cantharophily to anemophily. At that time there probably did not exist other pollination agents besides beetles and the wind. Some of the ancestors of the *Winteraceae* radiated into anemophily with a subsequent change-over to dicliny, as seen in *Tasmannia*. From a certain point of specialization on, such a radiation towards anemophily does not seem to have occurred anymore. I am not aware of any example of pure anemophily in the *Magnoliaceae, Annonaceae, Calcyanthaceae, Eupomatiaceae* or *Nymphaeaceae*, although there exist tendencies towards dicliny (for *Annonaceae* see FRIES 1959), supposedly here more an attempt to escape autogamy. I shall come back to the question of an early radiation into anemophily when I present the conclusions at the end of this paper.

Finishing the account of beetle pollination syndromes and presumable implications in the flower structures of some *Magnoliidae*, I admit that my reasoning is somewhat oversimplified, but I think it necessary to show some basic trends likely to have occurred during the evolution of the ancestors of these groups. In reality I am aware that reversals of trends may also have frequently occurred, which finally brought out the tremendous diversity in this subclass. Unless we have a more complete knowledge of how so called primitive flowers function, we will not be able to explain their characters in a more satisfying way. For the time being my contribution is meant to open discussion and also to show where more observation is imperative. An understanding of cantharophily in the *Magnoliidae* may be essential also for an understanding of what this mode of pollination may have caused in the flower structures of other subclasses.

The theory to be presented now is that the secondary numerical increase of stamens in other subclasses may represent an adaptation to pollination by beetles and other pollen-eaters. Such an interpretation certainly needs substantiation to be reliable. Therefore, let me first present the facts.

I was brought to this idea when I studied the pollination mechanisms of some of the polyandrous members of the *Dilleniidae*. In the *Dilleniaceae* I have observations for *Davilla* and *Doliocarpus*. *Davilla elliptica*, a shrub in the Brazilian cerrado vegetation opens its yellow-petaled flowers at dawn, which was at 6 : 20 in the morning in June. The two innermost sepals, enlarged and scoop-like, separate and the petals and stamens unfold. Immediately there was flower visiting by the introduced *Apis mellifera* and later on also by the native *Meliponinae*-bees, which all collected pollen. During the whole morning there was also frequent approach by many beetle species, which cut the stamens and fed on the filaments, anthers and pollen. From about 9 : 00 in the morning on, individuals of another species of a curculionid beetle arrived, usually crawling directly to the ovary. This beetle was seen to bite holes into the ovary's interior, afterwards depositing an egg in one of the locules and than changing to another flower. At about 11 A.M. the petals fall, and the two large sepals start to fold together, during which time there are still visits by the last mentioned beetles. At about 2 P.M. and 3 P.M. of the same day the ovary of flowers is completely enclosed by the two inner sepals. Thus, the flowers are pollinated by bees and beetles, of which the last are quite destructive. Some of them gnaw on the stamens and others use the ovary as a breeding place. The folding of the sepals after flowering seems to be an adaptation to save some of the ovaries from destruction. From three different counts it was seen that about 25% to 30% of ovaries are destroyed by the beetle's larvae. From another count in *Davilla rugosa* it became evident that the destruction sometimes attains nearly 100%. Unfortunately, when I made these observations some years ago, I did not pay special attention to the flower odour of *Davilla*. In *Doliocarpus schottianus* the flower odour has a heavy fruit-like component; its stamens and petals are also cut or gnawed by beetles.

In the *Clusiaceae* (*Theales*) beetles are involved in the pollination of several species of *Kielmeyera*. *K. variabilis* has large white-petaled, slightly protogynous flowers, which emit a strong fruit like odour. Besides some bees, numerous beetles are regularly attracted by the flowers and are co-pollinators. They crawl over all flower parts and do much destruction in gnawing petals and cutting stamens. Older flowers are often without any stamens because the beetles have rasped them all at the base. For some beetles there is already proof that they are true fruit eaters; they were found also on ripe, odoriferous fruits of *Campomanesia* (*Myrtaceae*).

In species of *Clusia*, beetle pollination is almost an exclusive feature. Pollination in these dioecious species is attained quite precisely because the strong odour, with a fruit-like component, is localized only on the massive stamens and the large stigmata. Here also there occurs some gnawing of flower parts.

Species of *Paeonia* (*Paeoniaceae*) have long been known as having beetle flowers with cantharophilous odours (DELPINO 1873; DIELS 1916; PORSCH 1950; GRANT 1950 b; VAN DER PIJL 1960; LEPPIK 1964; etc.). There is further suspicion of regular participation of beetles in the pollination of some *Cistaceae* (e.g. KNOLL 1914) and *Tiliaceae* (e.g. species of *Luehea*, BAKER et al. 1973; GOTTSBERGER, unpublished). Also here, destruction of stamens commonly occurs.

For the basic polyandrous group of the *Rosidae* data are most instructive also. A very large number of species of *Rosaceae* belonging to the genera *Spiraea*, *Aruncus*, *Physocarpus*, *Sorbus*, *Pyrus*, *Amelanchier*, *Crataegus*, *Rubus*, *Filipendula*, *Rosa*, *Sorbaria*, *Prunus*, *Chaenomeles*, *Mespilus*, *Potentilla*, *Geum*, etc. are known to be pollinated by beetles. Usually beetles are co-pollinators together with other insects; flower destruction may vary from population to population but occurs frequently (KNUTH 1904; PORSCH 1950; GRANT 1950b; KNOLL 1956; PROCTOR & YEO 1973; etc.). It is remarkable that a large number of these beetle pollinated flowers emit cantharophilous odours and are protogynous. Some beetles on *Rosaceae* even belong to species found otherwise in *Magnolia* flowers (KNUTH 1904: 334; DAUMANN 1930: 111).

For the polyandrous *Caryophyllidae* similar observations exist. *Mesembryanthemum* of the *Aizoaceae* frequently has beetles as co-pollinators of the flowers (KNUTH 1904: 278—279) and in the *Cactaceae* beetles are nearly omnipresent flower visitors (KNUTH 1904: 518–519; PORSCH 1939; VAN DER PIJL 1961: 45). One only needs to check the list of flower odours given by PORSCH (1939) to see how commonly fruit-like or aminoid odours occur in this family.

These are the flower-biological data, some of which have been known for decades. Analyzing the facts, my interpretation tends in the direction of seeing in the polyandry of the basic groups in *Rosidae*, *Dilleniidae* and *Caryophyllidae* an adaptive character that provides an excess of stamens and pollen for foraging beetles and other pollen-eaters. This adaptive character of many stamens is manifested also in some more advanced members with more advanced pollination syndromes.

One argument against my interpretation might be the belief that beetles are better attracted to these polyandrous flowers, because they have many stamens and therefore provide larger amounts of food than do flowers with few stamens. This argument is, however, easy to invalidate by observations. In the polyandrous members of more advanced families of these subclasses, as for example, *Caryocaraceae*, *Cochlospermaceae*, *Begoniaceae*, *Capparaceae*, *Chrysobalanaceae* and others, I never saw the flowers visited regularly by beetles and also did not often see damaged flowers. They have advanced pollination systems, being visited by bats, bees or moths, and there is no regular utilization of their polyandrous flowers by beetles. There may be casual visits by beetles, as occur on every plant, and there may be also a casual damage, but there is not usually any frequent visiting. Another good argument for beetles as the real co-pollinators of *Dilleniaceae, Rosaceae, Clusiaceae, Cactaceae*, etc. is the widespread occurrence of the imitative fruit-like or aminoid odours, which deceive beetles precisely; this character is known already from the beetle-pollinated *Magnoliidae*. Protogyny concludes the picture of this syndrome.

In the last few years it has become evident that the primary polyandry of *Magnoliidae* cannot be directly compared with the so called secondary one in other subclasses (LEINS 1964, 1971; KUBITZKI 1973; etc.). However, it is certainly right to say that if we assume that there existed an evolutionary trend within the *Magnoliidae* towards a numerical increase of stamens as a response to selection pressure, than this advanced condition might be called secondary polyandry also (STEBBINS 1974: 226). But as the numerical increase of stamens in *Magnoliidae* follows the spiral arrangement and in other subclasses it usually does not, the problem remains the same. It is still difficult to imagine how any form of secondary polyandry in *Rosidae, Dilleniidae* and *Caryophyllidae* may have derived directly from the spiral one of the *Magnoliidae*. There must have been first oligomerisation with a cyclic arrangement of stamens and from this condition on a new initiation of stamen primordia.

As for the *Caryophyllidae*, it is commonly assumed that they derived via the *Phytolaccaceae* from some *Ranunculanae* ancestors. In the *Magnoliidae*, besides the initially mentioned two trends of radiation into anemophily and specialized cantharophily, obviously at least a third trend has existed towards oligomerisation and fixation of numbers of flower parts. The last cited trend might have started from a time when, besides the beetles, other, probably nectar-sucking, visitors had attained enough frequency to exert selection pressure on flowers. The derivation of the *Caryopyllidae* via the *Magnoliidae* with oligomerised stamens is therefore at least theoretically possible. For the polyandrous *Rosidae* and *Dilleniidae* an origin via the woody *Saxifragales* is often assumed, in which group also flowers with oligomerised stamens predominate.

One reason why such an adaptation to beetle pollination has occurred may have been the presence of cantharophilous odours in flowers of the assumed ancestors. Cantharophilous odours together with the protogynous condition are indeed quite frequent in present-day *Magnoliidae* (even in those with a reduced number of stamens) and in *Saxifragales*. Such an imitative flower odour could have been a kind of preadaptational character, based on which new waves of radiation towards beetle pollination have occurred.

Is there any way to explain why there is a centrifugal stamen development in the polyandrous *Caryophyllidae* and *Dilleniidae* but a centri-

petal one in the *Rosidae*? Are these characters evolved by chance or are
these opposite directions of stamen development explainable? Let us
try to use the immensely diversified *Saxifragales* as a model for deriving
Rosidae and *Dilleniidae*. A large number of *Saxifragales* are characterized by a nectariferous disc in an intrastaminal position, thus separating the stamens and the ovary. The stamens are inserted at the periphery
of the flowers close to the perianth and there is space between them and
the gynoecium. If we assume that the polyandrous *Rosidae* are derived
via the *Saxifragales* with intrastaminal nectariferous discs, then in
response to a selection pressure for numerical increase of stamens a
centripetal development in direction of the flower-centre may have
occurred, filling the space of the former nectariferous disc, now partly or
totally functionless. That this idea may have some truth is shown by the
fact that the whole *Rosidae* are characterized by an intra- or also inter-
staminal disc which appears everywhere within this group, especially in
the more derived members (with nectar sucking pollinators!) but also in
more primitive ones. Other present-day *Saxifragales* are without such
a disc and their stamens and ovary are close together. The *Dilleniidae*
may have been derived from *Saxifragales*-like (or other) ancestors
without an intrastaminal disc. In response to selection for numerical
increase of stamens meristems are more likely to have increased towards
the periphery of flowers, because development towards the centre was
spatially limited by the gynoecium. The tendency to form a disc is
indeed much less pronounced in the *Dilleniidae*. Polyandrous *Caryophyllidae* may be comparable with the *Dilleniidae*, in being derived from
ancestors also without an intrastaminal disc, therefore appearing now
also with centrifugal stamen development. The evolution of secondary
polyandry, based on these reflections, may well have followed the
direction of least resistance.

To complete the picture of evolution of dicotyledons influenced by
beetle pollination, I wish finally to present some selected examples in
which another trend, that towards aggregation of small flowers into
inflorescences becomes evident. There are virtually hundreds of cases
but I shall restrict myself to those appearing in *Polemoniaceae*, *Proteaceae*
and *Asteraceae*. These three families have been well studied so that we
are aware of their major flower-biological radiation.

The *Polemoniaceae*, a basically bee-pollinated group, show occasional
radiation into beetle pollination (GRANT and GRANT 1965). In canthero-
philous *Linanthus parryae* the flowers are large and cup shaped, but in
Ipomopsis congesta small flowers are aggregated into terminal capitate
heads.

In the *Proteaceae* there is "a 'retrograde' development of pollination
syndromes from the brush blossom type back to the more primitive

bowl-shaped one concomitant with the progressive morphological development of the blossom. This development is apparently accompanied by a similar 'retrograde' development amongst the pollinators, from pure or predominating ornithophily back to the assumed most primitive stage of cantharophily" (FAEGRI 1965). In *Leucadendron discolor* the inflorescence is even condensed into a structure which forms a perfect parallel with *Magnolia* (VAN DER PIJL 1969: 92). Cantharophily is evidently an end line in both families. There is no doubt that this is to be considered a secondary specialization for beetle pollination.

In the *Asteraceae*, the aggregation of small flowers into flower heads seems to be an original adaptation to beetles (see KNUTH 1905; GRANT 1950 b; VAN DER PIJL 1960; LEPPIK 1960, 1970; etc.), from which a radiation towards more specialized pollinators, like Hymenoptera, Lepidoptera, birds, etc. seems to have occurred. The fact that *Compositae* and other groups presumably follow an evolutionary sequence from the more primitive pollination syndrome towards a more advanced one, has led LEPPIK (1970: 327) to put forward a theory which says that there is a general trend of floral evolution in various groups with a recapitulation of the historical appearance of pollinators and pollination syndromes. This theory, however, does not work for all cases. In the *Polemoniaceae* and *Proteaceae* an opposite development has in fact occurred.

The basic condition of cantharophily in the *Compositae* led several authors to think that they must be a very old group having originated in the early beetle-era. This assumption is not likely and in my opinion is not even necessary. From the origin of angiosperms up to the present, beetles were and are potential flower visitors and pollinators. There has been a continuous trend of adaptation towards cantharophily in the most diverse groups of most diverse evolutionary level. Therefore, the basic adaptation to beetle pollination in *Asteraceae* might well have occurred more recently. Also, the palaeontological data plead for a relatively late appearance of this family. Only in mid-Tertiary does the *Compositae* began to play a major role in the vegetation (WAGENITZ 1976).

Conclusion

Beetles probably influenced the origin of the angiosperms which led to the formation of a basically entomophilous, hermaphrodite flower. The original flowers must have had very primitive structures and cantharophily supposedly was of an open unspecialized type. Present-day members which still seem to show some of these original features are some *Winteraceae*, like *Drimys*. The presence of anemophily in this family probably indicates that there already existed a strong tendency to radiate into anemophily in the ancestral complex now extinct. It is likely that this line or these lines led to entomophilous-anemophilous and/or pure anemophilous groups of the *Magnoliidae* and some *Hamamelididae*. From the ancestral complex there might have been several waves of radiation towards specialized cantharophily concomitant with a development of specialized cantharophilous flower characters, as for example a numerical increase of sexual organs and their flattening (e.g. *Magnoliaceae, Annonaceae*, etc.). Another, but probably somewhat later trend, presumably connected with the increased influence of sucking insects, led to groups with oligomerized flower parts. From such groups with a reduced number of stamens, several waves of radiation towards cantharophily seem to have occurred again, concomitant with a secondary numerical increase of stamens. These radiation waves may have occurred at a time when other flower visitors besides the beetles were already in existence. Flowers of primitive members of secondarily polyandrous groups show a mixed pollination syndrome with beetles and other insects as co-pollinators. A nearly exclusive beetle pollination was reached by some *Magnoliidae* in earlier times when beetles probably were the predominant visitors. Beetles continued as potential pollinators up to the present, so that there was a continuous tendency of radiation into beetle pollination. In more modern groups, however, instead of an increase in size and number of flower parts, there is a new trend towards condensation of flowers into inflorescences.

This is my actual picture of a presumed evolution of dicotyledons regarding, however, only cantharophily. We now only need to fill the space with wind and water, bees, flies, moths, butterflies, birds and bats to get a more complete account.

My thanks go to Prof. Dr. L. van der Pijl, The Hague, who gave valuable suggestions on the content of the paper and to Dr. G. Eiten, Brasília, and Mr. B. L. Burtt, Edinburgh, who corrected the language and the style.

References

Baker, H. G., Baker, I., and Opler, P. A., 1973: Stigmatic exudates and pollination. In: Pollination and Dispersal (Brantjes, N. B. M., and Linskens, H. F., eds.), 47—60. Nijmegen: Dept. Bot., Kath. Universiteit.

Carlquist, S., 1969: Toward acceptable evolutionary interpretations of floral anatomy. Phytomorphology **19**, 332—362.

Corner, E. J. H., 1940: Wayside trees of Malaya. Vol. 1. Singapore: Government Printing Office.

Daumann, E., 1930: Das Blütennektarium von *Magnolia* und die Futterkörper von *Calycanthus*. Planta **11**, 108—116.

Delpino, F., 1873: Ulteriori osservazioni sulla dicogamia nel regno vegetale. Milano: G. Bernardoni.

DIELS, L., 1916: Käferblumen bei den *Ranales* und ihre Bedeutung für die Phylogenie der Angiospermen. Ber. deutsch. bot. Ges. **34**, 758—774.

EHRENDORFER, F., 1976: Evolutionary significance of chromosomal differentiation patterns in gymnosperms and primitive angiosperms. In: Origin and early evolution of angiosperms (BECK, C. B., ed.), 220—240. New York: Columbia University Press.

— KRENDL, F., HABELER, E., and SAUER, W., 1968: Chromosome numbers and evolution in primitive angiosperms. Taxon **17**, 337—353.

FAEGRI, K., 1965: Reflections on the development of pollination systems in African *Proteaceae*. J. South Afr. Bot. **31**, 133—136.

— and PIJL, L. VAN DER, 1971: The principles of pollination ecology. Ed. 2. Oxford: Pergamon Press.

FRIES, R. E., 1959: *Annonaceae*. In: Die natürlichen Pflanzenfamilien (ENGLER, A., and PRANTL, K., eds.), Ed. 2, **17 a II**, 1—171. Berlin: Duncker & Humblot.

GOTTSBERGER, G., 1970: Beiträge zur Biologie von Annonaceen-Blüten. Österr. bot. Z. **118**, 237—279.

— 1974: The structure and function of the primitive angiosperm flower—a discussion. Acta bot. Neerl. **23**, 461—471.

GRANT, V., 1950 a: The pollination of *Calycanthus occidentalis*. Amer. J. Bot. **37**, 294—297.

— 1950 b: The protection of the ovules in flowering plants. Evolution **4**, 179—201.

— and GRANT, K., 1965: Flower pollination in the *Phlox* family. New York-London: Columbia Univ. Press.

HAMILTON, A. G., 1897: On the fertilisation of *Eupomatia laurina* R. BR. Proc. Linn. Soc. New South Wales **22**, 48—55.

HEISER, CH. B., 1962: Some observations on pollination and compatibility in *Magnolia*. Proc. Indiana Acad. Sci. **72**, 259—266.

KNOLL, F., 1914: Zur Ökologie und Reizphysiologie des Andröceums von *Cistus salvifolius* L. Jahrb. wiss. Bot. **54**, 498—527.

— 1956: Die Biologie der Blüte. Berlin-Göttingen-Heidelberg: Springer.

KNUTH, P., 1904—1905: Handbuch der Blütenbiologie. 3. Band, 1. und 2. Teil (LOEW, E., ed.). Leipzig: Verlag W. Engelmann.

KRAL, R., 1960: A revision of *Asimina* and *Deeringothamnus* (*Annonaceae*). Brittonia **12**, 233—278.

KUBITZKI, K., 1973: Probleme der Großsystematik der Blütenpflanzen. Ber. dtsch. bot. Ges. **85**, 259—277.

— and VINK, W., 1967: Flavonoid-Muster der *Polycarpicae* als systematisches Merkmal. II. Untersuchungen an der Gattung *Drimys*. Bot. Jb. **87**, 1—16.

LEINS, P., 1964: Das zentripetale und zentrifugale Androecium. Ber. dtsch. bot. Ges. **77**, 22—26.

— 1971: Das Androecium der Dikotylen. Ber. dtsch. bot. Ges. **84**, 191—193.

LEPPIK, E. E., 1960: Early evolution of flower types. Lloydia **23**, 72—92.

— 1964: Floral evolution in the *Ranunculaceae*. Iowa State J. Sci. **39**, 1—101.

— 1970: Evolutionary differentiation of the flower head of the *Compositae*. II. Ann. Bot. Fennici **7**, 325—352.

PERVUKHINA, N. V., 1967: The pollination of the primary angiosperms and the evolution of modes of pollination (Russ.). Bot. Zh. (Moskow) 52, 157—188.

PIJL, L. VAN DER, 1953: On the flower biology of some plants from Java. Ann. Bogor. 1, 77—99.

— 1960: Ecological aspects of flower evolution. I. Phyletic evolution. Evolution 14, 403—416.

— 1961: Ecological aspects of flower evolution. II. Zoophilous flower classes. Evolution 15, 44—59.

— 1969: Evolutionary action of tropical animals on the reproduction of plants. Biol. J. Linn. Soc. 1, 85—96.

PORSCH, O., 1939: Das Bestäubungsleben der Kakteenblüte. Cactaceae, Jahrb. deutsch. Kakteen-Ges., 1—142.

— 1950: Geschichtliche Lebenswertung der Kastanienblüte. Österr. bot. Z. 97, 269—321.

PRANCE, G. T., and ARIAS, J. R., 1975: A study of the floral biology of Victoria amazonica (POEPP.) SOWERBY (Nymphaeaceae). Acta Amaz. 5, 109—139.

PROCTOR, M., and YEO, P., 1973: The pollination of flowers. Glasgow: W. Collins Sons and Co. Ltd.

STEBBINS, G. L., 1974: Flowering plants. Evolution above the species level. Cambridge, Mass.: Belknap Press.

THIEN, L. B., 1974: Floral biology of Magnolia. Amer. J. Bot. 61, 1037—1045.

— HEIMERMANN, W. H., and HOLMAN, R. T., 1975: Floral odors and quantitative taxonomy of Magnolia and Liriodendron. Taxon 24, 557—568.

UPHOF, J. C. TH., 1933: Die nordamerikanischen Arten der Gattung Asimina. Mitt. deutsch. dendrol. Ges. 45, 61—76.

WAGENITZ, G., 1976: Systematics and phylogeny of the Compositae (Asteraceae). Plant Syst. Evol. 125, 29—46.

WESTER, P. J., 1910: Pollination experiments with Anonas. Bull. Torrey Bot. Club 37, 529—539.

ZIMMERMAN, G. A., 1941: Hybrids of the American papaw. J. Hered. 32, 83—91.

Address of the author: Univ.-Doz. Dr. G. GOTTSBERGER, Departamento de Botânica, Faculdade de Ciências Médicas e Biológicas de Botucatu, Universidade Estadual Paulista "Julio de Mesquita Filho", BR-18600 Botucatu, S. P. Brazil.

Plant Syst. Evol., Suppl. 1, 227—234 (1977)

Botanisches Institut der Universität Wien, Austria

New Ideas About the Early Differentiation of Angiosperms

By

Friedrich Ehrendorfer, Wien

Abstract: Three premises are introduced: Origin of angiosperms from one common ancestral group, stamens and carpels as appendicular (phyllomic) sphorangiophores, and adaptive nature of basic angiosperm characters. The widely accepted assumptions that *Magnoliidae* link up directly to the *Rosidae-Dilleniidae* alliance, and that *Hamamelididae* are heterogeneous and have originated from several *Rosidae-Dilleniidae* groups are hardly tenable: There are drastic differences between the *Magnoliidae* and *Rosidae-Dilleniidae*, in the morphology of their functionally similar entomophilous flower types, and in many other respects; the *Hamamelididae* demonstrate considerable internal coherence together with ambivalent affinities to both, *Magnoliidae* and *Rosidae-Dilleniidae*. To solve this dilemma, the hypothesis is discussed that the extant *Hamamelididae* mark the remnants and descendants of an ancient "transitional field" from the *Magnoliidae* to the *Rosidae-Dilleniidae* (Fig. 1). The *Hamamelididae* would thereby correspond to an early phase of flower reduction and trends towards anemophily, the *Rosidae-Dilleniidae* to a subsequent phase of floral elaboration and intensified adaption towards zoophily.

There are three premises (1–3) which are essential to the following discussion; let me start with them without going into too many details:

1. The angiosperms have originated from one common ancestral group. There is growing evidence from many fields to back such an assumption in spite of continued scepticism (e.g. MEEUSE in this symposium): Paleobotany demonstrates the simultaneous appearance of typical monosulcate and later of tricolpate angiosperm pollen grains, leaves, etc. during the Lower Cretaceous (Barremian to Albian) on a world wide scale (cf. BRENNER 1976, DOYLE and HICKEY 1976, HUGHES 1976, DOYLE 1977). These data clearly suggest an origin of the group in the tropical zones and a spread towards the poles, paralleled by increasing diversification. Among the very large array of morphological, anatomical, palynological, biochemical, and other characters connecting all angiosperms, but lacking in other spermatophytes, the embryological similarities always have been particularly impressive. They are not

limited to the well known and unique structure of the embryo sack, the peculiar double fertilization, and the subsequent secondary endosperm formation, but include unexpected details concerning the entrance of the pollen tube through the degenerating synergide and the remarkable ultrastructure of the "filiform apparatus". As a taxon the *Angiospermae* are particularly coherent. This is most evident from the fact that none of the very numerous taxonomic subdivisions proposed has been generally accepted (not even in regard to *Monocotyledoneae* and *Dicotyledoneae*; cf. Huber in this symposium). On the other hand, the separation from all other *Spermatophyta* (including the *Gnetatae*—obviously the most closely related extant *Gymnospermae*), never has been a problem.

2. The stamens and carpels of angiosperms are sporangiophores of an appendicular (phyllomic) nature, homologous throughout the group. The increasing knowledge of pollen sac- and ovule-bearing organs in pteridosperms and their descendants hints to their morphological interpretation along modified classical concepts without taking refuge to unnecessarily complex auxiliary hypotheses (cf. Ehrendorfer 1971). From such a view-point the flowers of all angiosperms appear comparable (as would have to be expected from their common origin).

3. The origin and successful diversification of angiosperms must have been linked, just as in all other groups of organisms, to their basic characters, giving them adaptive superiority over their predecessors and allowing them to enter a new ecological niche. Their primarily hermaphrodite flowers with enclosed ovules evidently have made possible wind-independent, more precise, and more economic insect pollination. (Size, ornamentation, and ultrastructure of earliest known angiosperm monosulcate pollen suggest entomophily and germination on the stigma of closed carpels! Doyle 1977, J. Muller, pers. comm.) Neotenic gametophytes, double fertilization and secondary endosperm must have allowed angiosperms faster and more economic reproduction (Stebbins 1974), their improved conducting system and increased flexibility of vegetative growth clearly have stimulated ecological expansion. Consequently, early angiosperms may have grown as small, subordinate and scattered woody plants in the understory and along the margins of unstable successional gymnosperm forests in ± humid seasonal tropical climates (cf. the thin pollen walls of earliest angiosperm pollen; Doyle and Hickey 1976).

If these three premises are accepted, we can suggest on a comparative basis a likely basic flower model for the angiosperms (Fig. 1-I): generalized, entomophilous, rather large, with spiral arrangement of numerous flower elements, undifferentiated perianth (perigon), primary polyandry, medium-sized, monosulcate, psilate to reticulate pollen, carpels with stigmatic margins but without styles, and a trend to staminal nectaries.

Among extant angiosperms certain *Magnoliidae* seem to have preserved such a flower type rather closely.

If we want to interpret other angiosperm flower types with the help of our *Magnoliidae* model, and if we try to speculate about possible phylogenetic links between extant *Magnoliidae* (s. lat.) and the *Rosidae-Dilleniidae* alliance, we encounter two major obstacles, to be outlined in the following paragraphs (A and B).

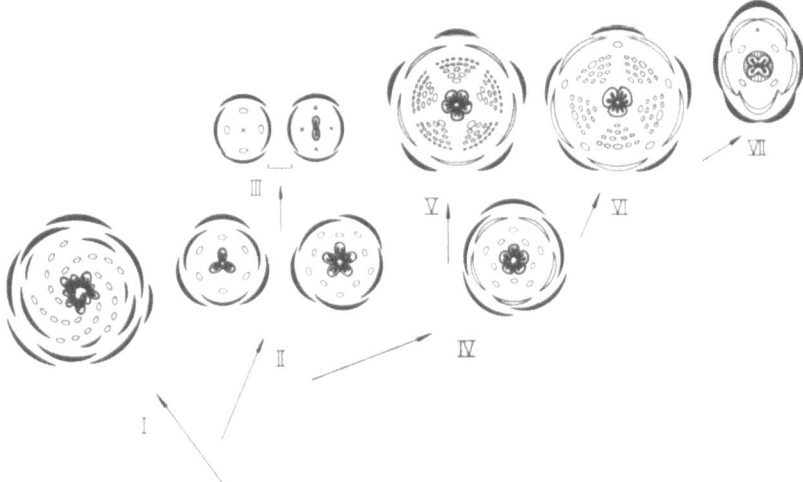

Fig. 1. Proposed relationships between several major dicotyledonous angiosperm groups and their flower types. I basic *Magnoliidae* (polymerous, perigon, primary polyandry); II basic *Hamamelididae* (and some basic *Rosidae-Dilleniidae*) (oligomerous, hermaphrodite); III advanced *Hamamelididae* (strongly oligomerous, unisexual); IV basic *Rosidae-Dilleniidae* (calyx and corolla, androecium 1-2 cyclic); V some *Dilleniidae* (secondary polyandry: centrifugal); VI some *Rosidae* (secondary polyandry: centripetal), VII advanced *Asteridae* (synsepaly, sympetaly, dorsiventrality, oligomery, axial nectary)

A. Among the *Rosidae-Dilleniidae* alliance we also find entomophilous and rather large, polyandrous flower types but they differ from our *Magnoliidae* model (Fig. 1-I) in some important aspects: (spiro)cyclic arrangement of fewer flower elements, differentiated perianth (calyx related to perigon, corolla related to stamens), secondary polyandry (usually through centripetal or centrifugal elaboration: LEINS 1975), rather large, tricolpate and often more strongly sculptured pollen, carpels with stigmas on styles, and a trend towards axial nectaries (Fig. 1-V, VI). A direct derivation of the entomophilous flower models

I → V, VI is hardly possible, if functional aspects are considered: Why are there new semaphylls (i.e. petals) in V + VI, if a showy pergion was already present in I? Why is there secondary polyandry in V + VI, as primary polyandry had already been established in I? Why the different types of nectaries in I and V + VI? It is these questions, so far not answered, together with the obvious differences between *Magnoliidae* and *Rosidae-Dilleniidae* in chemistry (Kubitzki 1969, 1973) and other characters which have supported ideas about a pleiophyletic origin of *Dicotyledoneae*.

B. The *Hamamelididae* (in the sense of Takhtajan 1969 and Cronquist 1968) are nowadays usually regarded as secondarily simplified in flower structure as a consequence of adaptation towards wind pollination (e.g. Endress in this symposium; Fig. 1-II, III). This and the recognition of affinities of *Hamamelididae* to several groups of the *Rosidae-Dilleniidae* alliance has recently led to a total demolition of the group (cf. Thorne 1974 or Dahlgren in this symposium: his *Hamamelidanae* include *Cunoniales* besides *Fagales, Casuarinales, Hamamelidales, Trochodendrales*, etc.; *Juglandales* and *Myricales* are placed in *Rutanae, Didymelaceae* (= *Didymelales*) and *Barbeyaceae* (= *Barbeyales*) in *Celastranae, Eucommiales* in *Cornanae, Urticales* in *Dillenianae*). Such a procedure is based on the opinion that the *Hamamelididae* are a polyphyletic assemblage, that their orders are ± independent anemophilous derivatives from several zoophilous *Rosidae-Dilleniidae* groups, and that therefore they should be regarded as "dead ends" in a morphological and evolutionary sense. This interpretation obscures important relationships and fails to explain several relevant phenomena: Why are the various orders of *Hamamelididae* connected by palynological, anatomical (wood, leaf venation, etc.), biochemical, and other affinities which can not be explained as accidental or as a result of parallel adaption to secondary anemophily? How is it possible that there are not only the unquestionable affinities of *Hamamelididae* to various groups of *Rosidae-Dilleniidae* but at the same time to the *Magnoliidae*, if they are to be derived separately from the former only? Let us remember in connection with the last question: The wide-spread occurrence of 3-merous flowers in *Magnoliidae* and in *Hamamelididae* (e.g. *Fagaceae, Platanaceae, Daphniphyllaceae, Barbeyaceae* etc.); the remarkable similarities in the pattern of chromosomal differentiation within *Hamamelididae* and between *Hamamelididae* and woody *Magnoliidae* (s. str.) (Ehrendorfer 1976a), the obviously ambivalent affinities of *Trochodendrales, Cercidiphyllales* and *Eupteleales* in regard to partly homoxylous wood, chloranthoid and platanoid leaf dentation (Hickey and Wolfe 1975), ultrastructure of 3-colpoidate or 3-colpate pollen (Walker 1976a, b), and chemistry: lack of benzyl-isoquinoline alkaloids (as in *Magnoliidae* s. str.), but presence of polyphenolics (like leuco-delphinidine) and ellagic acid (as in *Rosidae-Dilleniidae*); the broadly documented biochemical, anatomical, chromosomal and other affinities (benzyl-isoquinoline alkaloids, bitter substances, idioblasts, chromosomal patterns, choricarpy, etc.) which connect *Magnoliidae* and the primitive *Rutaceae-Zanthoxyleae* and *-Toddalioideae* (FISH and WATERMAN 1973, EHRENDORFER 1976a), which in turn are clearly linked through the *Rhoipteleaceae* to the *Juglandaceae* and *Myricaceae*. Why do evidently wind-transported and *Hamamelididae*-like pollen appear so remarkably early in the fossil record (Albian-Cenomanian), and why are there well documented records of several *Hamamelididae* (reminiscent of *Cercidiphyllaceae*, *Platanaceae*, *Fagaceae*, *Juglandaceae*, etc.) (MULLER 1970, BRENNER 1976, DOYLE and HICKEY 1976) which antedate their presumed zoophilous *Rosidae-Dilleniidae* predecessors? The widely accepted but evidently over-simplified reductional interpretation of the *Hamamelididae* has no answer to these questions. This deficiency again has provided arguments for ideas about primarily simple flowers in *Hamamelididae* and their direct gymnospermic origin, independent from *Magnoliidae* and the *Rosidae-Dilleniidae* alliance.

Is there a solution to the problems concerning the two basic zoophilous flower types of dicotyledons, i.e. *Magnoliidae* and *Rosidae-Dilleniidae* (A), and the ambivalent affinities of the evidently very ancient *Hamamelididae* (B)? To this end I would like first to discuss the flower types illustrated in Fig. 1. This diagram indicates that the primarily polyandrous *Magnoliidae* type (I) and the secondarily polyandrous *Rosidae-Dilleniidae* types (II) can be linked through types II and IV. Among less advanced *Hamamelididae* we encounter the flower types II which tend towards anemophily (but often still attract animal pollinators): small size, (spiro)-cyclic arrangement of a reduced number of flower elements, simple and inconspicuous perigon, oligandry, carpels ± free, with stigmas exposed on styles, small and smooth tricolpate to triporate pollen, and lack of nectaries. The development from II to III illustrates extreme reduction and unisexual differentiation of flowers typical for many exclusively anemophilous *Hamamelididae*, often reaching a stage which appears hardly reversible. On the other hand, flower type IV indicates an adaptive reversal to a new level of zoophily; it is typical for many *Rosidae-Dilleniidae*, and can be characterized by increasing flower size, origin of a new type of semaphylls, i.e. petals from stamens, larger pollen grains with variously sculptured exine and elaborated apertures. There are trends towards secondary polyandry, particularly in beetle pollinated groups (cf. GOTTSBERGER in this symposium), but also trends towards the formation of axial nectaries for the attraction of more advanced flower visitors with sucking mouth parts. Finally, flower model VII depicts well-known further steps in the specialization of sympetalous dicotyledons.—If we accept the sequence of changes in flower structure and function as outlined in this paragraph and in Fig. 1, the question listed under A) can be answered.

In terms of phylogenetic relationships the data presented suggest that the entomophilous *Magnoliidae* and *Rosidae-Dilleniidae* are not of obscure parallel origin, but that the extant *Hamamelididae* mark the remnants and descendants of an ancient (Lower to Upper Cretaceous "transitional field" from the *Magnoliidae* to the *Rosidae-Dilleniidae*. This has already been suggested, particularly on the basis of palynological evidence, by WALKER (1976: Fig. 4 and pers. comm.). We therefore come to the unorthodox view that the *Hamamelididae* are not altogether "dead ends" but indicators for an early phase of flower reduction and adaptation towards anemophily in the evolution and differentiation of the angiosperms. It is obvious that such an interpretation solves the problems raised under B), particularly those concerning the ambivalent affinities of *Hamamelididae* to *Magnoliidae* and *Rosidae-Dilleniidae*.

As an ecological and functional corollary to the hypothesis presented, one has to assume that early entomophilous angiosperms, after a phase of establishment and coexistence with anemophilous gymnosperms expanded into more northern and more xeric habitats, not yet colonized by seed plants. A tendency towards anemophily must have been an adaptive advantage under such circumstances, with more uniform pioneer populations in an open environment, and with the obligate lag of animals (including potential pollinators) following the plants in the conquest of new ecological niches. It could have been that these developments have contributed to the increase in pollen wall thickness and the origin of 3-colpate pollen (for better accomodation to volume changes due to varying humidity and/or better germination chances?) in the early angiosperms (DOYLE, VAN CAMPO and LUGARDON 1975).

From the considerable amount of evidence available to support the hypothesis discussed, only little can be added at this occasion to the paleobotanical, palynological, anatomical, morphological and biochemical data already briefly mentioned. The suggestion that petals, particularly in the *Rosidae*, *Dilleniidae* and *Asteridae*, are of androecial origin, dates back to C. DE CANDOLLE and ČELAKOVSKÝ, and has been well substantiated (BAUM 1950). Crucial instances include families, where apetalous and petaliferous taxa coexist, and where the origin of diplochlamydeous from monochlamideous flowers can be inferred, as in *Hamamelidaceae* (J. WALKER, pers. comm., but cf. ENDRESS in this symposium), *Cunoniaceae* (also with trends towards numerical increase of stamens), *Rutaceae-Zanthoxyleae/-Toddalioideae*, *Buxaceae-Euphorbiaceae*, etc. The ancient *Proteaceae* have remained on a monochlamydeous level but have compensated this in their return to elaborate zoophily through the formation of $\pm$ efficient pseudanthia, just as numerous *Hamamelidaceae* and *Moraceae* (cf. ENDRESS and BERG in this symposium), *Euphorbiaceae*, etc. Finally, a note on remarkable cases of parallelism in the *Caryophyllidae* and the *Monocotyledoneae*, groups which appear directly linked to the *Magnoliidae:* For both groups one can suggest—just as for the *Hamamelididae* and *Rosidae-Dilleniidae*—an early phase of flower reduction and trends towards anemophily (e.g. in primitive members of *Molluginaceae*, *Aizoaceae*, *Phytolaccaceae* or of *Dioscoreaceae*, *Arecaceae*, etc.), followed by a phase of intensified adaptation towards zoophily, with development of showy tepals or petals, nectaries (axial in *Caryophyllidae*, gynoecial-septal in *Monocotyledoneae*), and occasional increase of stamen numbers (cf. EHRENDORFER 1976).

Let me finish by saying that I hope that the facts presented, and the suggestions and hypotheses based upon them will stimulate further research on the fascinating and wide-open problem of the early differentiation of the angiosperms; to the benefit of the facts, and at the expense of the hypotheses.

The author is very grateful to Dr. J. WALKER (Amherst) for stimulating discussions and personal communications during his stay as Visiting Professor at the University of Massachusetts.

References

BAUM, H., 1950: Unifaziale und subunifaziale Strukturen im Bereich der Blütenhülle und ihre Verwendbarkeit für die Homologisierung der Kelch- und Kronblätter. Österr. Bot. Z. **97**, 1—43.

BRENNER, G. J., 1976: Middle cretaceous floral provinces and early migrations of angiosperms. In: Origin and early evolution of angiosperms (BECK, C. B., ed.), 23—47. New York and London: Columbia Univ. Press.

CRONQUIST, A., 1968: The evolution and classification of flowering plants. London: Nelson & Sons Ltd.

DOYLE, J. A., 1977: Patterns of evolution in early angiosperms. In: Patterns of evolution (HALLAM, A., ed.), 501—546. Amsterdam: Elsevier.

- - HICKEY, L. J., 1976: Pollen and leaves from the mid-cretaceous Potomac group and their bearing on early angiosperm evolution. In: Origin and early evolution of angiosperms (BECK, C. B., ed.), 139—206. New York and London: Columbia Univ. Press.

---- VAN CAMPO, M., LUGARDON, B., 1975: Observations on exine structure of *Eucommiidites* and lower cretaceous angiosperm pollen. Pollen et Spores **17**, 429—486.

EHRENDORFER, F., 1971: *Spermatophyta*. In: Lehrbuch der Botanik für Hochschulen, 30. Aufl., 586—741. Stuttgart: G. Fischer.

—— 1976a: Evolutionary significance of chromosomal differentiation patterns in gymnosperms and primitive angiosperms. In: Origin and early evolution of angiosperms (BECK, C. B., ed.), 220—240. New York and London: Columbia Univ. Press.

Ehrendorfer, F., 1976 b: Closing remarks: Systematics and evolution of centrospermous families. Plant Syst. Evol. **126**, 99—105.

Fish, F., Waterman, G., 1973: Chemosystematics in the *Rutaceae* II. The chemosystematics of the *Zanthoxylum/Fagara* complex. Taxon **22**, 177—203.

Hickey, L. J., Wolfe, J. A., 1975: The base of angiosperm phylogeny: Vegetative morphology. Ann. Missouri Bot. Gard. **62**, 538—589.

Hughes, N. F., 1976: Palaeobiology of angiosperm origins. Cambridge: Univ. Press.

Kubitzki, K., 1969: Chemosystematische Betrachtungen zur Großgliederung der Dicotylen. Taxon **18**, 360—386.

— 1972: Probleme der Großsystematik der Blütenpflanzen. Ber. dtsch. bot. Ges. **85**, 259—277.

Leins, P., 1975: Die Beziehungen zwischen multistaminaten und einfachen Androeceen. Bot. Jahrb. Syst. **96**, 231—237.

Muller, J., 1970: Palynological evidence on early differentiation of angiosperms. Biol. Rev. **45**, 417—450.

Stebbins, G. L., 1974: Flowering plants. Evolution above the species level. Cambridge: Harvard Univ. Press.

Takhtajan, A., 1969: Flowering plants. Origin and dispersal. Edinburgh: Oliver & Boyd.

Thorne, R. F., 1974: The *"Amentiferae"* or *Hamamelidae* as an artificial group: A summary statement. Brittonia **25**, 395—405.

Walker, J. W., 1976 a: Comparative pollen morphology and phylogeny of the Ranalean complex. In: Origin and early evolution of angiosperms (Beck, C. B., ed.), 241—299. New York and London: Columbia Univ. Press.

— 1976 b: Evolutionary significance of the exine in the pollen of primitive angiosperms. Linn. Soc. Symp. Ser. **1**, 251—308.

Address of the author: Prof. Dr. Friedrich Ehrendorfer, Botanisches Institut der Universität Wien, Rennweg 14, A-1030 Wien, Austria.

Plant Syst. Evol., Suppl. 1, 235—251 (1977)

Department of Botany, University of Singapore

Phyllocladus and Its Bearing on the Systematics of Conifers

By

Hsuan Keng, Singapore

Abstract: The so-called phylloclade—the flattened, simple or compound, leaf-like organ which emerges from the axil of an acicular leaf—of *Phyllocladus* is a complicated lateral branch system. It possibly represents a relic structure resembling the photosynthetic organ of the advanced members of *Progymnosperms*. It is therefore suggested to uphold the erection of a monogeneric family *Phyllocladaceae* as proposed by CORE, and to accept the basic scheme of classification of *Coniferales* as proposed by BUCHHOLZ.

Introduction

My previous studies on the genus *Phyllocladus*, one of the most curious members of conifers, were based on a single Malesian species, *Ph. hypophyllus* HOOK. f. (KENG 1963a, 1963b). From April to August, 1976, I visited Tasmania and New Zealand and had a good opportunity to examine living and preserved material and herbarium specimens of the other four known species, to have discussion with many specialists in different fields, and to have access to much literature. It is therefore necessary to further elaborate the theoretical conclusions reached earlier (KENG 1974, 1975) and bring them into a more definite form. This is presented in the following three sections: 1. On *Phyllocladus*; 2. Comparison between *Phyllocladus* and the *Archaeopteris*-like progymnosperms, and 3. *Phyllocladus* and the systematics of conifers.

On *Phyllocladus*

The unique feature of the genus *Phyllocladus* L. C. & A. RICHARD ex MIRBEL, as indicated in the generic name, is the presence of the so-called phylloclade or cladode because it is emerged from the axil of an acicular structure which is generally regarded as the true leaf (GOEBEL 1933). Phylloclades are extremely variable in shape and size, as they are highly subjected to environmental changes. Broadly speaking there are two types of phylloclades, simple and pinnately compound. Simple phylloclades are found in *Ph. alpinus* HOOK. f. (Fig. 21) and *Ph. aspleniifolius* (LABILL.) HOOK. f. (Fig. 22), while pinnately compound ones are characteristic of *Ph. glaucus* CARR. (Fig. 19), *Ph. hypophyllus* (Fig. 20) and *Ph. trichomanoides* D. DON (Fig. 18).

Simple phylloclades are 1–4 cm long, mostly rhomboid in outline. They have finely toothed or nearly entire margin, but sometimes shallowly or deeply 2–3-lobed on one or both sides below. In still other cases, they are pinnatifid but very rarely also found to be pinnately compound. In general, only some, not all, of the pinnatifid phylloclades bear a terminal bud which is covered with awl-shaped scales. These buds either remain dormant, or give rise to a short shoot of usually 1–4 cm long, coated spirally over its length by lanceolate scale leaves which are deciduous. The short shoot is normally ended up in a new terminal bud which is surrounded by a crop of 3–5 simple or sometimes pinnatifid phylloclades which are usually arranged in a false whorl.

The pinnately compound phylloclades can reach 10–15 cm long, consist of 5–10 (–12) segments. These segments, in taxonomic literature (e.g. ALLAN 1961, LAUBENFELS 1969), are often conveniently designated as phylloclades or cladodes. The pinnate phylloclades, like the simple ones, may or may not possess a terminal bud. In the latter case, there is always a large prominent terminal segment which is often partly lobed or partly fused with one or both lateral segments beneath. Further development of the terminal bud into a short shoot is quite similar to that of the simple phylloclades just described, except the short shoot of the pinnate phylloclades is usually much longer and stouter, and the new crop of phylloclades (3–5, up to 10) thus produced are normally pinnately compound, very rarely simple.

A complete and comprehensive taxonomic account of the genus *Phyllocladus* was prepared by PILGER (1903). The latest taxonomic descriptions of the species from various geographical regions are scattered in local floras and other sources (e.g. ALLAN 1961, CURTIS & MORRIS 1975, LAUBENFELS 1969, etc.).

Morphological studies of some *Phyllocladus* species, principally on *Ph. alpinus*, and sometimes supplemented by *Ph. glaucus* or *Ph. trichomanoides*, were carried out by ROBERTSON (1906), KILDAHL (1908a), YOUNG (1910), SINNOT (1913), HOLLOWAY (1937), BUCHHOLZ (1941) and others.

SINNOT (1913), for example, discussed the morphological significance of the strobilar anatomy and compared the development of the male and female gametophytes and of the embryo with other members of the *Podocarpaceae*. He stressed that the structure of the female gametophyte and embryo of *Phyllocladus* agrees more closely with that of *Dacrydium*, especially in the reduced number of archegonia and the structure of the neck, than with that of any other members of the family *Podocarpaceae*.

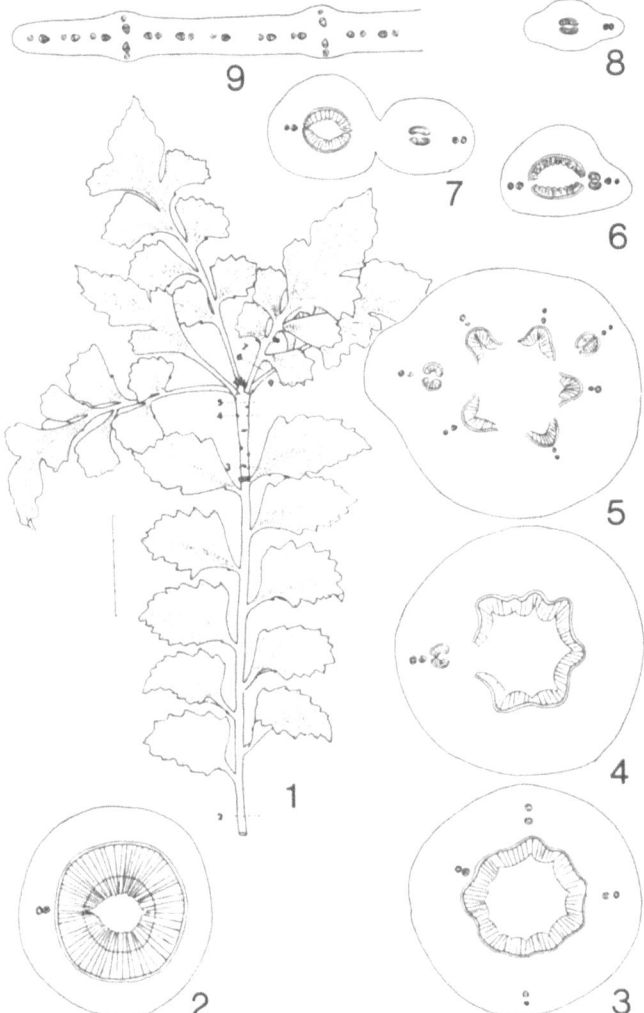

Fig. 1. An old pinnate phylloclade of *Ph. glaucus* giving rise to a short
shoot with a crown of new pinnate phylloclades. (Scale = 4 cm)

Figs. 2–9. Diagramatic transverse sections made from different parts as
indicated in Fig. 1

Vasculature of the segments of a pinnate phylloclade of *Ph. hypo-phyllus* has been reported in a previous paper (KENG 1963 b). Fresh material of *Ph. glaucus* (Fig. 1) and *Ph. trichomanoides* were sectioned with the aid of a Hooker Plant microtome and a freezing microtome. Their vasculatures are very similar and are briefly described below.

Transverse sections (Fig. 2) of the stalk of last year's pinnate phylloclade indicate that 1. in the centre a cylinder of secondary xylem envelops the primary wood, and 2. in the cortex a leaf trace enters the leaf subtending the lowermost segment of the pinnate phylloclade.

As a result of activation of the terminal bud of the pinnate phylloclade, a new shoot is formed. This new shoot is covered with spirally arranged, single-veined, deciduous, lanceolate leaves and is crowned with a crop of young pinnate phylloclades (Fig. 1). Transverse sections from the lower level of the shoot (Fig. 3) show an undulating central cylinder of primary growth with leaf-traces in the cortex. Sections from a higher level indicate that the undulation becomes more pronounced, finally five large V-shaped ribs alternating with four or five smaller ones gradually become evident. Further upwards (Fig. 4), a single bundle, associated with a resin canal, diverges from the pole of one of the ribs to supply the leaf which is subtending the entire pinnate phylloclade. This is followed by the departure of two blocks of vascular tissues which constitute the rest of the rib. These two blocks of vascular tissues form the basic vasculature of the lowermost pinnate phylloclade. Subsequently the departure of another single bundle, followed by two blocks of vascular tissues which form the basic vasculature of a second pinnate phylloclade (Fig. 5). Eventually, a crown of 3–5, up to 8–10 new pinnate phylloclades can be formed.

Sections of the stalk of a young pinnate phylloclade show a central vascular system consisting of two closely associated blocks of vascular tissues. The divergence of a single bundle from one side followed by two small blocks of vascular tissue from the same side constitutes the vasculature of the lowermost segment and its subtending leaves (Figs. 6, 7). As soon as these two small blocks of vascular tissues enter the base of a segment, divergence begins (Fig. 8). A single trace enters into the leaf-like structure of the segment, and two smaller patches of vascular tissues follow to form the basis of the vasculature of one of the several small branches within a segment. Eventually, these smaller branches result in five to seven leaf-like structures with their xylem portions oriented towards the centres (Fig. 9).

The wood structure of *Phyllocladus* species has been described by Greguss (1955), Patel (1967), and others. The latter author remarked that the wood of *Phyllocladus* species can be separated from that of other New Zealand members of the *Podocarpaceae* by its distinct rings, large cross-field pits, abundant tangential pits and lack of axial parenchyma. He further emphasized that the presence of abundant tangential pits in *Phyllocladus* can be regarded as a useful diagnostic feature.

Haploid chromosome numbers of three New Zealand species of *Phyllocladus* were reported by HAIR & BEUZENBERG (1958) which are uniformly 9, the lowest among the living conifers. Those of the Tasmanian and Malesian species remain unknown.

At the present time the genus *Phyllocladus* is confined to the Australasian and Malesian regions (Fig. 10). Fossil records, however, reveal its former

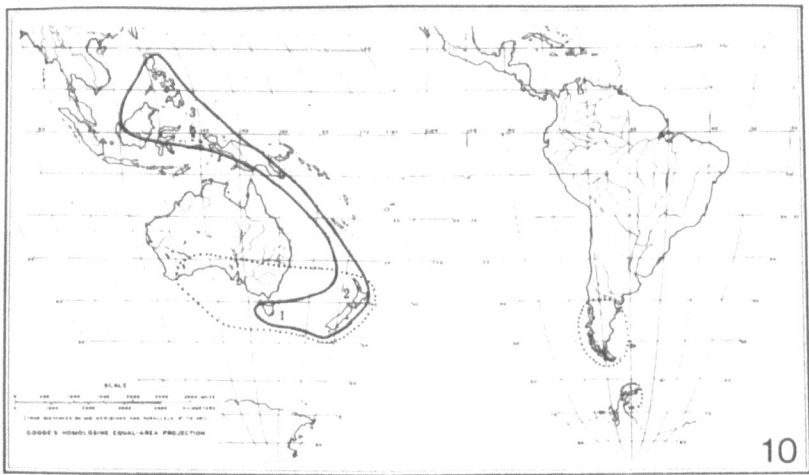

Fig. 10. Present (in solid lines) and past (in broken lines) distribution of *Phyllocladus*. Five living species are found in 1. Tasmania (with one species), 2. New Zealand (with three spp.) and 3. Malesia (with one sp.). (Base map copyright by the University of Chicago Press)

much wider range of distribution in geological time. After enumerating and evaluating the palaeo-botanical data, COUPER (1960) reaches the following conclusions (Fig. 10): 1. Macrofossils identified as *Phyllocladus* on cuticular structure and morphology of the cladodes are recorded from the Oligocene of continental Australia and the Oligocene, later Tertiary and Quarternary of New Zealand. 2. Microfossils (pollens) are found sporadically in Tertiary sediments in Australia, Tasmania, and New Zealand. They are also found in the Lower Tertiary of Seymour Island in western Antarctica. 3. No fossils of *Phyllocladus* are known from the Cretaceous and Tertiary of northern South America, New Guinea, Borneo or India. A phyllocladean affinity has been suggested for some fossil woods from southern South America, but these identifications are inconclusive.

The absence of fossils in the tropics and in the northern Hemisphere deserves special attention. It supports the view that the present extension of *Phyllocladus* into eastern Malesia was a relatively recent development (HAIR 1963, LAUBENFELS 1969).

Comparison Between *Phyllocladus* and the *Archaeopteris*-like Progymnosperms

The photosynthetic organs of *Phyllocladus* are generally known as phylloclades or cladodes. The prevailing view to-day in interpreting the origin and nature of these organs is expressed in the following statement made by Robertson (1906: 259) over half a century ago: "*Phyllocladus* is ... characterized by the reduction of its true leaves to pointed scales, and the expansion of certain of its stem-branches into flattened leaf-like structures". A simple photosynthetic organ as commonly found in *Ph. alpinus* and *Ph. aspleniifolius* (the type species of the genus) (Figs. 21, 22) could be summarily described as a phylloclade, without paying much attention to its detailed internal structure. Superficially it is a flattened green organ arising from the axil of a true leaf, the latter often being reduced. In the case of the pinnately compound photosynthetic organs as found in *Ph. glaucus*, *Ph. hypophyllus* and *Ph. trichomanoides* (Figs. 18, 19, 20), however, such terms as phylloclade or cladode are hardly applicable. Observations made on the Malesian species, *Ph. hypophyllus* (Keng 1963 b) reveal that the whole pinnate photosynthetic organ is subtended by a leaf-like structure, and that it consists of usually 6–8 segments. Each segment is also situated in the axil of a leaf-like structure and it is composed of a number of smaller branches. These smaller branches, moreover, are individually subtended by a leaf-like structure but they are fused together laterally. Examinations of the pinnate photosynthetic organs of the mature plants of *Ph. glaucus* and *Ph. trichomanoides* (Figs. 14, 15) show basically the same configuration as that of the Malesian species, but the bipinnate nature becomes much more pronounced in the photosynthetic organs of the seedlings of *Ph. trichomanoides* (Figs. 16, 17 and 28).

Transverse sections of a mature segment of the pinnate photosynthetic organs also indicate some unusual features (Keng 1963 b). The upper and lower epidermis are well defined. Mesophyll is more or less differentiated into two parts comparable to palisade and spongy tissues. The deeply sunken stomata possess guard cells almost indistinguishable from those found in the juvenile leaves. Furthermore, many of these photosynthetic organs, simple or pinnate, possess a terminal bud, which upon activation, gives rise to a short shoot with a number of spirally arranged leaves on the axis together with a crown of new photosynthetic organs (Fig. 1).

The photosynthetic organ of *Phyllocladus* is thus much too complicated to be designated as a phylloclade in its original sense as referring to the specialized branch of an angiospermous plant. Or in other words, phylloclade is a misnomer when applied to the photosynthetic organ of *Phyllocladus*. It becomes apparent that the photosynthetic organ of *Phyllocladus* represents a unique structure among gymnosperms. Gymnosperms were plausibly derived from the *Psilophyton*-like earlier vascular plants through progymnosperms (BECK 1962, 1966; BANKS 1968; and others). It is therefore probably not too far-fetched to attempt to compare the photosynthetic organs of *Phyllocladus* with those of *Archaeopteris*, one of the most advanced progymnosperms.

The photosynthetic organs of *Archaeopteris* (Fig. 11) was thought at one time to be the ancestral form of ferns. They were thus described as a bipinnate "frond" (e.g. BECK 1962). Each "frond" is composed of pinnae and pinnules arranged on the rachis and its branches respectively. This plant has been specially noted for the possession of a large stipule-like structure at the base of the "frond" and for the presence of the so-called rachial pinnules. CARLUCCIO et al. (1966) found the so-called "rachis" has a radially symmetrical vascular system like a stem and that the axes of the lateral pinnas are characterized by a similar type of vasculature. Based on these facts, they concluded that the foliar organ of *Archaeopteris* is not a bipinnate "frond" but rather should be regarded as a planated lateral branch system. According to this interpretation, the actual leaves are helically arranged and comprise 1. the pairs of "rachial pinnules", and 2. the fertile and sterile appendages (pinnules) borne on the lateral axes which had previously been regarded as pinnae. BECK (1970) later accepted this view and further noted that the entire lateral branch system appears to be subtended by a leaf which he had earlier (1962) designated as a "stipule".

Photosynthetic organs of *Archaeopteris macilenta* LESQ. and *Phyllocladus trichomanoides* are presented in Figs. 11 and 14, and for convenience of comparison and discussion, a set of new, simple, lucid terms are proposed in brackets along with the terms used in the previous sections of this paper on *Phyllocladus* on the one hand, and the terms used by BECK in his 1971 paper on *Archaeopteris* on the other.

The lateral branch system (primary branch) of *Archaeopteris* is composed of a number of distichously disposed lateral branches (secondary branches). The large notched foliar structure (primary laminar appendage) has only recently been recognized as a leaf (BECK 1970). The leaves, or formerly the rachial pinnules (secondary laminar appendages), along the main axis are probably totally different from those on the lateral branches. The latter are possibly referable both to the tertiary branches and the tertiary laminar appendages (Fig. 12): those distichously disposed along the secondary branches might represent the tertiary branches, while those inserted between the distichous leaves represent the tertiary laminar appendages. Another possibility would be that these tertiary laminar appendages (like the secondary laminar appendages of *Phyllocladus*), might be of deciduous nature,

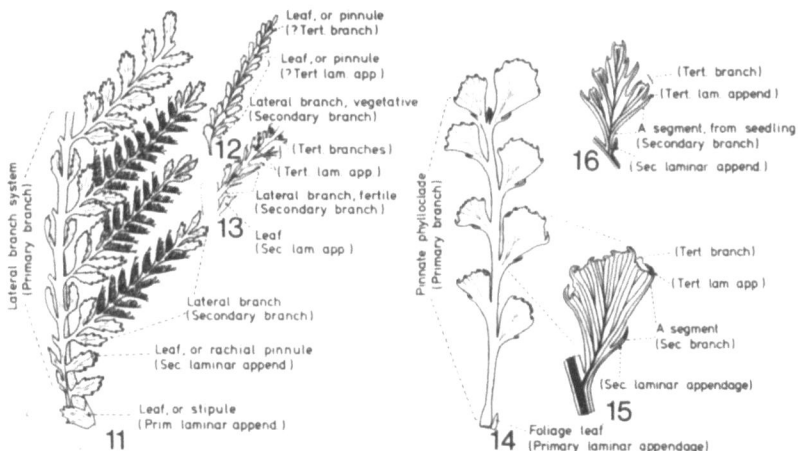

Figs. 11–16. Comparison of a lateral branch system of *Archaeopteris macilenta* (Fig. 11) with a young pinnate phylloclade of *Ph. trichomanoides* (Fig. 14). Figs. 12 and 13, sterile and fertile lateral branches of *Archaeopteris*. Figs. 15 and 16, cleared segments showing the vasculatures from an adult plant and a seedling of *Phyllocladus*, respectively. (Fig. 11 from BECK 1962, Figs. 12 and 13, from BECK 1971)

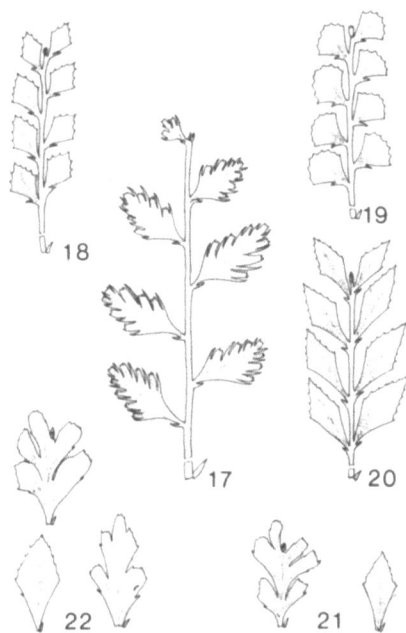

Figs. 17–22. Photosynthetic organs of *Phyllocladus*, (a) from the seedling (Fig. 17) and adult plant (Fig. 18) of *Ph. trichomanoides*, and (b) from the adult plants of *Ph. glaucus* (Fig. 19), *Ph. hypophyllus* (Fig. 20), *Ph. alpinus* (Fig. 21) and *Ph. aspleniifolius* (Fig. 22). (All schematic)

thus the persistent ones are all tertiary branches. These tertiary branches and their laminar appendages become extremely clearly differentiated in the fertile lateral branches (Fig. 13, redrawn from BECK 1971, or as reconstructed by PHILLIPS et al 1972).

The putative ancestor of *Phyllocladus* probably possessed a large photosynthetic organ resembling closely that of *Archaeopteris*. It probably

Figs. 23–27. Photosynthetic organs ($\pm$ reduced) of *Phyllocladus* bearing ovulate strobili: *Ph. trichomanoides* (Fig. 23), *Ph. glaucus* (Fig. 24), *Ph. hypophyllus* (Fig. 25), *Ph. alpinus* (Fig. 26) and *Ph. aspleniifolius* (Fig. 27). (All schematic)

consisted of a twice-divided primary branch, and several of these branches were probably arranged on the main axis forming a false whorl (cf. the reconstruction of *Tetraxylopteris* by BECK 1957, in his Fig. 32). Each primary branch probably bore a large terminal bud. At present the photosynthetic organs found in the seedlings of *Ph. trichomanoides* (which has conspicuous tertiary branches with subtending tertiary laminar appendages as well as a large terminal bud) is probably the nearest to the putative ancestral form (Figs. 17 and 28).

The simple photosynthetic organs (of which further ramification has been arrested) as found in *Ph. aspleniifolius* (Fig. 22) and *Ph. alpinus* (Fig. 21), probably represent a more specialized form.

With regard to the reproductive structures, among *Phyllocladus* species, *Ph. glaucus* (Fig. 24) possesses the most well-organized ovulate cones which consist of a large number of cone-scales. *Ph. hypophyllus* (Fig. 25), has less well-organized cones and the latter are generally seated on the notch of a secondary branch (segment). The combination of vegetative and reproductive functions performed by a single photo-synthetic organ is highly remarkable among the living seed plants. The ovulate reproductive structure of a putative ancestor of *Phyllo-cladus* probably possessed a highly well-organized cone seated on the upper part of a secondary branch.

The above comparison indicates some similarity between the photo-synthetic and reproductive organs of progymnospermous *Archaeopteris* and those of coniferous *Phyllocladus*, yet the geological and morpho-logical gaps between them are enormous. For instance, *Archaeopteris* is found in Upper Devonian to the beginning of Carboniferous, over 340 million years ago (BANKS 1968), whereas *Phyllocladus* fossils are known only from the Oligocene onwards, less than 40 million years ago (COUPER 1960). Morphological gaps include the arrangement of pits in the radial walls of the tracheids which in *Archaeopteris* are in groups separated by spaces free of pits (ARNOLD 1930) and the hetero-sporous sporangia which stand erect in one or more rows on adaxial surface of fertile ultimate appendages (BECK 1976). In *Phyllocladus* pits occur on radial walls of tracheids in one or rarely in two rows (PATEL 1967), the staminate and ovulate strobili are well organized, and the ovule and seeds are of highly evolved form.

Therefore any attempt to postulate a connection between them needs to be extremely cautious. Without further evidence, it can only be said that the photosynthetic and reproductive organs of *Phyllocladus* might have been derived from a putative ancestor which had general features resembling *Archaeopteris* or its related genera.

Phyllocladus and the Systematics of Conifers

It is not intended to give an exhaustive historical sketch of classi-fication systems of conifers here, as this has been adequately done by PILGER (1926), FLORIN (1955) and others. For the convenience of discussion, the following brief account will be sufficient.

In the first edition of ENGLER & PRANTL's "Die natürlichen Pflan-zenfamilien", EICHLER (1887) divided *Coniferae* into two subfamilies: *Pinoideae* (with two tribes, *Abietineae* and *Cupressineae*) and *Taxoideae* (also with two tribes, *Podocarpeae* and *Taxeae*—the genus *Phyllocladus* was included in the latter). In the 9th and 10th edition of ENGLER's Syllabus (ENGLER & GILG 1924), the two subfamilies were elevated

Fig. 28. A seedling of *Phyllocladus trichomanoides* showing the bipinnate photosynthetic organs. Inset, a bipinnate photosynthetic organ. (Photographed by Mr. C. J. MILES; material supplied by Dr. B. P. J. MOLLOY, Botany Division, DSIR, Christchurch, New Zealand)

to two families, *Taxaceae* and *Pinaceae*, with the former further divided
into three subfamilies (namely, *Podocarpoideae, Phyllocladoideae* and
Taxoideae) and the latter, four tribes (*Araucarieae, Abieteae, Taxodieae*
and *Cupresseae*). All the subfamilies and tribes, except the *Phyllocladoideae* (which were retained as a subfamily in *Podocarpaceae*) were
promoted to family rank by PILGER (1926), plus a new family *Cephalotaxaceae*, formerly under the *Taxaceae*. Thus the *Coniferales* constitutes
the following seven families: *Taxaceae, Podocarpaceae, Cephalotaxaceae,
Araucariaceae, Pinaceae, Taxodiaceae,* and *Cupressaceae*.

The composition and classification of the Conifers were significantly
altered in the 12th edition of ENGLER's Syllabus (MELCHIOR & WER-
DERMANN 1954), in which PILGER and MELCHIOR proposed the following scheme:

Class *Coniferopsida*
    Order *Cordaitales*
    Order *Coniferales* (*Lebachiaceae, Voltziaceae, Cheirolepidaceae, Proto-
    pinaceae, Pinaceae, Taxodiaceae, Cupressaceae, Podocarpaceae, Ce-
    phalotaxaceae, Araucariaceae*)
Class *Taxopsida*
    Order *Taxales* (*Taxaceae*)

Following this brief account of the historical background, discussion
will be concentrated on three essential problems, namely, 1. the inclusion of the *Taxaceae* in conifers, 2. the establishment of the *Phyllocladaceae*, and 3. the systematics of conifers.

The Inclusion of the *Taxaceae* in Conifers

SAHNI (1920) first noted that the taxads (including *Cephalotaxus*)
were so different from the other conifers that they deserved the rank
of a separate order, *Taxales*. FLORIN (1938–1945, summarized in 1948,
1951) studied the reproductive organs of fossil taxads—*Palaeotaxus*
of Triassic and *Taxus* (or *Marskea*, cf. HARRIS 1976) of Jurassic—and
found that they were clearly different from those of true conifers.
Therefore he accepted the separation of the taxads—including the
following five genera: *Taxus, Amentotaxus, Torreya, Austrotaxus* and
Pseudotaxus, but not *Cephalotaxus*—from the *Coniferae* into a separate
order, *Taxales*, and even a new class, *Taxopsida*. Many authors, such
as CHAMBERLAIN (1935: 439), PULLE (1937) and others, expressed a
different view, namely that the single ovulate strobilus of taxads is
most probably derived from the multiovulate cone. The present writer
(KENG 1963 b, 1969) pointed out that the development and evolution
of the ovulate strobili in *Phyllocladus* might indicate the possible way
in which the single, seemingly terminal ovule of taxads could have been

achieved. He also drew attention to the fact that *Phyllocladus* is closely related to the *Taxaceae* and *Podocarpaceae* on the one hand, and *Amento-taxus*, to the *Taxaceae* and *Cephalotaxaceae* on the other. Therefore he suggested that the *Taxaceae* should be retained in the *Coniferales*.

From palaeobotanical studies, HARRIS (1976) recently observed that "there is something in common between the ovuliferous short shoot of *Taxus* and the fertile shoot of *Lebachia*", and suggested that early fossil conifers like *Walchia* could "shift" their ovules from a lateral to a terminal position. He therefore implied that the class *Taxopsida* is unnecessary. (Both *Lebachia* and *Walchia* belong to the family *Lebachiaceae* which is the earliest member of the *Coniferales*, cf. MELCHIOR & WERDERMANN 1954.)

The Establishment of the *Phyllocladaceae*

The somewhat intermediate morphological features of *Phyllocladus* between the *Taxaceae* and *Podocarpaceae* lead to quite diverse taxonomic treatment of the genus, depending on the relative weight assigned to these features. For example, such characters as the leafy microsporophyll bearing two microsporangia at the base and the winged microspores resemble those of the *Podocarpaceae*; the erect ovules and the arillate seed appear to be similar to those of the *Taxaceae*. For this reason, three different schemes of classification of this genus have been proposed. The first includes it in the *Podocarpaceae* (or formerly *Podocarpoideae* or *Podocarpineae*) (e.g. PILGER 1926); the second includes it in the *Taxaceae* (or formerly *Taxoideae, Taxineae*) (e.g. EICHLER 1887); and the third is to consider *Phyllocladus* as representing a separate taxon coordinate with and intermediate between the *Podocarpaceae* and *Taxaceae* (PILGER 1903, ENGLER & GILG 1924, CORE 1955).

Morphological studies of *Phyllocladus* by many authors (e.g. KILDAHL 1908b) led to more or less the same conclusions as ROBERTSON's (1906: 264): "*Phyllocladus* occupies an intermediate position between *Podocarpoideae* and *Taxoideae*, but with greater affinities for the former". YOUNG's view (1910: 91) slightly deviates from this, she observed that *Phyllocladus* has primitive characters of the *Taxineae* which become eliminated in the *Podocarpineae*, and it also has primitive characters of the *Podocarpineae* which have been eliminated in the *Taxineae*.

The present writer (KENG 1973, 1975) emphasized the fact that the photosynthetic organs of *Phyllocladus* commonly known as phyllo-clades are probably a very ancient structure, and that they could have been misinterpreted morphologically. It may represent a relic structure of which the distinction between leaf and branch has not been sharply differentiated. Therefore the present writer supported the proposition made by CORE (1955) to establish the family *Phyllocladaceae*. He further considered this monogeneric family to be probably the most primitive members of the *Coniferae*. The fact that this family has external and internal morphological characters in common with the *Taxaceae* on the one hand, and with the *Podocarpaceae* on the other could be explained by assuming that this family inherited these traits directly from the ancestral stock from which both *Taxaceae* and *Podocarpaceae* were subsequently evolved.

The Systematics of Conifers

With the re-instatement of the *Taxaceae* in the *Coniferales*, and with the establishment of the family *Phyllocladaceae*, the living members of this order thus include the following eight families: *Phyllocladaceae, Taxaceae, Podocarpaceae, Cephalotaxaceae, Pinaceae, Taxodiaceae, Cupressaceae*, and *Araucariaceae*. This treatment resembles closely the basic scheme proposed by PILGER (1926) with the addition of the *Phyllocladaceae* which was formerly a sub-family of the *Podocarpaceae*.

Attemps have been made, from time to time, to subdivide this order into two to several smaller groups. For examples, GAUSSEN (1944—1952, cited in FLORIN 1955) recognized three sub-orders: (a) *Taxineae*, with the *Taxaceae*, (b) *Podocarpineae*, with the *Podocarpaceae*, and (c) *Pinoidineae*, with the remaining families. PULLE (1937) created five orders: *Araucariales, Podocarpales, Pinales, Cupressales* (with *Taxodiaceae* and *Cupressaceae*) and *Taxales* (with *Cephalotaxaceae* and *Taxaceae*). The fragmentation of this order culminated in a system of FLORIN (in ERDTMAN 1952) in which two classes and six orders were proposed. They are (a) the Taxad Class or *Taxopsida*, with the order *Taxales*, and (b) the Conifer Class, or *Coniferopsida*, with five orders: *Araucariales, Podocarpales, Pinales, Cupressales*, and *Cephalotaxales*.

BUCHHOLZ's (1934) proposal to divide the order *Coniferales* into two sub-orders, which he called *Phanerostrobilares* and *Aphanostrobilares*, appears to be the most reasonable one and thus deserves special attention. It adheres in principle, to the original schemes of EICHLER (1887) and ENGLER & GILG (1924). This general scheme was later adopted by CHAMBERLAIN (1935, as *Taxares* and *Pinares*), JANCHEN (1949, as *Taxales* and *Pinales*) and others. The main difference between these two sub-groups, in the words of CHAMBERLAIN (1935: 229, 230) is the one sub-group (called *Phanerostrobilares, Pinares, Pinineae* or *Pinales* by different authors) "with an obvious (ovulate) cone", including *Pinaceae, Araucariaceae, Taxodiaceae* and *Cupressaceae*; and the other sub-group (called *Aphanerostrobilares, Taxares, Taxineae* or *Taxales*) "without such an obvious cone", including *Podocarpaceae, Taxaceae* and *Cephalotaxaceae*, and to be added here, the *Phyllocladaceae*.

At the first glance, the validity of this distinction seems to be dubious, as it is ambiguous. Members of the first sub-group do possess very obvious cones, although most of the members of the second sub-group lack an obvious cone, yet in some cases such as *Cephalotaxus* (*Cephalotaxaceae*), *Microcachrys* (*Podocarpaceae*) and *Phyllocladus*, especially *Ph. glaucus* (Fig. 24) (*Phyllocladaceae*), "obvious" ovuliferous cones are in fact present.

FLORIN's monumental work (1938–1945, summarized in 1951) has almost conclusively demonstrated that the seemingly simple ovuliferous cones of *Pinus* and possibly other members of one sub-group (*Pinineae*) have evolved from a much more complicated structure such as those found in the palaeozoic fossil groups *Lebachia*, *Ernestiodendron*, *Walchia*, and *Pseudovoltzia*. But whether the same conclusion could be extended to explain the cones of the members of the other sub-group (*Taxineae*) is highly debatable. Without further evidence, it is thought better to keep these two subgroups separate.

For this reason, the following scheme of classification which is a modification of BUCHHOLZ's (1934), is presented below.

Order *Coniferales*
    Sub-order *Taxineae*
        Family 1. *Phyllocladaceae*
        Family 2. *Podocarpaceae*
        Family 3. *Taxaceae*
        Family 4. *Cephalotaxaceae*
    Sub-order *Pinineae*
        Family 5. *Pinaceae*
        Family 6. *Araucariaceae*
        Family 7. *Taxodiaceae*
        Family 8. *Cupressaceae*

This study was carried out during a Sabbatical Leave from the University of Singapore. An Inter-University Exchange Scheme grant was awarded by the British Council. Professors L. CONSTANCE and W. R. PHILIPSON and Dr. K. R. SPORNE kindly criticized the manuscript. Professors W. R. PHILIPSON (University of Canterbury, Christchurch, New Zealand) and W. D. JACKSON (University of Tasmania, Hobart, Australia) kindly provided laboratory and other facilities, while many individuals, including B. C. ARNOLD, G. BROWNLIE, R. K. CROWDEN, W. M. CURTIS, J. W. DAWSON, H. K. MAHANTY, B. J. MOLLOY, L. G. MOORE, A. E. ORCHARD, P. WARDLE and many others provided helpful discussions. My wife, Mrs. R. S. KENG prepared several drawings for illustration.

References

ALLAN, H. H., 1961: Flora of New Zealand, Vol. 1. Wellington: R. E. Owen.

ARNOLD, C. A., 1930: The genus *Callixylon* from the Upper Devonian of Central and Western New York. Pap. Mich. Acad. Sc. **11**, 1—50.

BANKS, H. P., 1968: The early history of land plants. In: Evolution and Environment (DRAKE, E. T., ed.), 73—107. New Haven: Yale Univ. Press.

BECK, C. B., 1957: *Tetraxylopteris schmidtii* gen. et sp. nov., a probably pteridosperm precursor from the Devonian of New York. Amer. J. Bot. **44**, 350—367.
— 1962: Reconstructions of *Archaeopteris* and further consideration of its phylogenetic position. Amer. J. Bot. **49**, 373—382.
— 1966: On the origin of gymnosperms. Taxon **15**, 337—339.
— 1970: The appearance of gymnospermous structure. Biol. Rev. **43**, 379—400.
— 1971: On the anatomy and morphology of lateral branch systems of *Archaeopteris*. Amer. J. Bot. **58**, 758—784.
— 1976: Current status of the *Progymnospermopsida*. Rev. Palaeobot. Palyn. **21**, 5—23.
BUCHHOLZ, J. T., 1934: The classification of *Coniferales*. Trans. Illinois Acad. Sci. **25**, 112—113.
— 1941: Embryogeny of the *Podocarpaceae*. Bot. Gaz. **103**, 1—37.
CARLUCCIO, L. W., HUEBER, F. M., and BANKS, H. P., 1966: *Archaeopteris macilenta*, anatomy and morphology of its frond. Amer. J. Bot. **53**, 719—730.
CHAMBERLAIN, C. J., 1935: Gymnosperms, structure and evolution. Chicago: Univ. Press.
CORE, E. L., 1955: Plant taxonomy. Englewood: Prentice-Hall.
COUPER, R. A., 1960: Southern Hemisphere Mesozoic and Tertiary *Podocarpaceae* and *Fagaceae* and their palaeogeographic significance. Proc. Roy. Soc. B. **152**, 491—500.
CURTIS, W. M., and MORRIS, D. I., 1975: The Student's Flora of Tasmania. 2nd ed. pt. 1. Hobart: Government Printer.
EICHLER, A. W., 1887: *Coniferae*. In: Die nat. Pflanzenfamilien, Teil II, 1 (ENGLER, A., and PRANTL, K.), 28—116. Leipzig: W. Engelmann.
ENGLER, A., and GILG, E., 1924: Syllabus der Pflanzenfamilien. Ed. 9 & 10. Berlin: Gebr. Borntraeger.
ERDTMAN, H., 1952: Chemistry of some heartwood constituents of conifers and their physiological and taxonomic significance. In: Progress in Organic Chemistry, 22—63. London: Butterworths Sc. Publs.
FLORIN, R., 1948: On the morphology and relationships of the *Taxaceae*. Bot. Gaz. **110**, 31—39.
— 1951: Evolution in Cordaites and Conifers. Acta Hort. Berg. **15**, 285—388.
— 1955: The systematics of the Gymnosperms, In: A century of progress in the natural sciences 1853-1953, 233—403. San Francisco: Calif. Acad. Sc.
GOEBEL, K., 1933: Organographie der Pflanzen. 3. Aufl. Jena: G. Fischer.
GREGUSS, P., 1955: Identification of living Gymnosperms on the basis of Xylotomy. Budapest: Akad. Kiado.
HAIR, J. B., 1963: Cytogeographical relationships of the southern podocarps. In: Pacific Basin Biogeography (GRESSITT, J. L., ed.), 401—414. Honolulu: Bishop Museum Press.
— and BEUZENBERG, E. J., 1958: Chromosomal evolution in the *Podocarpaceae*. Nature (London) **181**, 1584—1586.
HARRIS, T. M., 1976: The Mesozoic Gymnosperms. Rev. Palaeobot. Palyn. **21**, 119—134.
HOLLOWAY, J. T., 1937: Ovule anatomy and development and embryogeny in *Phyllocladus alpinus* HOOK. and in *P. glaucus* CARR., Trans. R. Soc. N. Z. **82**, 329—410.

JANCHEN, E., 1949: Das System der Koniferen, Sitzungsber. Oest. Akad. Wiss. math.-naturw. Kl. (div. 1) **158**, 155—162.

KENG, H., 1963a: *Phyllocladus hypophyllus* HOOK. f. Gard. Bull. Singapore **20**, 123—126.

— 1963b: Aspects of morphology of *Phyllocladus hypophyllus*. Ann. Bot. n.s. **27**, 69—78.

— 1969: Aspects of morphology of *Amentotaxus formosana* with a note on the taxonomic position of the genus. J. Arnold Arb. **50**, 432—446.

— 1973: On the family *Phyllocladaceae*. Taiwania **18**, 142—145.

— 1974: The phylloclade of *Phyllocladus* and its possible bearing on the branch systems of progymnosperms. Ann. Bot. n.s. **38**, 757—764.

— 1975: A new scheme of classification of the *Coniferales*. Taxon **24**, 289—292.

KILDAHL, N. J., 1908a: The morphology of *Phyllocladus alpinus*. Bot. Gaz. **46**, 339—348.

— 1908b: Affinities of *Phyllocladus*. Ibid. **46**, 464—465.

LAUBENFELS, D. J. DE, 1969: A revision of the Malesian and Pacific Rain Forest Conifers. I. *Podocarpaceae*, in part. J. Arnold Arb. **50**, 274—369.

MELCHIOR, H., and WERDERMANN, E., 1954: A. ENGLER's Syllabus der Pflanzenfamilien. Ed. 12, **I**. Berlin: Gebrüder Borntraeger.

PATEL, R. N., 1967: Woody anatomy of *Podocarpaceae* indigenous to New Zealand. 3. *Phyllocladus*. N. Z. J. Bot. **6**, 3—8.

PHILLIPS, T. L., ANDREWS, H. N., and GENSEL, P. G., 1972: Two heterosporous species of *Archaeopteris* from the Upper Devonian of West Virginia. Palaeontographica **139 B**, 47—71.

PILGER, R., 1903: *Taxaceae*. In: Das Pflanzenreich, IV. (ENGLER, A.), **5**. Leipzig: W. Engelmann.

— 1926: *Coniferae*. In: Die nat. Pflanzenfamilien, 2. Aufl. (ENGLER, A., and PRANTL, K.), **13**, 121—407. Leipzig: W. Engelmann.

PULLE, A., 1937: Remarks on the system of the Spermatophytes. Med. Bot. Mus. Utrecht **43**, 1—17.

ROBERTSON, A., 1906: Some points in the morphology of *Phyllocladus alpinus* HOOK. Ann. Bot. **20**, 259—265.

SAHNI, B., 1920: On certain archaic features in the seed of *Taxus baccata*, with remarks on the antiquity of the *Taxineae*. Ann. Bot. **34**, 117—133.

SINNOT, E. W., 1913: The morphology of the reproductive structures in the *Podocarpinae*. Ann. Bot. **27**, 39—82.

YOUNG, M. S., 1910: The morphology of *Podocarpineae*. Bot. Gaz. **50**, 82—100.

Address of the author: Prof. Dr. H. KENG, Department of Botany, Bukit Timah Road, Singapore 10.

Plant Syst. Evol., Suppl. 1, 253—283 (1977)

Botanical Museum of the University of Copenhagen, Denmark

A Commentary on a Diagrammatic Presentation of the Angiosperms in Relation to the Distribution of Character States

By

Rolf Dahlgren, Copenhagen

Abstract: The difficulty, not to say impossibility, of presenting groups of orders as variously shaped figures in a two-dimensional diagram to indicate relationships is first emphasized. However, the method can be used to demonstrate the distribution of character states. The distributional patterns of a number of character states are presented with brief comments, viz. sympetaly, epigyny and perigyny, centrifugal succession in multi-staminate androecia, pollen grains with one aperture, the successive type of microsporogenesis, apocarpy, three-nucleate pollen grains (the "dry" stigmas and the C_4 pathways being largely correlated with this attribute), unitegmic and tenuinucellate ovules and cellular and helobial types of endosperm formation. In addition, the distribution in dicotyledons is outlined for the synthesis of ellagic acid, iridoids, benzylisoquinoline alkaloids, glucosinolates and polyacetylenes derived from fatty acids. In monocotyledons other characters prove useful in gross taxonomy, for instance the presence of vessels in the stem, the presence of starch in the endosperm and the presence of oxalate raphides and silica bodies, as well as more conventional characters such as floral symmetry, epigyny versus hypogyny, stamen number and fruit type. Comments on the possible taxonomic significance of the characters shown are included. An apparent connection between certain constellations of orders (superorders) is pointed out. Finally, an appeal is made for further cooperation in presenting the distribution of other character states which may be of taxonomic importance at the higher levels.

Introduction

The method of illustrating an angiosperm system as an imaginary phylogenetic tree in transection was first used by the author and his colleagues in a textbook of angiosperm taxonomy (DAHLGREN et al. 1974, 1975a and b, 1976a). Some 25 different characters, most of which were considered to be of importance in gross taxonomy, were shown in this diagram. The details of the system and the method of presentation needed further elaboration, however. A revised system with a diagram was presented in 1975 (DAHLGREN 1975a). Diagrams showing

the distribution of some embryological features were presented in
DAHLGREN (1975 b).

As in DAHLGREN (1975 a) reservations must be made when presenting
a two-dimensional diagram to illustrate phylogenetic relationships.
The attempt has been made to place the variously shaped figures

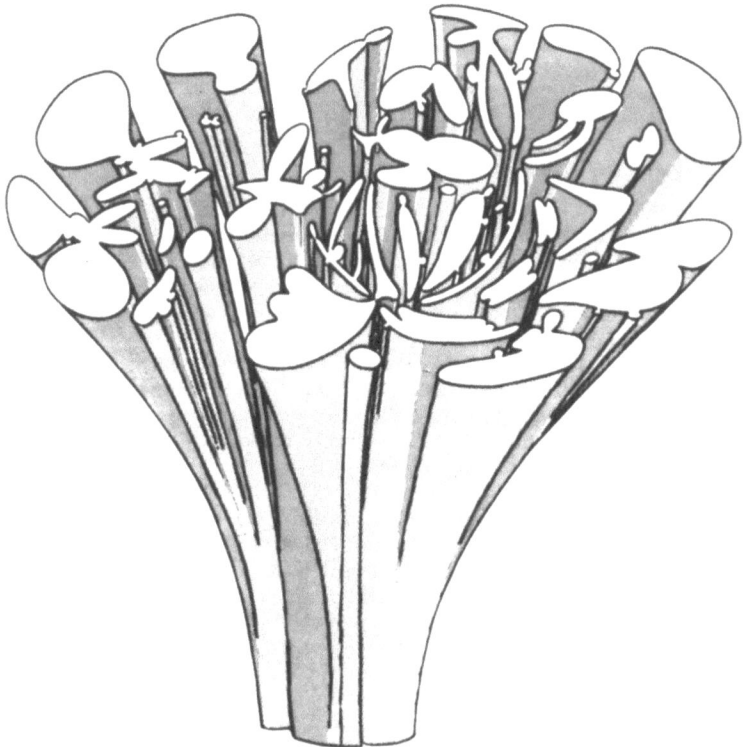

Fig. 1. The diagrams used in this article presented as the transection of an
imaginary phylogenetic "shrub"

representing the orders close to one another when members of the
orders are similar in several characters that reflect presumably phylogenetic relationships. The size of these figures is as far as possible an
expression of the number of species in the orders. Where members of
an order closely resemble members of more than one other order the
figure representing the order has been extended in two or more directions. Other figures have been rounded.

Differentiation within the ancestral forms of angiosperm orders
naturally was extensive and affected the appearance of different parts

of the plant. While some characters underwent considerable evolutionary change others remained at a level that most of us would regard as primitive. Greater importance is often attached to the primitive state of pollen grains (viz. the monosulcate type) or to that of the gynoecium (viz. the apocarpous) than to the primitivity of other characters.

Earlier evolutionary trends may have continued to develop in the same direction, but in many cases evolution may have been reversed or at least taken a different course. Whereas there may be a conspicuous similarity between major taxa of plants with regard to certain characters this may be counteracted by equally conspicuous differences in other characters. Thus a diagram showing the variation pattern of one character will seldom or never coincide with the pattern of variation of another independent character. Affinities as judged by the sum of character similarities will therefore always be relative. Moreover, no such diagram can be completely accurate, since the position of a group as determined by similarities to two or three other groups in a two-dimensional diagram will inevitably correspond poorly with its position in relation to a fourth of fifth group whose positions are determined by similarities to still other groups.

The probability of a polyphyletic origin for the angiosperms has been stressed, for example, in several papers by MEEUSE. But I do not consider it likely that the angiosperms have evolved from widely different groups of gymnosperms. The combination of several very characteristic attributes (secondary endosperm, 8-nucleate embryo sac, companion cells in the phloem, etc.) would hardly have evolved independently in different gymnosperm groups.

Is there anything to be gained from a diagrammatic presentation of the angiosperms? Firstly, I am convinced that for numerous features their distributional patterns can be shown in the diagram presented here. Secondly, the evaluation of distributional patterns of different categories of characters will doubtless cast light on phylogenetic relationships between groups of orders.

The distributional patterns of a number of characters will be shown together with brief comments on the following pages. Some major features of the system presented will also be pointed out together with a discussion on the subclasses of CRONQUIST (1968) and TAKHTAJAN (1969).

Distributional Patterns of Some Characters Within the Angiosperms

Some of the following characters will either be dealt with in greater detail in forthcoming papers or have already been presented (DAHLGREN 1975b, JENSEN et al. 1975, BEHNKE and DAHLGREN 1976).

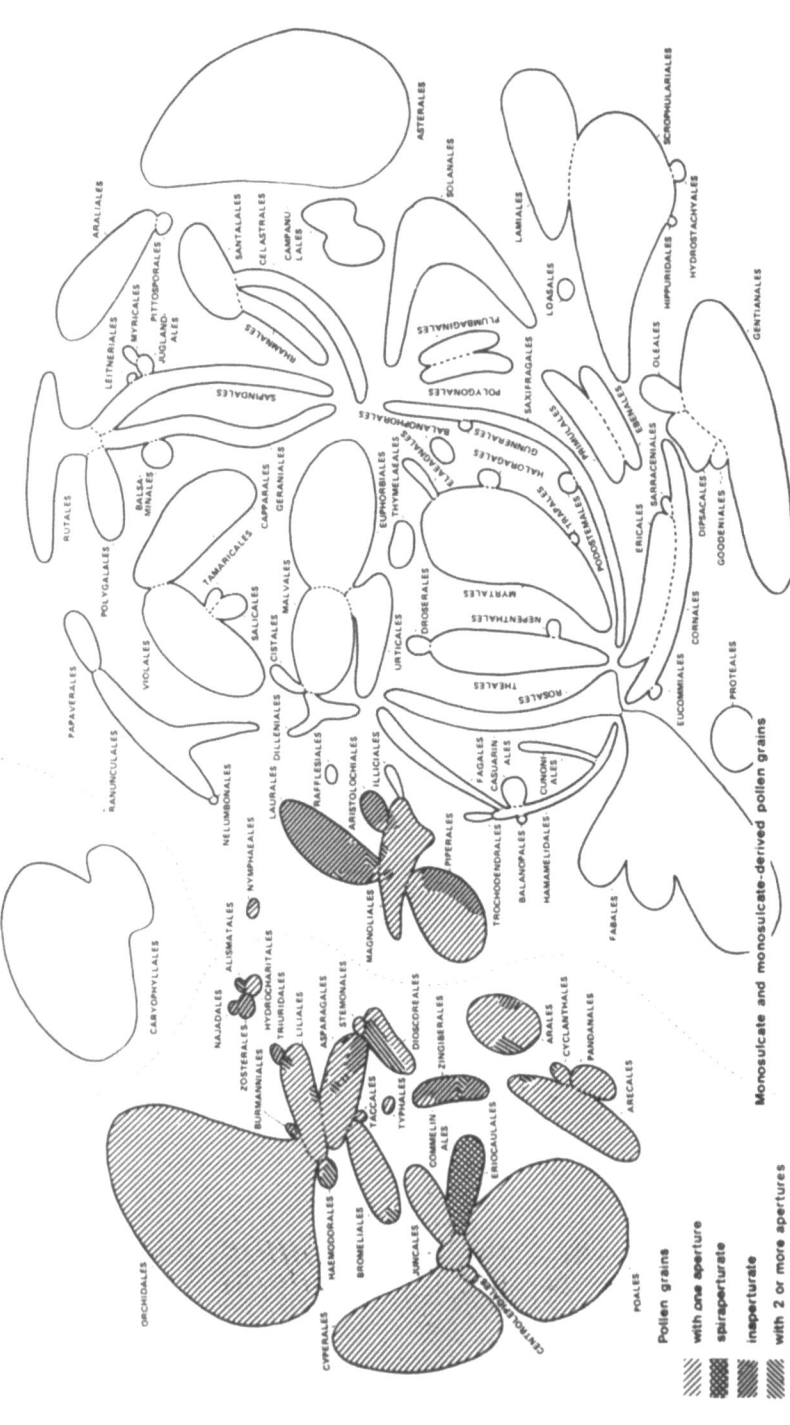

Fig. 2. Distribution of pollen grains with one aperture or of pollen grains without or with more than one aperture considered to be derived from types with one aperture. In *Magnolianae* the apertures do not lie in the equatorial plane except in *Illiciales* and some *Chloranthaceae*. Inaperturate pollen grains are found in many *Laurales*, *Aristolochiales* and *Piperales* and some *Nymphaeales* in the dicotyledons and in *Zosterales* and *Najadales*, many *Zingiberales*, *Smilacaceae*, etc., in the monocotyledons. Grains with 2 to many apertures occur sporadically, the most conspicuous group being perhaps *Alismatales*. Greater detail is found in works of Walker and collaborators

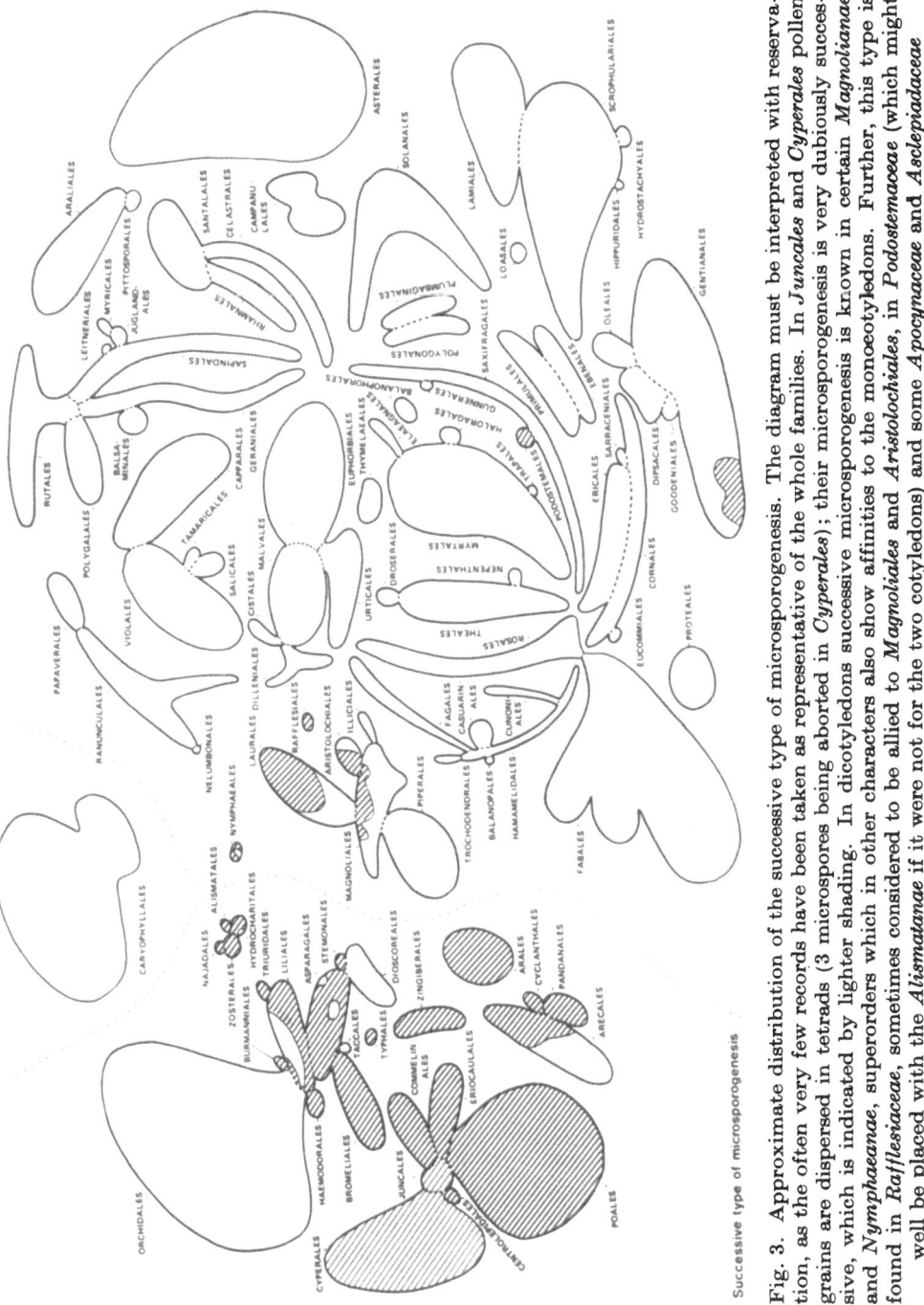

Successive type of microsporogenesis

Fig. 3. Approximate distribution of the successive type of microsporogenesis. The diagram must be interpreted with reservation, as the often very few records have been taken as representative of the whole families. In *Juncales* and *Cyperales* pollen grains are dispersed in tetrads (3 microspores being aborted in *Cyperales*); their microsporogenesis is very dubiously successive, which is indicated by lighter shading. In dicotyledons successive microsporogenesis is known in certain *Magnolianae* and *Nymphaeanae*, superorders which in other characters also show affinities to the monocotyledons. Further, this type is found in *Rafflesiaceae*, sometimes considered to be allied to *Magnoliales* and *Aristolochiales*, in *Podostemaceae* (which might well be placed with the *Alismatanae* if it were not for the two cotyledons) and some *Apocynaceae* and *Asclepiadaceae*

1. The distribution of monoaperturate pollen grains or of in-aperturate or bi- to polyaperturate grains derived presumably from monoaperturate types is hown in Fig. 2. It can be seen that they are found in (a) the monocotyledons, (b) the superorder *Magnolianae* and (c) in *Nymphaeanae* (= *Nymphaeales*). Apertures in these groups are rarely in the equatorial plane (except in *Illiciales* and a few *Chloranthaceae*). For details of Fig. 2, see for example WALKER & DOYLE (1976) and WALKER (1974, 1976).

2. A connection between monocotyledons and certain *Magnolianae* is also shown in the distribution of protein-accumulating (P-type) sieve-tube plastids (BEHNKE & DAHLGREN 1976). All groups of monocotyledons have P-type plastids of a rather distinctive type with cuneate protein bodies matched interestingly enough by those in *Asarum*, *Aristolochiaceae* (BEHNKE 1971, 1975: 93). The sieve-tube plastids in *Caryophyllanae* differ from those in other angiosperms except some *Magnoliales* but resemble those of certain *Pinatae* and contribute to separate out *Caryophyllanae* as a very distinct group.

3. The successive type of microsporogenesis (Fig. 3) indicates a connection between the majority of the monocotyledons and certain *Magnolianae* and *Nymphaeanae*, for example in members of *Myristicaceae*, *Lauraceae*, *Aristolochiaceae*, *Cabombaceae* and *Ceratophyllaceae*. The successive type of microsporogenesis is also found in many *Apocynaceae* and *Asclepiadaceae*, where it has doubtless arisen independently, and in *Rafflesiaceae* and *Podostemaceae*, where the character has not yet been decisively evaluated. The basic data on microsporogenesis is very limited and the diagram is thus provisional.

4. Apocarpy (Fig. 4) is regarded as a primitive state compared with syncarpy (incl. paracarpy). It has a wider distribution in dicotyledons than the character states previously shown. Thus it is found not only in *Magnolianae* but is dominant in the *Ranunculanae* and *Rosanae*. Its distribution in the monocotyledons is relatively restricted, however, occurring in three or four independent groups, viz. in a number of palm genera, in most *Alismatanae* and in the *Triuridales* and a few *Melanthiaceous* genera. A "disjunctive" distribution can be excpeted for certain so-called primitive states. They may have been conserved in some disparate groups, whereas in most groups the characters are advanced, the gynoecium in this case being variously syncarpous. When the carpel is solitary syncarpy is naturally an impossibility, as in *Amygdalaceae* (in *Rosales*), *Fabales*, *Proteales* and *Elaeagnales*. This condition is not to be equated with apocarpy. Nor must the parietal placentation in monomerous gynoecia be compared to that of the syncarpous gynoecia. A transition from apocarpous to syncarpous gynoecia is found in several orders (*Magnoliales*, *Ranunculales*,

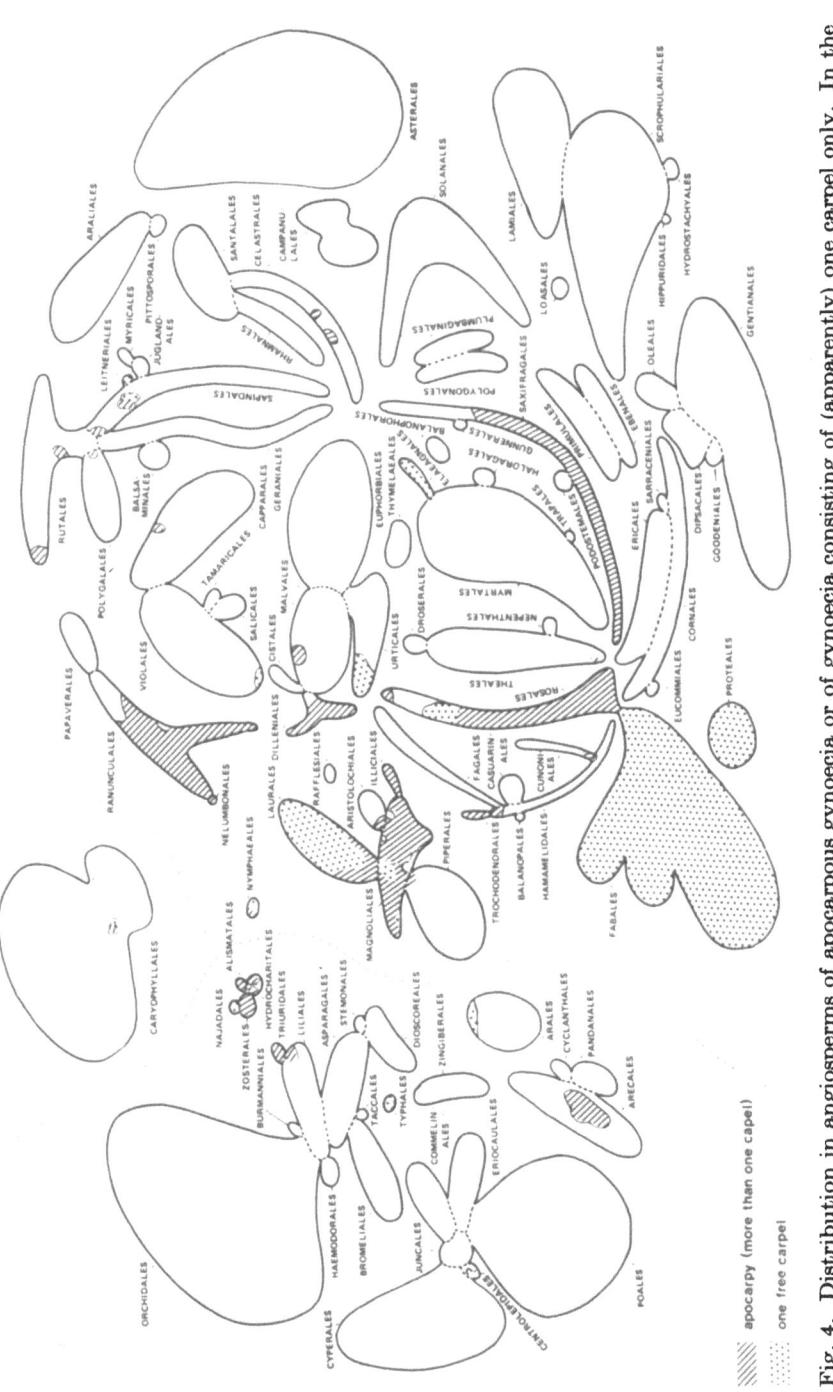

Fig. 4. Distribution in angiosperms of apocarpous gynoecia or of gynoecia consisting of (apparently) one carpel only. In the monocotyledons apocarpy is found in most families of *Alismatanae*, in *Triuridales*, in some *Melanthiaceae* (*Liliales*) and in certain palms. In the dicotyledons apocarpy is concentrated to *Magnolianae*, *Nymphaeanae*, *Rosanae* and *Saxifraganae* (*Saxifragales*). Outside these groups free carpels occur occasionally in *Hamamelidanae*, *Rutanae*, *Celastranae* and *Caryophyllanae*. Doubts are often connected with some groups having so-called monomerous gynoecia. They are found in approximately the same parts of the diagram as the apocarpous groups, culminating in the *Fabales*, *Proteales* and *Elaeagnales*

17*

Rosales, Rutales, Sapindales, Dilleniales, Saxifragales). A case of presumably secondary apocarpy is met with in *Resedaceae* (e.g. *Caylusea*).

5. Endosperm formation (DAHLGREN 1975b: 189–193) is cellular
in at least some family in each of all the orders of *Magnolianae* and
Nymphaeanae. Cellular endosperm is also found in members of *Nelumbonaceae, Lardizabalaceae* and *Circaeasteraceae* in *Ranunculanae*
and in *Mitrastemon* (*Rafflesiaceae*) in *Rafflesianae*. This type also
occurs in *Trochodendraceae* and *Cercidiphyllaceae* here placed in *Hamamelidanae*. All these groups are usually classified as primitive. In
addition, the ab initio cellular endosperm is found in many *Saxifraganae* and *Gentiananae* (though *Gentianales* generally has the nuclear
type), in most *Cornanae*, in *Lamianae, Loasanae, Solananae* and *Campanulanae* and in many *Asteranae* and *Celastranae*. This feature is
thus common to some of the most "primitive" and some of the most
"advanced" groups. That the cellular type is the original state was
probably first stated by WUNDERLICH (1959: 274). In the "advanced"
orders, according to her, it can have been retained as a consequence
of the restricted nucellus size. However, the endosperm is nuclear
ab initio in for example the small nucelli of *Theales*.

6. The distribution of benzylisoquinoline alkaloids (Fig. 5)
covers some of the often apocarpous groups (see also KUBITZKI 1969).
They are found in many taxa of *Magnolianae*, though so far there is
only a couple of records in *Piperales* (see, e.g., HÄNSEL et al. 1975)
and none as yet in *Illiciales*. Further, the benzylisoquinoline alkaloids
are known in many *Ranunculanae*, reaching a maximum of complexity
and variation in *Papaverales*. Complex benzylisoquinoline alkaloids
are also found in certain genera of *Rutaceae* (FISH & WATERMAN 1973)
and a new find was recently reported in *Heracleum* (*Apiaceae*) in
Aralianae by GUPTA et al. (1976). The presence of benzylisoquinoline
alkaloids in some genera of *Rhamnaceae* (*Celastranae*) indicates that
this family may have some affinity to *Ranunculanae* and *Magnolianae*, although this is dubious. Benzylisoquinoline alkaloids are also
found in *Croton* (*Euphorbiaceae*).

7. The distribution of sesquiterpene lactones and of polyacetylenes derived from fatty acids (Fig. 6) supports a connection between
the *Asteranae, Campanulanae, Aralianae, Celastranae* (*Santalales*),
Rutanae (*Rutales*) and possibly also *Ranunculanae* and *Magnolianae*.
Thus polyacetylenes are chiefly found in *Asteranae* and *Campanulanae* but
also in *Apiaceae, Araliaceae* and *Pittosporaceae* (*Aralianae*), in *Olacaceae, Opiliaceae, Santalaceae, Viscaceae* and *Loranthaceae* (*Santalales,
Celastranae*), in *Simarubaceae* (*Rutanae*) and in *Lauraceae* and *Annonaceae* (*Magnolianae*). Further, it is also known in isolated members
of *Sterculiaceae* (*Dillenianae*) and *Fabaceae* (*Rosanae*). Reports of their

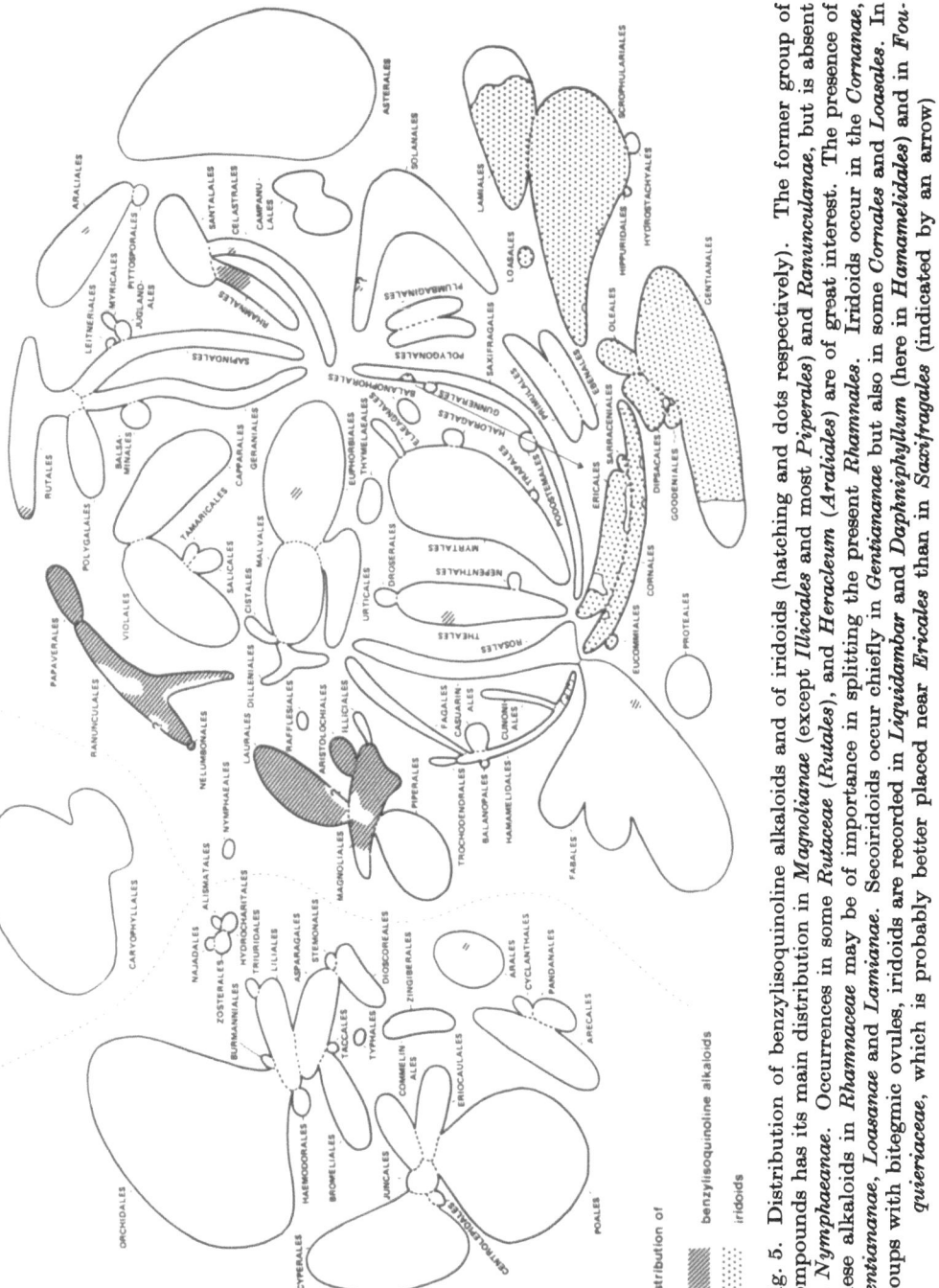

Fig. 5. Distribution of benzylisoquinoline alkaloids and of iridoids (hatching and dots respectively). The former group of compounds has its main distribution in *Magnolianae* (except *Illiciales* and most *Piperales*) and *Ranunculanae*, but is absent in *Nymphaeanae*. Occurrences in some *Rutaceae* (*Rutales*), and *Heracleum* (*Araliales*) are of great interest. The presence of these alkaloids in *Rhamnaceae* may be of importance in splitting the present *Rhamnales*. Iridoids occur in the *Cornanae*, *Gentiananae*, *Loasanae* and *Lamianae*. Secoiridoids occur chiefly in *Gentiananae* but also in some *Cornales* and *Loasales*. In groups with bitegmic ovules, iridoids are recorded in *Liquidambar* and *Daphniphyllum* (here in *Hamamelidales*) and in *Fouquieriaceae*, which is probably better placed near *Ericales* than in *Saxifragales* (indicated by an arrow)

Distribution of polyacetylenes

Fig. 6. Distribution of polyacetylenes derived from fatty acids. The greatest concentration is found in *Asteranae*, *Aralianae* (incl. *Pittosporales*), *Campanulanae* and *Santalales* (*Celastranae*). Scattered occurrences are reported in *Simaroubaceae* (*Rutales*), *Lauraceae* (*Laurales*), *Annonaceae* (*Magnoliales*), *Fabaceae* (*Fabales*) and *Sterculiaceae* (*Malvales*). The records in *Euphorbiaceae* (*Euphorbiales*) and *Valerianaceae* (*Dipsacales*) are very dubious, and a find in *Scaevola* in *Goodeniaceae* deserves particular mention as it is a deviation from the pattern

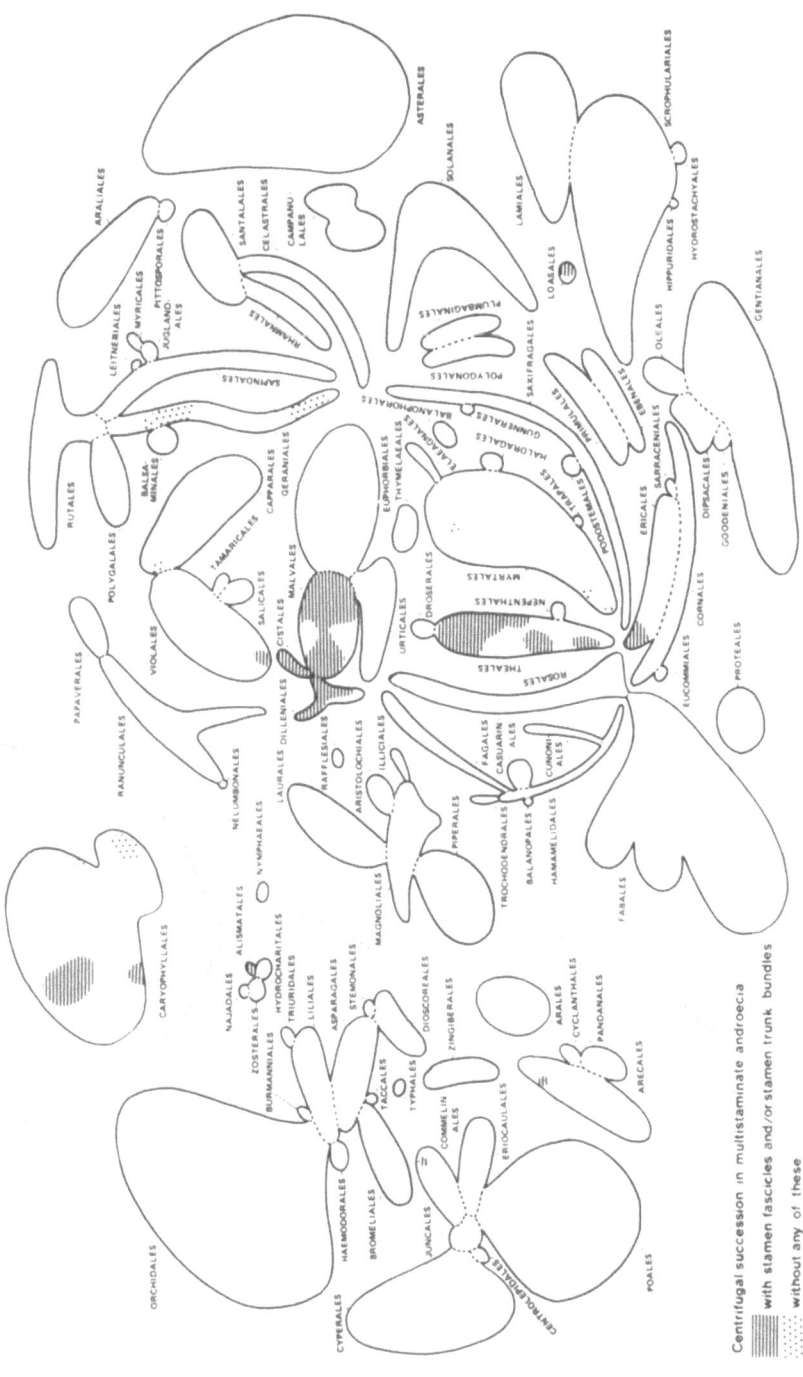

Fig. 7. Distribution of centrifugal succession in multistaminate androecia. Hatching represents groups with stamen fascicles and/or stamen trunk bundels, the dots groups with none of these structures; sometimes as in *Geraniaceae*, with centrifugal obdiplostemony. Most data from TUCKER (1972)

Centrifugal succession in multistaminate androecia

▨ with stamen fascicles and/or stamen trunk bundles

⋯ without any of these

presence in *Ricinus* and *Valeriana* are dubious. One case, *Scaevola* (*Goodeniaceae, Gentiananae*), an iridoid-containing genus, does not conform to the expected pattern.

8. Androecial characters, particularly centrifugal v. centripetal succession in multistaminate androecia (Fig. 7), have been dealt with by MERXMÜLLER (1972), KUBITZKI (1973), TUCKER (1972) and LEINS (1972, 1975) and are of importance in distinguishing the subclasses *Rosidae* and *Dilleniidae* in TAKHTAJAN's and CRONQUIST's systems. Taxa with multistaminate androecia displaying centrifugal succession and usually fascicular stamens occur in a number of groups, most of which are not closely related:

Certain *Caryophyllanae* (in *Aizoaceae* and *Cactaceae*).
A great many *Dillenianae* (*Dilleniales, Cistales* and *Malvales*).
Certain *Violanae* (*Violales, Capparales*), some of which may be closely related to certain *Dillenianae*.
Many *Theanae*: mainly *Theales* (incl. *Lecythidaceae*).
Certain *Ericales* (*Cornanae*), e.g. *Actinidiaceae*.
Genera in the apparently isolated *Loasaceae*, here treated as a separate superorder, *Loasanae*.
Some *Butomales* and *Hydrocharitales, Alismatanae*.
Some genera of palms (*Arecanae*).

The obdiplostemonous taxa have not been included here.

Facts seem to indicate that in many (or all?) cases the multistaminate-fascicular androecia with centrifugal succession are derived from androecia with one or two whorls of stamens. The character is of taxonomic value but should not be over-estimated as it has arisen within several lines of evolution. There are also obvious inconsistencies as in *Begoniaceae* (*Violales*), where the stamens develop centripetally in contrast to other multistaminate *Violales*; and as in *Punicaceae* (*Myrtales*) which has centrifugal succession in contrast to most other *Myrtales*. The consistently spiral organization of the stamens in *Magnoliales* does not hold either: stamen trace bundles are found for example in *Degeneria* (STEBBINS 1974: 221).

9. The presence of glucosinolates (ETTLINGER, unpubl. ms.) or similar isothiocyanate precursors characterizes *Tovariaceae, Resedaceae, Capparaceae* and *Brassicaceae*, which make up to the nucleus of the order *Capparales*, but also a number of small families: *Tropaeolaceae, Limnanthaceae, Bretschneideraceae, Moringaceae, Pentadiplandraceae, Caricaceae, Salvadoraceae, Gyrostemonaceae* and *Bataceae*, most of which are of uncertain taxonomic position. In addition, glucosinolates are found in *Drypetes* (incl. *Putranjiva*), usually placed in *Euphorbiaceae*, but possibly meriting distinction as a separate family, *Putranjivaceae*. The question of whether all or most of the glucosinolate

families are at all closely related or whether they form several independent aggregates has not yet been settled. Research is being done
on several of the families. Typical so-called "myrosin cells" are found
in most but not all of the glucosinolate families. Glucosinolates and
cyanohydrins are formed from amino acids by the same initial reaction.
Several of the glucosinolate families also seem to be allied to *Flacourtiaceae* and *Passifloraceae*, which contain cyanogenic compounds (ETT-
LINGER, op. cit.).

10. Ellagic acid and ellagitannins are of taxonomic importance.
Much has been written on this by BATE-SMITH (e.g. 1973), who has
kindly supplied me with the data presented in Fig. 8. The absence
of ellagitannins in minor groups is of restricted taxonomic value. For
example they tend to be lacking in herbaceous plants, whereas they
may be present in related woody plants. There is a noticeable absence
of ellagitannins in all monocotyledons and in the sequence of groups
Magnolianae—Ranunculanae—Rutales (of *Rutanae*)—*Aralianae—Asteranae—Campanulanae—Solananae*, many orders of which contain either
benzylisoquinoline alkaloids or polyacetylenes or both (see above).
Ellagitannins are also (generally) lacking in another sequence of superorders, *Cornanae—Gentiananae—Lamianae—Loasanae*, i.e. those in which
iridoids are produced, although there are some exceptions where both kinds
of compounds occur together, viz. in some *Ericales*, some *Cornales* and
the monogeneric *Fouquieriaceae*. Other superorders where ellagitannins are of very restricted occurrence are *Violanae* (except *Tamaricales*), *Dillenianae* (except *Dilleniales* and certain *Malvales-Euphor-
biales*), *Thymelaeanae* and *Rutanae* (except some *Geraniales* and *Juglandales*). Individual orders where they are usually or consistently
lacking are *Fabales* and *Polygonales*, a fact that should be taken into
consideration in the future.

The presence of ellagic acid in some members of *Nymphaeanae*,
viz. in *Nymphaeaceae* and *Cabombaceae*, is unique in taxa with monoaperturate pollen grains. *Nelumbo* is lacking in ellagic acid and contains benzylisoquinoline alkaloids. This together with its tricolpate
pollen grains suggests a position near *Ranunculales* and at some distance
from *Nymphaeales*.

11. Epigynous and perigynous flowers (Fig. 9) characterize
taxa at different levels. Epigyny can with certainty be classified as a
derived condition not likely to revert to hypogyny. Where epigyny
has appeared early it is likely to characterize larger categories: *Aralianae*; *Asteranae*; most *Myrtanae*; *Gentiananae*: *Rubiaceae*; *Orchidales*;
and some other *Lilianae*. Concrescence between a well-developed
floral axis or perianth and the carpel walls is likely to arise within
different groups and thus to be scattered in the system. The distribution

Distribution of ellagic acid

Fig. 8. Preliminary survey of the distribution of ellagic acid and/or ellagitannins in the angiosperms. Families with records of ellagic acid are shaded; partially shade only in certain larger families where their presence is known to be sporadic (as in *Caesalpiniaceae, Fabales*). Ellagitannins are concentrated mainly to the *Hamamelidanae—Rosanae* (*Rosales*)—*Theanae—Dilleenianae* (in part)—*Myrtanae—Saxifraganae—Cornanae* (*Ericales*)—*Primulanae* (in part) complexes, being conspicuously absent or rare in most other parts of the system. Striking occurrences outside the complexes mentioned are the orders *Nymphaeales, Tamaricales, Santalales* and *Juglandales* and, e.g., the families *Geraniaceae, Aceraceae, Simarubaceae* and *Limnanthaceae*. The absence of ellagitannins in all monocotyledons in the *Magnolianae—Ranunculanae—Rutanae—Violanae—Aravianae—Asteranae—Campanulanae—Solananae* complexes and in *Gentiananae—Loasanae—Lamianae* is conspicuous (see also Bate-Smith 1973). The data, kindly supplied by E. C. Bate-Smith, are preliminary

Fig. 9. Distribution of epigyny and perigyny in angiosperms. Epigyny has developed independently in many groups. The largest of these are the *Lilianae* (especially *Orchidales*), the *Myrtales*, the *Araliales—Asterales(—Campanulales)* complexes and the *Rubiaceae* in *Gentianales*. In some of these it has doubtless appeared in more than one evolutionary branch. Epigyny is less frequent in most of the so-called "primitive" groups such as *Magnolianae*, *Nymphaeanae* and *Ranunculanae*, and is likewise rather uncommon in *Rutanae*, *Dillenianae*, *Lamianae*, *Primulanae*, *Solananae*, etc. In monocotyledons epigyny is absent in *Commelinanae*, *Arecanae*, *Aranae* and *Typhanae*, all groups where the tepals are usually reduced or glume-like. Perigyny is a somewhat ill-defined concept, and in a slightly broader sense has a more extensive distribution than is shown here

of epigyny, perigyny and hypogyny should be considered in relation to position of nectaries, function of fruit dehiscence and other biological phenomena of importance. In monocotyledons the total absence of epigyny in *Commelinanae*, *Aranae* and *Arecanae* can be explained by the early reduction of the perianth to hyaline scales, bristles or hairs (or lack of perianth) in most of the families.

12. Sympetaly (Fig. 10) may well be compared with epigyny and can be expected to appear in widely different groups as a consequence of intercalary concrescence (STEBBINS 1974: 111). A distinction can be made between the "main sympetalous groups" or the *Sympetalae*, (most of) which are often considered to be closely related, and the other sympetalous groups. The "*Sympetalae*" include *Asteranae*, *Campanulanae*, *Solananae*, *Lamianae*, *Primulanae*, *Gentiananae*, many *Cornanae*, and *Plumbaginales* (*Plumbaginanae*) in the present system. It is generally agreed that sympetaly in other (minor) groups has evolved along a number of separate lines and in many cases perhaps at a late stage of evolution. The problem as to whether the "main sympetalous groups" form one, two or several natural complexes is often discussed. As indicated above (under polyacetylenes, sesquiterpene lactones) as well as below (under iridoids) it seems that the *Sympetalae* (excl. *Primulales* and *Plumbaginales*) form at least two main branches: the iridoid-bearing branch (lacking polyacetylenes, sesquiterpene lactones and usually inulin) and the iridoid-lacking sympetalous branch. The former, comprising *Cornanae* (often choripetalous; including *Ericales*)— *Gentiananae—Lamianae* seems to approach *Theanae* and/or *Saxifraganae—Hamamelidanae*, while the latter, comprising *Asteranae—Campanulanae—Solananae*, seems to approach *Aralianae*(*—Rutanae*) and *Santalales* of the *Celastranae*. *Primulanae* and *Plumbaginales* (*Plumbaginanae*) probably fall outside both of these groups and are perhaps best placed near *Theanae*. Moreover the ovules of *Primulanae* and *Plumbaginanae* are usually bitegmic, whereas in the other large sympetalous groups they are unitegmic.

13. Iridoids (Fig. 5) are mostly confined to sympetalous plants (JENSEN et al. 1975), but only in the superorders *Cornanae*, *Gentiananae* and *Lamianae*. As far as is known they are absent in *Solananae*, *Campanulanae* and *Asteranae*. With the exception of a few genera, *Daphniphyllum*, *Fouquieria* (DAHLGREN et al. 1976 b) and *Liquidambar* all the iridoid-bearing taxa have unitegmic, generally tenuinucellate ovules. With the chief exception of most *Gentianales* their endosperm is usually cellular ab initio. Terminal endosperm haustoria are common in the iridoid groups. An unexpected occurrence of this combination of features is found in *Loasaceae* which is possibly best placed near *Cornales* or the orders of *Gentiananae*.

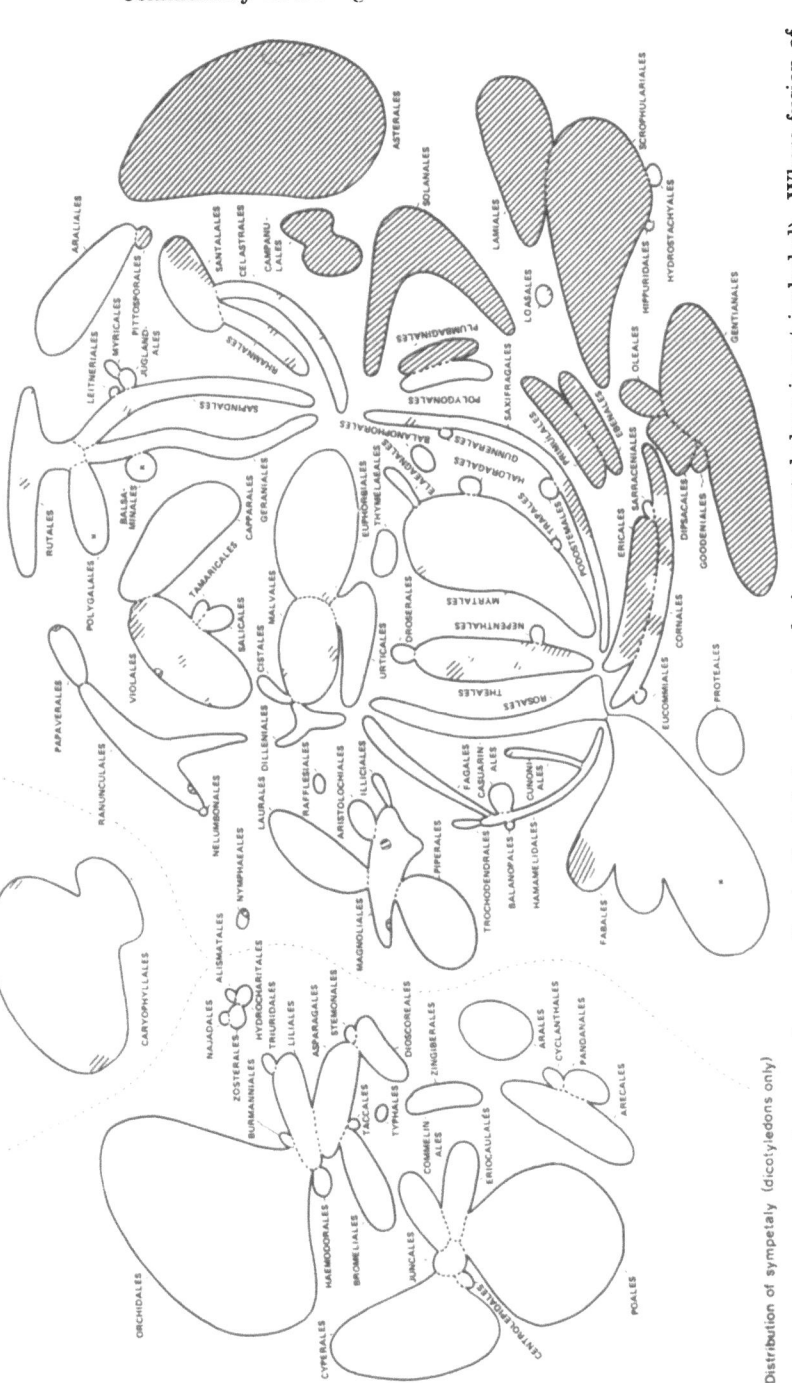

Distribution of sympetaly (dicotyledons only)

Fig. 10. Distribution of sympetalous corollas in dicotyledons (syntepaly in monocotyledons is not included). Where fusion of pairs or of three of the petals has occurred (such as in *Balsaminaceae* or *Polygalaceae*) or where two petals of five are united (such as in *Fabaceae*) these are not included but are indicated by asterisks. Some problems arise where there is doubt as to whether calyx or corolla is represented in monochlamydeous flowers: in *Santalales Loranthaceae* is shaded but not *Viscaceae*. The perianth in *Thymelaeaceae* and *Proteaceae* is interpreted as a calyx. Sometimes the petals are connate apically to form a calyp.ra (*Marcgraviaceae*, some *Vitaceae*). The polyphyletic origin of sympetaly is obvious. However, the assemblage of super-orders in the right-hand part of the diagram (*Primulanae—Plumbaginanae: Plumbaginales—Cornanae—Gentiananae—La-mianae—Solananae—Campanulanae—Asteranae*) comprising the "Sympetalae" in the broad sense is worthy of note although it probably does not represent a natural group

The importance of iridoids as an indication of relationships was not accepted by STEBBINS (1974: 135). Nor is it acknowledged by THORNE in his system of angiosperms (1968; pers. comm.), where iridoid-containing families are widely scattered. The concentration of seco-iridoids and indole alkaloids chiefly to the orders of *Gentiananae* as well as the high correlation with embryological characters support their taxonomic value, however.

14–15. The distribution of **unitegmic** and of **tenuinucellate ovules** and its implications have been stressed in particular by PHILIPSON (1974), and have also been discussed by DAHLGREN (1975 b) largely on the basis of an extensive study by WUNDERLICH (1959). Unitegmic ovules characterize *Cornanae, Gentiananae, Lamianae, Loasanae, Solananae, Campanulanae, Asteranae* and *Aralianae*, but also occur in a considerable number of other groups. The presence of both bitegmic and unitegmic ovules in, for example, *Ranunculaceae, Rosaceae, Piperaceae* and *Salicaceae* prove that the unitegmic condition is highly polyphyletic. The tenuinucellate and unitegmic conditions usually occur together (especially in sympetalous groups), but there are notable exceptions. In a number of usually restricted, isolated groups with unitegmic ovules (such as in the families just mentioned) these are crassinucellate. In most *Theales* and *Primulales* the ovules are bitegmic and tenuinucellate.

16. The distribution of **2- and 3-nucleate pollen grains** has been presented by BREWBAKER (1967) and outlined in the present system by DAHLGREN (1975 b). Certain major groups of dicotyledons are consistently 3-nucleate viz. *Caryophyllanae, Aralianae, Asteranae, Plumbaginanae, Thymelaeanae (Dichapetalaceae* excluded) and *Brassicaceae (Violanae)*. BREWBAKER pointed out the coincidence in dicotyledons between 3-nucleate pollen grains and sporophytic self-incompatibility and between 2-nucleate pollen grains and gametophytic self-incompatibility. Recent studies show further correlation. Thus almost all groups with sporophytic self-incompatibility (HESLOP-HARRISON 1975: 413) and thus also 3-nucleate pollen grains (but also other groups) have the "dry" type of papillate stigma. The fact that 3-nucleate pollen grains and "dry" (smooth in SEM micrographs) stigmas and 2-nucleate and "wet" (secretory) stigmas occur together even down to specific level is shown by HANSEN (1976: 343) for members of *Balanophoraceae*. The basic studies of dry/wet stigma distribution by K. R. SHIVANNA at Kew have not yet been published (HESLOP-HARRISON 1976: 143). In the case of the 3-nucleate pollen grains it seems that the active materials for pollen tube growth are conveyed in the exine of the pollen grain wall and are originally derived from the tapetum (thus being of sporophytic origin); further, there

appears to be no synthesis in them at all (HESLOP-HARRISON 1968: 90). Thus we have here a syndrome of interdependent characters of profound biological importance. The taxa photosynthesizing over the C_4 pathway are also mainly restricted to groups with 3-nucleate pollen grains.

Some Distributional Patterns in Monocotyledons

The taxonomy of the monocotyledons does not present problems of the same magnitude as in the dicotyledons. The superorders *Alistanae*, *Lilianae*, *Commelinanae*, *Zingiberanae*, *Typhanae*, *Arecanae* and *Aranae* are those which most taxonomists tend to agree upon, *Aranae* and *Zingiberanae* perhaps being placed with *Arecanae* and *Lilianae* respectively. Whether *Bromeliales* and *Haemodorales* are best placed with *Lilianae* or *Commelinanae*, and whether *Triuridales* is best placed with *Lilianae* or *Alismatanae* depends on the evaluation of characters. *Typhales* (= *Typhanae*) may be closest to *Juncales*. The fact that they can be attacked by the same species of *Uromyces* as *Acorus* (*Araceae*) (NANNFELDT 1968: 94) supports another affinity, but *Typhales* differs from *Arales* in having helobial endosperm formation. It may on the other hand be closest to families in *Lilianae*, as supported by the serological investigations of LEE and FAIRBROTHERS (1972).

Some criteria that contribute to distinguish between *Lilianae* and *Commelinanae* are shown here. They have been presented with comments in DAHLGREN et al. (1976a).

The presence or absence of vessels (Fig. 11) in different parts of the plant presents an interesting pattern of variation (WAGNER, in·prep.). *Commelinanae*, *Arecanae* and *Typhanae* consistently have vessels in the stem, *Aranae* and *Alismatanae* lacking them in the stems (*Alismatanae* usually also in the roots). *Zingiberanae* and *Lilianae* are variable in this respect. *Lilianae* generally lacks vessels in the stems, exceptions being, for example, *Dioscoreales*, many *Bromeliales* and certain *Asparagales*, particularly some of those that are bacciferous and/or have secondary thickening growth.

Epigyny is common in *Lilianae* and particularly so in *Zingiberanae*. It is also found in *Hydrocharitaceae* (*Alismatanae*). All these groups have a well-developed perianth. Reduction of the perianth is extensive in *Commelinanae*, which has consistently hypogynous flowers. Likewise, zygomorphic and asymmetric flowers are concentrated to *Lilianae* and *Zingiberanae*. In the monocotyledons tri-nucleate pollen grains are concentrated to *Commelinanae* (except most *Commelinales*, *Eriocaulales* and some *Cyperales*), *Alismatanae*, some saprophytic *Lilianae* and certain *Aranae* (BREWBAKER 1967).

The endosperm where present (it is absent in the seeds of *Or-*

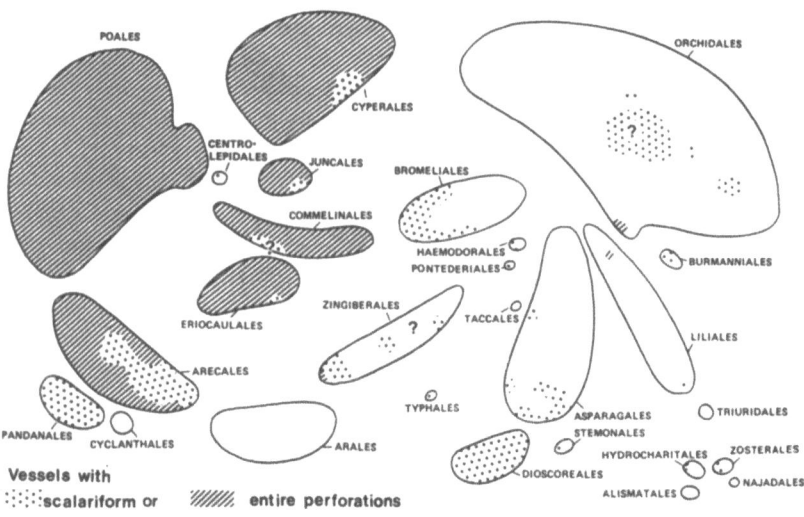

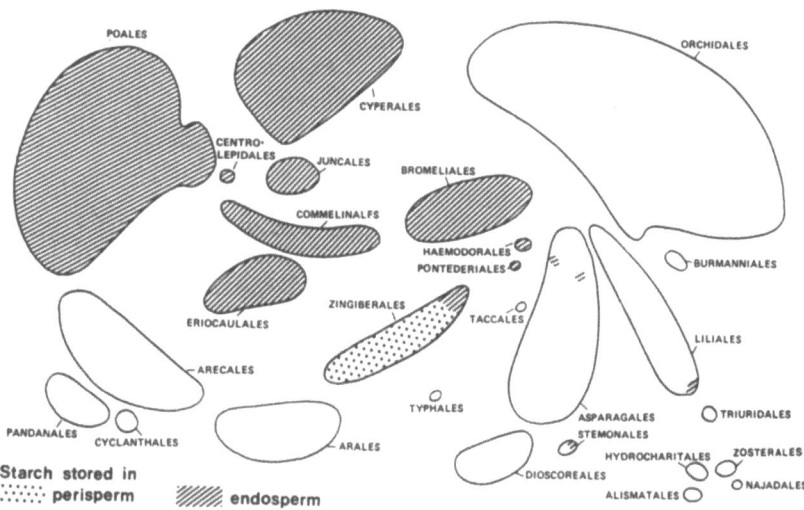

Fig. 11. Top: Approximate distribution of vessels with scalariform perforation or with entire perforation (sometimes also with scalariform perforation)
in monocotyledons.   Bottom: Approximate distribution of seeds with
starchy endosperm or starchy perisperm.—Both characters serve to some
extent to illustrate the distinctness of *Commelinanae* and *Lilianae*. The
diagram corresponds to the monocotyledonous sector of the main
diagram although the orders have been separated from one another and
their positions slightly alterered (moreover *Pontederiaceae* is treated as a
separate order)

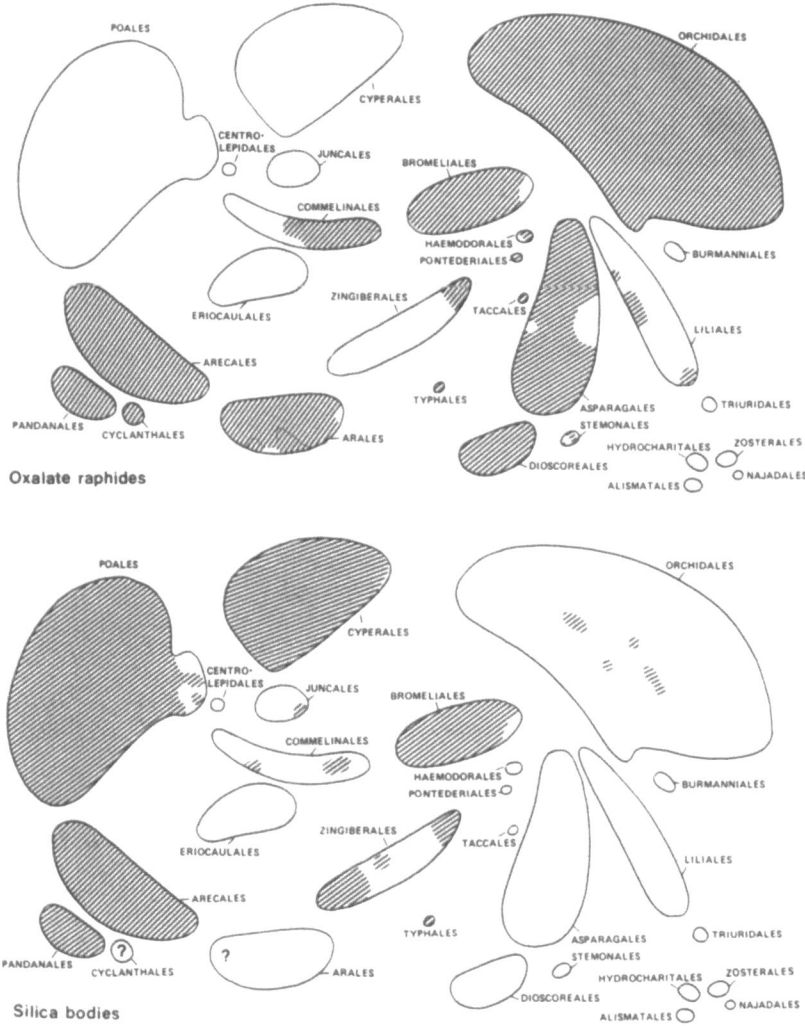

Fig. 12. Top: Approximate distribution of oxalate raphides in monocotyledons. *Alismatanae* and *Commelinanae* except most *Commelinaceae* are characterized by the absence of oxalate raphides. In *Lilianae* they are much commoner in *Asparagales* than in *Liliales*, where they are scattered in *Alstroemeriaceae* and *Melanthiaceae*. Bottom: Approximate distribution of silica bodies in monocotyledons. The shape of the silica bodies characterizes families or minor taxa. They are typical of most *Commelinanae* and *Arecanae* but are lacking in most *Lilianae*

Table 1. Classification of the families of angiosperms according to DAHL-GREN (1975a), where further information is obtainable. An asterisk (*) marks the families with weakly based or dubious position. Alternative positions are given for many of these in the publication mentioned

Magnolianae

Magnoliales: Winteraceae, Degeneriaceae, Himantandraceae, Magnoliaceae, Annonaceae, Canellaceae, Myristicaceae, Eupomatiaceae

Laurales: Monimiaceae, Trimeniaceae, Lauraceae, Idiospermaceae, Austrobaileyaceae, Gomortegaceae, Amborellaceae, Calycanthaceae, Hernandiaceae, Lactoridaceae*, Chloranthaceae*

Aristolochiales: Aristolochiaceae

Piperales: Saururaceae, Piperaceae

Illiciales: Illiciaceae, Schisandraceae

Rafflesianae

Rafflesiales: Rafflesiaceae, Hydnoraceae

Ranunculanae

Nelumbonales: Nelumbonaceae

Ranunculales: Lardizabalaceae, Menispermaceae, Sargentodoxaceae, Kingdoniaceae, Ranunculaceae, Circaeasteraceae, Hydrastidaceae, Glaucidiaceae, Podophyllaceae, Nandinaceae, Berberidaceae

Papaverales: Papaveraceae, Hypecoaceae, Fumariaceae

Nymphaeanae

Nymphaeales: Cabombaceae, Nymphaeaceae, Barclayaceae, Ceratophyllaceae

Rutanae

Rutales: Rutaceae, Cneoraceae, Surianaceae, Simaroubaceae, Kirkiaceae, Burseraceae, Meliaceae

Polygalales: Malpighiaceae, Trigoniaceae, Vochysiaceae, Xanthophyllaceae, Polygalaceae, Krameriaceae, Emblingiaceae*

Sapindales: Coriariaceae, Anacardiaceae, Podoaceae, Julianaceae, Akaniaceae, Uapacaceae*, Sapindaceae, Aitoniaceae, Aceraceae, Hippocastanaceae, Sabiaceae*, Meliosmaceae, Koeberliniaceae

Juglandales: Rhoipteleaceae, Juglandaceae

Myricales: Myricaceae

Leitneriales: Leitneriaceae

Geraniales: Zygophyllaceae, Nitrariaceae, Peganaceae, Balanitaceae, Erythroxylaceae, Dirachmaceae*, Geraniaceae, Ledocarpaceae, Vivianiaceae, Biebersteiniaceae, Ixonanthaceae, Humiriaceae, Hugoniaceae, Linaceae, Lepidobotryaceae, Averrhoaceae, Oxalidaceae, Hypseocharitaceae

Balsaminales: Balsaminaceae

Aralianae

Araliales: Araliaceae, Torricelliaceae, Apiaceae

Pittosporales: Pittosporaceae

Table 1 (continued)

Asteranae

   Asterales: Asteraceae

Dillenianae

   Dilleniales: Paeoniaceae, Dilleniaceae
   Cistales: Cistaceae, Bixaceae
   Malvales: Spaerosepalaceae, Cochlospermaceae*, Elaeocarpaceae*, Ster-
culiaceae, Huaceae*, Tiliaceae, Dipterocarpaceae, Bombacaceae, Malvaceae,
Neuradaceae*
   Urticales: Ulmaceae, Hymenocardiaceae, Moraceae, Cannabaceae, Ur-
ticaceae
   Euphorbiales: Euphorbiaceae, Pandaceae*, Aextoxicaceae*, Picro-
dendraceae

Thymelaeanae

   Thymelaeales: Dichapetalaceae*, Thymelaeaceae

Violanae

   Violales: Flacourtiaceae, Passifloraceae, Dipentodontaceae, Scyphoste-
giaceae, Violaceae, Turneraceae, Malesherbiaceae, Achariaceae, Cucurbitaceae, Begoniaceae, Datiscaceae*, Caricaceae
   Tamaricales: Tamaricaceae, Frankeniaceae
   Salicales: Salicaceae
   Capparales: Limnanthaceae, Tropaeolaceae, Bretschneideraceae*, Salvadoraceae*, Moringaceae, Resedaceae, Tovariaceae, Capparaceae, Pentadiplandraceae, Brassicaceae, Gyrostemonaceae*, Bataceae*

Celastranae

   Celastrales: Buxaceae*, Simmondsiaceae*, Stylocerataceae*, Didymelaceae*, Barbeyaceae*, Geissolomataceae*, Avicenniaceae*, Staphyleaceae,
Sphenostemonaceae, Aquifoliaceae, Celastraceae, Stackhousiaceae, Siphono-
dontaceae*, Goupiaceae*, Lophopyxidaceae*, Montiniaceae*
   Santales: Olacaceae, Opiliaceae, Loranthaceae, Misodendraceae, Santalaceae, Eremolepidaceae, Viscaceae
   Rhamnales: Rhamnaceae, Vitaceae, Leeaceae

Solananae

   Solanales: Solanaceae, Goetzeaceae, Nolanaceae, Convolvulaceae, Cuscu-
taceae, Cardiopterygiaceae, Cobaeaceae, Polemoniaceae, Hydrophyllaceae,
Ehretiaceae, Boraginaceae, Wellstediaceae, Lennoaceae*, Hoplestigmataceae*

Campanulanae

   Campanulales: Campanulaceae, Pentaphragmataceae, Lobeliaceae, Sphenocleaceae

Hamamelidanae

   Trochodendrales:* Trochodendraceae, Tetracentraceae, Eupteleaceae,
Cercidiphyllaceae

Table 1 (continued)

Hamamelidales: Myrothamnaceae, Hamamelidaceae, Platanaceae, Altingiaceae, Daphniphyllaceae, Rhodoleiaceae*
Casuarinales: Casuarinaceae
Fagales: Fagaceae, Corylaceae, Betulaceae
Balanopales*: Balanopaceae
Cunoniales: Cunoniaceae, Iteaceae, Brunelliaceae, Eucryphiaceae, Baueraceae, Bruniaceae

Rosanae

Rosales: Crossosomataceae*, Rosaceae, Malaceae, Amygdalaceae, Connaraceae*, Melianthaceae*, Chrysobalanaceae*
Fabales: Mimosaceae, Caesalpiniaceae, Fabaceae

Proteanae

Proteales: Proteaceae

Myrtanae

Myrtales: Lythraceae, Punicaceae, Rhizophoraceae, Dialypetalanthaceae*, Crypteroniaceae, Combretaceae, Oliniaceae, Melastomataceae, Penaeaceae, Myrtaceae, Onagraceae
Elaeagnales*: Elaeagnaceae
Trapales: Trapaceae
Haloragales: Haloragaceae

Saxifraganae

Saxifragales: Crassulaceae, Penthoraceae, Saxifragaceae, Fouquieriaceae*, Francoaceae*, Brexiaceae*, Cephalotaceae, Tremandraceae, Vahliaceae, Ribesiaceae*, Greyiaceae*
Podostemales*: Tristichaceae, Podostemaceae
Gunnerales: Gunneraceae

Balanophoranae

Balanophorales: Balanophoraceae, Cynomoriaceae

Plumbaginanae

Plumbaginales: Plumbaginaceae, Limoniaceae
Polygonales: Polygonaceae

Primulanae

Primulales: Myrsinaceae, Aegicerataceae, Theophrastaceae, Primulaceae, Cordidaceae
Ebenales: Ebenaceae, Sapotaceae, Lissocarpaceae, Styracaceae

Theanae

Theales: Stachyuraceae, Ochnaceae, Quiinaceae, Medusagynaceae, Scytopetalaceae*, Sarcolaenaceae, Strasburgeriaceae, Oncothecaceae, Theaceae, Pentaphylacaceae, Marcgraviaceae, Caryocaraceae, Pelliceriaceae, Napo-

Table 1 (continued)

leonaceae, Bonnetiaceae, Foetidiaceae, Lecythidaceae, Symplocaceae, Clusiaceae, Ancistrocladaceae*, Elatinaceae*
    Nepenthales: Nepenthaceae, Dioncophyllaceae*
    Droserales: Droseraceae, Lepuropetalaceae, Parnassiaceae

Cornanae

    Ericales: Actinidiaceae, Clethraceae, Cyrillaceae, Roridulaceae, Ericaceae, Monotropaceae, Pyrolaceae, Epacridaceae, Diapensiaceae, Byblidaceae*, Empetraceae, Grubbiaceae*
    Sarraceniales: Sarraceniaceae
    Eucommiales: Eucommiaceae
    Cornales: Garryaceae, Alangiaceae, Cornaceae, Davidiaceae, Nyssaceae, Icacinaceae, Escalloniaceae, Columelliaceae, Stylidiaceae, Hydrangeaceae, Alseuosmiaceae, Sambucaceae, Adoxaceae

Gentiananae

    Dipsacales: Caprifoliaceae, Valerianaceae, Triplostegiaceae, Dipsacaceae, Morinaceae*, Calyceraceae*
    Oleales: Oleaceae
    Goodeniales: Goodeniaceae.
    Gentianales: Loganiaceae, Buddlejaceae, Retziaceae, Rubiaceae, Menyanthaceae, Gentianaceae, Apocynaceae, Asclepiadaceae

Loasanae

    Loasales: Loasaceae

Lamianae

    Scrophulariales: Scrophulariaceae, Selaginaceae, Globulariaceae, Lentibulariaceae, Plantaginaceae, Pedaliaceae, Trapellaceae, Martyniaceae, Orobanchaceae, Gesneriaceae, Bignoniaceae, Henriqueziaceae, Myoporaceae, Acanthaceae
    Hippuridales: Hippuridaceae
    Hydrostachyales*: Hydrostachyaceae
    Lamiales: Verbenaceae, Callitrichaceae, Lamiaceae.

Caryophyllanae

    Caryophyllales: Phytolaccaceae, Agdestidaceae, Stegnospermataceae, Achatocarpaceae, Nyctaginaceae, Aizoaceae, Molluginaceae, Didiereaceae, Cactaceae, Portulacaceae, Hectorellaceae, Basellaceae, Chenopodiaceae, Dysphaniaceae, Halophytaceae, Amaranthaceae, Caryophyllaceae

Alismatanae

    Alismatales: Alismataceae, Limnocharitaceae
    Hydrocharitales: Butomaceae, Hydrocharitaceae, Aponogetonaceae
    Zosterales: Scheuchzeriaceae, Juncaginaceae, Potamogetonaceae, Zosteraceae, Posidoniaceae, Zannichelliaceae, Cymodoceaceae
    Najadales: Najadaceae

Table 1 (continued)

Lilianae

Dioscoreales: Dioscoreaceae
Stemonales: Stemonaceae, Trilliaceae
Asparagales: Smilacaceae, Philesiaceae, Ruscaceae, Convallariaceae, Asparagaceae, Dracaenaceae, Hypoxidaceae, Tecophileaceae, Phormiaceae, Xanthorrhoeaceae, Aphyllanthaceae, Asphodelaceae, Anthericaceae, Ixoliriaceae, Agavaceae, Hemerocallidaceae, Hyacinthaceae, Alliaceae, Amaryllidaceae.
Taccales: Taccaceae
Haemodorales: Haemodoraceae, Pontederiaceae, Philydraceae
Liliales: Colchicaceae, Iridaceae, Alstroemeriaceae, Liliaceae, Melanthiaceae
Triuridales: Triuridaceae
Burmanniales: Burmanniaceae, Corsiaceae, Thismiaceae
Orchidales: Apostasiaceae, Cypripediaceae, Orchidaceae
Bromeliales: Bromeliaceae, Velloziaceae *

Typhanae

Typhales: Sparganiaceae, Typhaceae

Zingiberanae

Zingiberales: Lowiaceae, Heliconiaceae, Musaceae, Strelitziaceae, Zingiberaceae, Costaceae, Cannaceae, Marantaceae

Commelinanae

Commelinales: Commelinaceae, Cartonemataceae, Mayacaceae, Xyridaceae, Abolbodaceae, Rapateaceae
Eriocaulales: Eriocaulaceae
Juncales: Juncaceae, Thurniaceae
Cyperales: Cyperaceae
Centrolepidales: Centrolepidaceae
Poales: Restionaceae, Ecdeiocoleaceae, Flagellariaceae, Joinvilleaceae, Poaceae

Arecanae

Arecales: Arecaceae
Pandanales: Pandanaceae
Cyclanthales: Cyclanthaceae

Aranae

Arales: Araceae, Lemnaceae

chidales, Alismatanae and many Aranae) contains starch (Fig. 11) in all Commelinanae but also in certain Lilianae such as Bromeliales, Haemodorales and some Asparagales and Liliales.

Oxalate raphides (Fig. 12) are common in monocotyledons except in Commelinanae where they seem to be confined to the (entomophilous) *Commelinaceae*; they are also absent in the *Alismatanae* and rare in *Zingiberanae*. Within *Lilianae* they are commoner in *Asparagales* than in *Liliales*.

Silica bodies (Fig. 12) in cells of the vegetative parts show an occurrence that is partly "vicarious" in relation to that of the oxalate raphides. They are present in most *Commelinanae* but also in *Arecanae*, while they are absent in *Alismatanae* and most *Lilianae* (except *Bromeliaceae*). *Zingiberanae* is variable in this respect. Finally, steroidal saponins, which are commoner in monocotyledons than in dicotyledons, are concentrated chiefly to *Lilianae* (incl. *Bromeliales*) and outside this superorder are of restricted occurrence.

Important Major Features Within This Angiosperm System in Relation to the Subclasses in TAKHTAJAN's and CRONQUIST's Systems

The main features of the angiosperm system as outlined by TAKHTA-JAN (1969) and CRONQUIST (1968) have received wide support. Like THORNE (1968) I have found it necessary, however, to refrain from using their subclasses. My reasons are as follows:

1. In the dicotyledons in particular certain orders placed in different subclasses display obvious similarities indicating close affinities.— Example: The families of *Ericales* (placed in *Dilleniidae*) display obvious similarities to those of *Cornales* (placed in *Rosidae*). These in their turn resemble families in *Dipsacales* (placed in *Asteridae*).

2. Relevant data justify excluding major groups from certain subclasses thus restricting their circumscription markedly.—Example: *Caryophyllales* is clearly distinct from *Polygonales* and *Plumbaginales*, which should both be placed in another part of the system. Further taxa must also be removed from the vicinity of *Caryophyllales*, for example the monogeneric *Theligonales* which is closely connected with *Rubiaceae*, and *Bataceae* and *Gyrostemonaceae* with other affinities. All these groups comprise the subclass *Caryophyllidae* which must thus be changed radically.

3. Entirely new constellations of orders can be discerned which do not correspond to the subclasses previously distinguished.—Example: The constellation comprising *Cornanae* (*Ericales*, *Cornales*, *Sarraceniales*, *?Eucommiales*)—*Gentiananae* (*Dipsacales*, *Goodeniales*, *Oleales*, *Gentianales*)—*Lamianae* (*Scrophulariales*, *Hippuridales*, *?Hydrostachyales*, *Lamiales*)—*Loasanae* (*Loasales*). This group is characterized by the presence of iridoids; unitegmic nearly always tenuinucellate ovules; usually cellular endosperm formation (exception most *Gentianales*); usual absence of ellagitannins and absence of polyacetylenes, sesquiter-

pene lactones and inulin. Another (similar) natural group is *Asteranae-Aralianae* with other characteristics.

4. The presence of orders that can only with difficulty be referred to any of the subclasses.—Example: *Nymphaeales*, which by virtue of its monoaperturate (or inaperturate) pollen grains agrees with the orders of *Magnolianae* but lacks their essential oils and benzylisoquinoline alkaloids. Ellagic acid is sometimes present in *Nymphaelaes* whereas it is totally lacking in *Magnolianae* and *Ranunculanae*. *Nymphaeales* exhibits conspicuous similarities to taxa in the monocotyledonous *Alismatanae*, especially *Hydrocharitales*. It is thus difficult to place in any subclass and it is best treated as a separate group between the *Magnolianae* and *Alismatanae*. Even more problematic is the position of *Rafflesiales*, *Balanophorales*, *Proteales*, *Elaeagnales* and *Podostemales*. *Eucommia*, placed in the subclass *Hamamelidae*, has important embryological and phytochemical features in common with the *Cornanae*.

Thus I consider it justified to split up the subclasses to facilitate the redistribution of their constituents. This will make it easier to take advantage of recent advances in, for example, phytochemistry and ultrastructural research.

Aims of the System and Diagram Used

A diagram such as that presented here can be used to survey the range of distribution, consistency and concentration of a particular character state. Where different attributes tend to occur in the same groups useful patterns of correlation and interdependence can thus be more easily discerned and assessed.

After the distributional patterns of a considerable number of characters have been illustrated and compared it will presumably be possible to deduce certain taxonomic conclusions. In the diagram the families have fixed positions within the orders; if they repeatedly prove to deviate in important characters from other families of the order or superorder where they have been placed they should probably be moved to another position in a revised version. There are several cases where it may also be justified to radically rearrange orders or superorders. It is hoped that in the course of time it will be possible to revise the system and diagram on the basis of a comparison of the diagrams in combination with new achievements presented in the literature.

Various suggestions for improvements have already begun to accumulate. The following can be mentioned as examples:

Fouquieriaceae should be removed from *Saxifragales* to the vicinity of *Ericales* (DAHLGREN et al. 1976b).
Cochlospermaceae should be included in *Bixaceae* and placed in *Cistales*.

Elaeocarpaceae should probably be removed from *Malvales*.

Dichapetalaceae is probably best placed in *Euphorbiales*.

Gyrostemonaceae and *Bataceae* should probably be removed from *Capparales*, into a separate order.

The position of several of the smaller families in *Celastrales* should be reconsidered; *Montiniaceae*, for example, should be placed in *Cornales*.

Rhamnales should be divided, *Vitaceae* and *Leeaceae* being placed in a separate order.

Rosales and *Fabales* are probably better treated in separate superorders, *Fabales* perhaps closer to *Sapindales*.

Connaraceae seems to be better placed in *Sapindales* than in *Rosales*.

A position for *Elaeagnaceae* closer to *Proteales* should be considered.

The superorders *Plumbaginanae* and *Primulanae* could perhaps be closer to *Theanae*.

Centrolepidaceae shobld be divided into two families.

Finally I should like to appeal to specialists in all relevant fields for contributions presenting the distribution of any attribute. Contributing information on affinity between groups of angiosperms at family level or higher will also be welcomed.

Several colleagues have contributed considerably to this article. The chemical data have been supplied by S. ROSENDAL JENSEN and B. JUHL NIELSEN, Technical University of Denmark, with the exception of those on ellagic acid which I have received from E. C. BATE-SMITH, Agricultural Research Council, Cambridge, England. P. WAGNER, Central Botanical Library, Copenhagen, has kindly supplied the data on vessels in monocotyledons. Mrs. MARGARET GREENWOOD-PETERSSON, Lund, Sweden, has revised the English text.

References

BATE-SMITH, E. C., 1973: Systematic distribution of ellagitannins in relation to the phylogeny and classification of the angiosperms. Nobel Symposium **25**, 93—102.

BEHNKE, H.-D., 1971: Zum Feinbau der Siebröhren-Plastiden von *Aristolochia* und *Asarum* (*Aristolochiaceae*). Planta **97**, 62—69.

— 1975: P-type sieve element plastids: a correlative ultrastructural and ultrahistological study on the diversity and uniformity of a new reliable character in seed plant systematics. Protoplasma **83**, 91—101.

— and DAHLGREN, R., 1976: The distribution of characters within an angiosperm system. 2. Sieve-element plastids. Bot. Notiser (Lund) **129**, 287—295.

BREWBAKER, J. L., 1967: The distribution and phylogenetic significance of binucleate and trinucleate pollen grains in the angiosperms. Amer. J. Bot. **54**, 1069—1083.

CRONQUIST, A., 1968: The evolution and classification of flowering plants. London-Edinburgh: Nelson.

DAHLGREN, R., 1975a: A system of classification of the angiosperms to be used to demonstrate the distribution of characters. Bot. Notiser (Lund) **128**, 119—147.

Dahlgren, R., 1975 b: Current topics. The distribution of characters within an angiosperm system. I. Some embryological characters. Bot. Notiser (Lund) **128**, 181—197.

— in cooperation with Hansen, B., Jakobsen, K., and Larsen, K., 1974: Angiospermernes taxonomi, 1. (In Danish.) København: Akademisk Forlag.

— 1975a: Ibid. 2.

— 1975b: Ibid. 3.

— 1976a: Ibid. 4.

— Jensen, S. R., and Nielsen, B. J., 1976 b: Iridoid compounds in *Fouquieriaceae* and notes on its possible affinities. Bot. Notiser (Lund) **129**, 207—212.

Ettlinger, M., (unpubl. ms.): Plant analysis by butterfly; the occurrence of benzylglucosinolate in *Batis maritima* and its context in systematics and chemical evolution.

Fish, F., and Waterman, P. G., 1973: Chemosystematics in the *Rutaceae* II. The chemosystematics of the *Zanthoxylum/Fagara* complex. Taxon **22**, 177—203.

Gupta, B. D., Banerjee, S. K., and Handa, K. L., 1976: Alkaloids and coumarins of *Heracleum wallichii*. Phytochemistry **15**, 576.

Hansen, B., 1976: Pollen and stigma conditions in *Balanophoraceae*. Bot. Notiser (Lund) **129**, 341—345.

Heslop-Harrison, J., 1968: Ribosome sites and S gene action. Nature **218**, 90—91.

— 1975: Incompatibility and the pollen-stigma interaction. Ann. Rev. Plant Physiol. **26**, 403—425.

— 1976: A new look at pollination. Rep. E. Malling Res. Stn. for 1975, 141—157.

Hänsel, R., Leuschke, A., and Gomez-Pompa, A., 1975: Aporphine-type alkaloids from *Piper auritum*. Lloydia **38**, 529—530.

Jensen, S. R., Nielsen, B. J., and Dahlgren, R., 1975: Iridoid compounds, their occurrence and systematic importance in the angiosperms. Bot. Notiser (Lund) **128**, 148—180.

Kubitzki, K., 1969: Chemosystematische Betrachtungen zur Großgliederung der Blütenpflanzen. Taxon **18**, 360—368.

— 1973: Probleme der Großsystematik der Blütenpflanzen. Ber. dtsch. bot. Ges. **85**, 259—277 (1972).

Lee, D. W., and Fairbrothers, D. E., 1972: Taxonomic placement of the *Typhales* within the monocotyledons: preliminary serological investigation. Taxon **21**, 39—44.

Leins, P., 1972: Das Androeceum der Dikotylen. Ber. Deutsch. Bot. Ges. **84**, 191—193.

— 1975: Die Beziehungen zwischen multistaminaten und einfachen Androeceen. Bot. Jahrb. Syst. **96**, 231—237.

Merxmüller, H., 1972: Systematic botany—an unachieved synthesis. Biol. J. Linn. Soc. **4**, 311—321.

Nannfeldt, J. A., 1968: Fungi as plant taxonomists. Festskrift till T. Segerstedt. Acta Univ. Uppsala, 85—95.

Philipson, W. R., 1974: Ovular morphology and major classification of the dicotyledons. Bot. J. Linn. Soc. **68**, 89—108.

Stebbins, G. L., 1974: Flowering plants. Evolution above the species level. Cambridge, Mass.: Harvard Univ. Press.

TAKHTAJAN, A., 1969: Flowering plants. Origin and dispersal. Edinburgh: Oliver & Boyd.

THORNE, R. F., 1968: Synopsis of a putative phylogenetic classification of the flowering plants. Aliso **6**, 57—66.

TUCKER, S. C., 1972: The role of ontogenetic evidence in floral morphology. Adv. Pl. Morph., 359—369.

WAGNER, P., in prep.: Distribution of vessels in roots, stems and leaves of the monocotyledons. Bot. Notiser **130**.

WALKER, J. W., 1974: Aperture evolution in the pollen of primitive angiosperms. Amer. J. Bot. **61**, 1112—1137.

— 1976: Comparative pollen morphology and phylogeny of the Ranalean complex. In: Origin and early evolution of angiosperms (BECK, C. B., ed.), 241—299. Columbia Univ. Press.

— and DOYLE, J. A., 1976: The bases of angiosperm phylogeny: Palynology. Ann. Missouri Bot. Gard. **62**, 664—723 (1975).

WUNDERLICH, R., 1959: Zur Frage der Phylogenie der Endospermtypen bei den Angiospermen. Österreich. Bot. Zeitschr. **106**, 203—293.

Address of the author: Professor ROLF DAHLGREN, Botanical Museum of the University of Copenhagen, Gothersgade 130, DK-1123 Copenhagen, Denmark.

Plant Syst. Evol., Suppl. 1, 285—298 (1977)
© by Springer-Verlag 1977

Institut für Allgemeine Botanik und Botanischer Garten,
Universität Hamburg, Federal Republic of Germany

The Treatment of the Monocotyledons in an Evolutionary System of Classification

By

Herbert Huber, Hamburg

Abstract: The more characters to distinguish monocotyledons and dicotyledons that are taken into consideration, the more numerous become the families which exhibit both mono- and dicotyledonous characters. In particular, monocotyledons and ranalean dicotyledons appear intimately related and the gaps between them do not allow a distinction into classes to be made. The author, therefore, considers the monocotyledons and the ranalean dicotyledons as two extreme wings of a single natural unit, with the *Annonaceae*, *Aristolochiaceae*, *Nymphaeaceae*, and *Piperaceae* as connecting links. Within the monocotyledonous wing, 12 natural units of higher than ordinal level (superorders) may be tentatively recognized and arranged according to the number of dicotyledonous features they possess and to the degree of their systematic isolation. Whereas the groups which present at least a few clearly dicotyledonous characters (like *Arales* and *Helobiae*) occupy a more isolated position, the reduction or absence of dicotyledonous characters (as in the anemophilous monocotyledons) is accompanied by decreasing isolation. This indicates that among monocotyledons which have attained a high evolutionary level, convergence largely camouflages relationship.

For more than half a century the fact has been acknowledged that certain characters, widely distributed among, and generally attributed to, the monocotyledons, occur sporadically in the dicotyledons as well, particularly within the ranalean orders. The occurrence of these mono-cotyledon-features outside the monocotyledons present a considerable problem in angiosperm classification, especially if one takes into account how few characters form the basis of our current definition of the monocotyledons. This overlap of characters is reflected in most modern treatments of angiosperm classification by the definition of monocotyledons as an early and strongly specialized offshoot of the ranalean group, an opinion against which there is little objection, provided it is admitted that differentiation took place before the recent orders and families had emerged.

A fundamental bipartition of angiosperms into monocotyledons and dicotyledons has nevertheless been unanimously maintained by practically all recent writers. This may serve convenience but, as I see the situation, it is not an appropriate way to demonstrate existing relationships. I should, therefore, like firstly to point out the problem of the delimitation of the monocotyledons from the ranalean dicotyledons, and secondly to illustrate what I consider to be the natural units above the rank of family within the monocotyledonous group.

Characters that obscure the delimitation between the monocotyledons and the dicotyledons may be roughly divided into two categories: the first which are centred within the dicotyledons, only extending casually or in very few groups into the monocotyledons, and the second, on the contrary, represented by typical monocotyledon-characters which are uncommon or rare within the dicotyledons. There is a striking difference in the overall distribution of these two sets of characters: Among the numerous features which belong to the first category there is not a single one which is exclusive to monocotyledons and the ranalean dicotyledons, however broadly one may construe the latter. All characters are commonly found in the non-ranalean dicotyledons as well. On the other hand, the typical monocotyledon characteristics which invade the dicotyledons are almost totally restricted to the ranalean group and many of them occur within the central orders distinguished by their synthesis of benzyltetrahydroisoquinoline alkaloids. I am consequently inclined to consider that the differentiation of the angiosperms into non-ranalean dicotyledons and the ranalean complex, including the monocotyledons, took place prior to that into the ranalean dicotyledons and the monocotyledons.

The characters of the second category are particularly important because they indicate how imperfect the differentiation into ranalean and monocotyledonous orders is still today. When present in dicotyledons, these characteristics tend to group into syndromes, as is well shown by *Annonaceae* and *Aristolochiaceae* which share several characters with monocotyledons. These characters are: (a) sieve tube plastids with protein crystalloids; (b) median position of prophyll; (c) primarily trimerous perianth; (d) nectar secretion from floral phyllomes instead from a disc; (e) unisulcate or inaperturate pollen grains. Among these characters, prophyll-position is perhaps the most remarkable, because this can hardly be considered primitive. According to FRIES (1959) the *Annonaceae* uniformly present single median prophylls within the floral region, and predominantly so but with numerous exceptions within their vegetative shoot, but normally with a dicotyledonous embryo. Apparently the tendency to produce single median prophylls arose peripherically in the plant body and extended proximally until it finally took over on the embryo whose cotyledons are equivalent to prophylls, thus giving rise to a monocotyledonous embryo.

Another example of concentration of monocotyledon-characters outside of the monocotyledons proper is in the *Piperaceae*. Again we find a primarily trimerous perianth, and pollen grains unisulcate or derived from the unisulcate type. In addition to this, the *Piperaceae* present medullary stem bundles; the stomata frequently surrounded by a rosette of subsidiary cells; and a tendency towards a completely closed reticulation of veinlets.

Benzyltetrahydroisoquinoline alkaloids have recently (HÄNSEL & al. 1975) been discovered in one species of *Piper*, but on the whole, arguments to include the *Piperaceae* and *Saururaceae*, in the ranalean group are weaker than for the *Annonaceae* and *Aristolochiaceae*. In particular, the *Piperaceae* have perisperm, otherwise known in the ranalean orders only from *Nymphaeales* (without *Nelumbo*).

This, however, is the third example of a major assemblage of monocotyledonous characters occurring outside monocotyledons proper and perhaps the most perplexing one. Apart from the medullary stem bundles and unisulcate pollen grains (at least in *Brasenia*, *Cabomba* and *Nuphar*), the order exhibits one most unexpected feature, not otherwise known outside typical monocotyledons, i.e., as HAINES & LYE (1975) have pointed out, an embryo possessing not only a single cotyledon but forming a coleoptile. Finally, the helobial formation of endosperm in *Cabomba* is worth mentioning. This clearly monocotyledonous assemblage of characters is counterbalanced by an apparently primarily polymerous, spirally constructed flower, which is quite out of place within the monocotyledons. The same is true of the ellagitannins, reported from *Nymphaeaceae* but absent from both monocotyledons and the ranalean dicotyledons. As to sieve tube plastids, the *Nymphaeales* and the *Piperales* form part of the ordinary dicotyledonous pattern and correspond neither with typical monocotyledons nor with the *Annonaceae*.

The *Nymphaeales* and *Piperales* are not closely related to each other nor to any particular group within or without the monocotyledons. Nevertheless, their general affinity strongly tends in the latter direction, far more than towards the ranalean orders. Apart from the characters already referred to, this fact is illustrated (Fig. 1) by the presence of laticiferous tubes in *Nymphaeaceae*, *Helobiae* (*Alismataceae*), *Araceae*, and *Scitamineae* (*Musa*); the occurrence of perispermous seeds in *Nymphaeales*, *Piperales* (*Chloranthaceae* excluded), and *Scitamineae*; and the cellular formation of endosperm in most *Nymphaeaceae*, many *Piperaceae* (not always in *Piper*), and in the *Arales* altogether. Generally speaking, the *Piperales*, *Nymphaeales*, *Arales*, *Helobiae*, and *Scitamineae* have in common a tendency not to employ endosperm as a principal storage tissue in their seeds but substitute it partly or completely, either by a storage nucellus commonly called perisperm, or by storage embryos. In the *Arales*, the latter is merely a tendency; in the *Helobiae* no trace of a storage tissue other than the embryo is left in the ripe seeds. I cannot pretend that perispermous seeds constitute a particularly derived character but storage embryos certainly do; in the monocotyledons, however, they do not indicate a very successful evolutionary tendency.

The five orders mentioned above are equally distinct from each other as from the rest of the monocotyledons. Accordingly I do not suggest any particularly close relationship of the *Piperales* with the *Arales* as suggested by EMBERGER (1960), although one must admit that *Arales* present certain piperalean features, which can not be rejected as casual.

The *Arales* agree well with the *Piperales* among other characters in leaf articulation and venation; stomata frequently with a rosette of subsidiary cells; cellular formation of endosperm; atropous ovules (in part of *Araceae*); and pseudomonomerous ovaries in *Peperomia* and a few *Araceae*. They differ fundamentally in the loss of vessels in the shoot, thus representing an outstanding monocotyledonous progression. Another highly evolved feature which the *Arales* have in common with various specialized but unrelated monocotyledons is an ameboid periplasmodium. The two last mentioned characters and to a certain degree the possession of subsidiary cells arranged in a rosette are shared by *Arales*, *Helobiae*, and *Scitamineae*. In these three groups the pollen is frequently inaperturate and trinucleate when shed. As to pollen apertures, unisulcate grains, supposed to be primitive in angiosperms, are common in the *Piperales* and *Nymphaeales*; they occur in few members of the *Arales* and *Helobiae*, but not in the *Scitamineae*. Inaperturate pollen, considered derived from unisulcate, occurs in the *Piperales* (*Peperomia*), *Nymphaeales* (*Ceratophyllum*), *Arales* (particularly abundant in the geophytic genera), and is frequent in the *Helobiae*. This may suggest an adaptation to aquatic pollination, but the predominance of inaperturate pollen in *Scitamineae* contradicts this explanation.

Abbreviations (Fig. 1):

Ara	*Arales*	Gram	*Graminales*	Phi	*Philydrales*
Arec	*Arecales*		= *Poales*	Pip	*Piperales*
Aspar	*Asparagales*	Haem	*Haemodorales*	Pont	*Pontederiales*
Bro	*Bromeliales*	Hel	*Helobiae*	Scit	*Scitamineae*
Cyp	*Cyperales*	Junc	*Juncales*	Ste	*Stemonales*
Dio	*Dioscoreales*	Lil	*Liliales*	Tac	*Taccales*
Far	*Farinosae*	Nym	*Nymphaeales*	Typh	*Typhales*
		Orch	*Orchidales*		
		Pand	*Pandanales*		

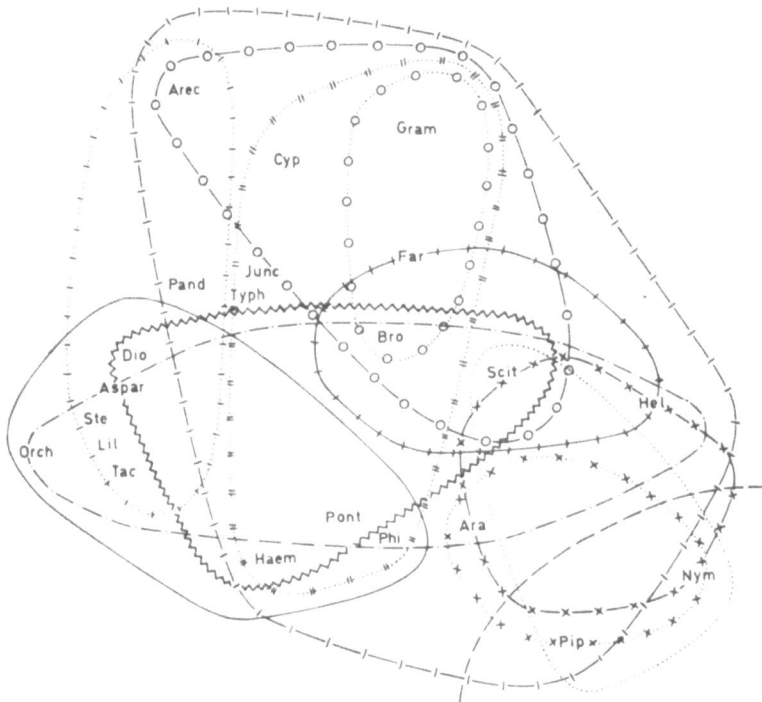

Fig. 1. Diagram showing specific superposition of characters, indicative of ordinal and higher level in the monocotyledons. Orders, or groups of orders, such as the *Farinosae* and *Helobiae*, surrounded by a signature, commonly possess the character indicated in the key, those placed outside do not. Names crossing a line indicate that the respective feature is present in a part of the order or orders only. Characters restricted to one order have been neglected, for which reason the *Taccales*, e.g., appear closer to the *Liliales* than they are; nor have primitive characters, such as apocarpy or presence of proanthocyanidins, been indicated as they tend to occur in conservative representatives of several evolutionary lines, nor have derived features obviously polyphyletic in origin, such as anemophilous and epigynous flowers (for abbreviations see p. 288)

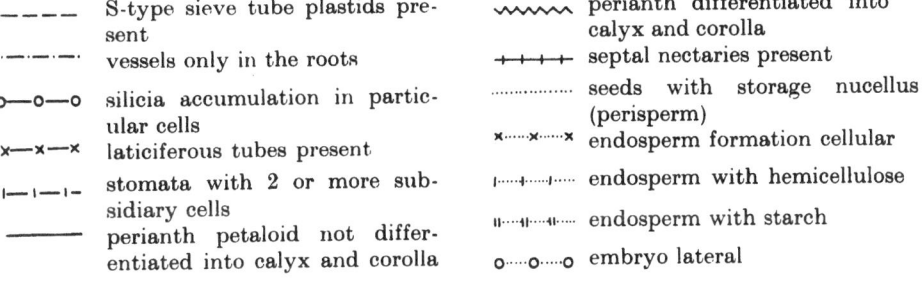

– – – –	S-type sieve tube plastids present	⋁⋀⋁⋀⋁⋀	perianth differentiated into calyx and corolla
·—·—·—	vessels only in the roots	++++	septal nectaries present
o—o—o	silicia accumulation in particular cells	················	seeds with storage nucellus (perisperm)
x—x—x	laticiferous tubes present	x·····x·····x	endosperm formation cellular
⊢—⊢—⊢	stomata with 2 or more subsidiary cells	⊢····+····⊣····	endosperm with hemicellulose
————	perianth petaloid not differentiated into calyx and corolla	‖····⊩····⊩····	endosperm with starch
		o·····o·····o	embryo lateral

Apart from the *Arales*, the *Helobiae* represent another example of extreme heterobathmy. In addition to the characters already mentioned, tHe *Helobiae* possess, insofar as they are zoophilous, heterochlamydeous flowers as found in the monocotyledons, e.g. in the *Scitamineae*, *Bromeliaceae*, and *Enantioblastae*. Secondarily polyandrous flowers mark another progression. On the other hand, apocarpy, laticiferous tubes, intrapetiolar stipules and the terminal plumule of *Ottelia* can be considered primitive or, more correctly, reminders of the dicotyledons.

If one disregards the reduced aquatic plants, such as the *Lemnaceae* and *Zostera*, the most derived group among the isolated evolutionary lines of monocotyledon affinity is the *Scitamineae*. This is evident from the following vegetative and reproductive characters: the flowers tend to become monosymmetric or asymmetric; the number of stamens becomes reduced; septal nectaries are universally present; and silica bodies or silica idioblasts are frequent in leaves and stems. The latter character is often correlated with the arborescent habit. Rosette trees indeed occur in *Scitamineae*. There is now a tendency to consider this habit rather primitive among monocotyledons, but at least as far as the *Scitamineae* are concerned I agree with STEBBINS (1974) in rejecting this opinion. As it seems, the arborescent genera *Ravenala* from Madagascar, *Strelitzia* from South Africa, and *Phenakospermum* from the Guayanas are pollinated either by bats or by birds (*Strelitzia* only). Disjunction in distribution indicates age, but pollinators do not: bats as well as *Passeriformes* date back, at most, into the lower Tertiary. Not much is known about the forest floor herb *Orchidantha*, but I think this would make a much better model of a primitive *Scitaminea*.

With respect to the somewhat marginal position occupied by *Piperales*, *Nymphaeales*, *Arales*, *Helobiae*, and *Scitamineae*, the "central" group of monocotyledons shows at least one character common to most of its members: this is, the presence of a well developed storage endosperm, whereas perisperm is scarce, or mostly absent, and storage embryos are almost lacking. Orchids, with mycotrophic germination, make the only considerable exception, but their seeds, which lack storage tissue throughout, have no equivalent in the previously mentioned groups either. Perhaps less important is the observation that, contrary to the marginal monocotyledonous orders, the tendency towards inaperturate pollen is much more weakly developed within the "central" group, a few *Asparagales*, *Iridaceae*, *Hanguana*, and part of *Xyridaceae* excepted.

Distinction of "marginal" and "central" groups does not imply that the latter are derived from the former nor does it imply the absence of primitive and dicotyledonous features from the central group. Within the latter, *Liliiflorae* can be called reasonably distinct, at least if one limits them to

·the orders *Dioscoreales*, *Stemonales*, *Asparagales*, and *Liliales*, as I have defined them elsewhere (HUBER 1969).

From an anatomical point of view, the *Liliiflorae* are almost sufficiently described by the possession of anomocytic stomata, and the absence of silica bodies or silica idioblasts. Vessels are mostly confined to the roots, but this is not so in *Dioscoreales* and *Smilacaceae*, which I consider, as to this character, to be the most primitive members of this assemblage. The outer integument consists commonly of more than two cell-layers and the inner integument is of two layers only. The endosperm accumulates fat oil, protein, and mostly hemicellulose deposited on the cell walls. The *Liliiflorae* primarily possess nectariferous tepals, but many of them replace this original condition by septal nectaries. Apart from this, the *Asparagales* produce rosette trees, with or without secondary thickening of the stem. An unusual character is the steroid saponins which abound in the *Dioscoreales*, *Stemonales*, and numerous *Asparagales*, particularly in their phanerophytic members. They have been found in smaller quantities in the *Liliales* and *Orchidales*, two orders presumably related to each other, although the *Orchidales* certainly did not descend from any recent member of *Liliales*.

The liliiflorous monocotyledons have some important characters in common with the *Scitamineae* (suppression of vessels in the shoot, septal nectaries, and presence of chelidonic acid being the most noteworthy) and the *Arales* (presence of steroid saponins and again suppression of vessels), but more pronounced is affinity with what I have previously called the palmaceous evolutionary line (*Arecanae* of DAHLGREN 1976), that is palms including the *Cyclanthaceae* and screw pines. The relationship of these three orders seems fairly well established.

As to anatomical characters, the palmaceous monocotyledons are distinguished by (a) stomata mostly surrounded by a rosette of subsidiary cells; (b) *Arecales*, but not *Cyclanthales* and *Pandanales*, by accumulation of silica; (c) *Arecales* and *Cyclanthales*, but not *Pandanales*, by postgenital partition of the leaves; (d) *Arecales* and *Pandanales* by presence of vessels throughout their vegetative organs; (e) *Arecales* and *Pandanales*, but not *Cyclanthales* (which are zoophilous), by a tendency towards reduction in number of ovules, a characteristic frequently correlated with anemophily; (f) all three orders by occurrence of polyandrous flowers. Pollen in the *Cyclanthaceae* and *Pandanaceae* often exhibits ulcoidate apertures which are almost absent from the *Liliiflorae*. Chelidonic acid, widely distributed in the *Liliiflorae*, has never been reported from palmaceous monocotyledons.

Most noteworthy is the arborescent habit of most palms and many

Pandanales. With respect to the association of this character with evidently derived features (silica accumulation in *Arecales*, septal nectaries in *Asparagales*, silica accumulation and septal nectaries in the *Scitamineae* and *Bromeliales*), I consider the pachycaul habit in the monocotyledons not as an archaic but a derived feature, and regard a plant like *Freycinetia* or a climbing *Cyclanthaceae* as a much better choice for a primitive palmaceous representative.

The affinity of the palmaceous monocotyledons with the *Liliiflorae* is stressed by presence of abundant endosperm storing hemicellulose, apart from fat oils and aleurone, whereas starch is mostly absent from the mature endosperm. Another seed-character in favour of an affinity of palmaceous with liliiflorous monocotyledons is ruminate endosperm as present in the *Trichopodaceae*, species of *Yucca* and many palms. This character, almost invariably correlated with woody habit throughout angiosperms (the *Trichopodaceae* are the only noteworthy exception) may indeed indicate primitiveness, as recently expressed by CORNER (1976). Steroid saponins are not common in palms, but occur in a few genera in large quantities (HEGNAUER 1963).

Faced with this list of characters that are shared by *Liliiflorae* and palmaceous monocotyledons, why not, one may wonder, attribute the *Dioscoreaceae* to the latter. The presence of vessels in axis and petioles, the tendency to reduce flowers (which are unisexual and anemophilous in *Dioscoreaceae* proper) and ruminate seeds might indicate this. Nevertheless, the structure, venation and vernation of the leaves, stomata (anomocytic in *Dioscoreales*), the tendency towards a geophytic habit, epigynous flowers, presence of chelidonic acid and absence of silica permit the *Dioscoreales* to be seen as undeniable members of the *Liliiflorae*.

Araceae, conventionally but wrongly associated with the palmaceous group, differ markedly by an opposite tendency in the evolution of the androecium, as the number of stamens tends to become reduced and not increased as in many palmaceous monocotyledons, quite apart from the vessel and endosperm characters already described.

The crucial areas in monocotyledon systematics are the families whose seeds present two commonly superposed specializations, i.e. both integuments consisting of two cell layers; and accumulation of starch instead of hemicellulose as the principal storage carbohydrate in the endosperm. The great diversity of families which exhibit these characteristics indicates that this syndrome must have arisen several times. There are examples of a positive correlation of this syndrome with almost every character reported above both for the liliiflorous and palmaceous monocotyledons, save three exceptions: anomocytic stomata; postgenital partition of leaf or compound or lobed leaves throughout; and ruminate seeds (there is, however, an indication of ruminate seeds in *Angiozanthos*, *Haemodoraceae*).

It is logical to compare the families with the petaloid perianth not differentiated into calyx and corolla with *Liliiflorae*. Additional liliiflorous characters found in the families which present seeds and perianth as described are: septal nectaries in *Haemodoraceae* and *Pontederiaceae*; and suppression of vessels in the aerial organs of *Philydraceae* and *Pontederiaceae*. As in the *Liliiflorae*, neither is silica accumulated nor has ulcerate or ulcoidate pollen been reported. Apart from this, the *Haemodoraceae* resemble the *Liliiflorae* in presence of chelidonic acid, although they differ from the majority by having vessels distributed throughout the vegetative organs. Information available at present does not allow to decide if the similarities between the *Haemodoraceae*, *Philydraceae*, and *Pontederiaceae* are due to affinity, as suggested by DAHLGREN (1974), or to convergence.

In my opinion, the *Haemodoraceae* are distantly related to the *Taccaceae*, a family which combines dicotyledonous features, one of them not otherwise encountered in monocotyledons (DAVIS 1966), with suppression of vessels in the shoot, both integuments two-layered and an endosperm, free from starch, storing hemicellulose although in modest quantities. A remote affinity may be deduced from the presence of a peculiar pigment in the seed coat both in the *Haemodoraceae* and the *Taccaceae* and other characteristics as well (HUBER 1969).

In contrast to the zoophilous and primarily homoiochlamydeous families, all but one of the families which combine a perianth either differentiated into calyx and corolla, or glumaceous, reduced and anemophilous with the above mentioned superposition of specialized seed characters, develop vessels in all their vegetative organs and they mostly accumulate silica in peculiar cells (except the *Sparganiaceae*, *Typhaceae*, and the greater part of the *Juncaceae*). In these families, silica accumulation is frequently associated with ulcerate or ulcoidate pollen (exceptions: *Commelinaceae*, *Eriocaulaceae*, *Mayacaceae*, *Rapateaceae*, *Typhaceae*). Endosperm formation, helobial throughout *Haemodoraceae*, *Philydraceae*, and *Pontederiaceae*, is nuclear in the calyciflorous and anemophilous families with exception of the *Juncaceae* (and possibly the *Sparganiaceae* and *Typhaceae* as well). These characters viewed together indicate affinity with the palmaceous evolutionary line and I would not hesitate to consider the familes in question en bloc as reduced offshoots of palmaceous ancestry, if these families were indeed uniformly anemophilous or achlamydeous and there were not conspicuous examples of highly evolved entomophily, presenting a perianth differentiated into calyx and corolla, as in the *Commelinaceae* and *Xyridaceae*. As a matter of fact there is no way in which one can imagine these calyciflorous families derived from, or in any definable way related to, the palmaceous orders. However, there seems to exist at least a chance of understanding the calyciflorous families better. It has long been known that both calyciflorous and certain anemophilous families (*Centrolepidaceae, Eriocaulaceae, Flagellariaceae, Gramineae, Rapateaceae,* and *Restionaceae*) possess an almost unique feature not otherwise observed in the angiosperms: that is the embryo removed from the axial position, which it originally helds, and attached laterally to the endosperm. This group of families corresponds with the *Farinosae* as defined by HAMANN (1961), including the *Gramineae*.

There is a strange resemblance of the *Enantioblastae*—the name I prefer for this assemblage—with the *Scitamineae*, with which they share, apart from heterochlamydeous flowers, silica accumulation, stomata with subsidiary cells arranged in a rosette (in *Enantioblastae* due to their mostly narrow leaves particularly in broad-leaved *Commelinaceae*) and, if one has the perfume grasses in mind, accumulation of essential oils in one-celled idioblasts. More stringent, however, is the affinity of the *Enantioblastae* with the *Bromeliaceae*. Among the monocotyledons with both two-layered integuments and farinose endosperm, the *Bromeliaceae* are the only family which combines differentiation of perianth into calyx and corolla with suppression of vessels in the shoot. The bromeliads indeed exhibit a most puzzling constellation of characters of different evolutionary lines. They agree with the *Enantioblastae* in their mostly two-layered outer integument, farinose endosperm, and commonly lateral position of the embryo; with the *Enantioblastae* and *Scitamineae*, in silica accumulation, stomata provided with subsidiary cells arranged in a rosette, and heterochlamydeous flowers; with the *Liliiflorae* in unisulcate or two- or multiaperturate but not ulcerate nor ulcoidate pollen grains; with the *Liliiflorae* and *Scitamineae*, in suppression of vessels in leaves and axis, occasionally arborescent habit (*Puya*), septal nectaries, and frequently helobial formation of the endosperm (in *Scitamineae, Zingiberaceae* only); with the *Scitamineae*, finally, in a decided tendency towards ornithophily combined with persistent petaloid bracts.

Leaves with a thorny toothed margin which are characteristic for the terrestrial bromeliads do not occur in either the *Enantioblastae* or the *Scitamineae* but are not infrequent in the *Liliiflorae* (particularly among *Asparagales*), in the *Juncales* (*Prionium* and *Thurnia*), and in the *Velloziaceae*, which seem to be the closest relatives of the *Bromeliaceae*. Steroid saponins, widely distributed in liliiflorous monocotyledons, occur in the *Bromeliaceae* in but small quantities, whereas chelidonic acid seems to be absent.

Generally speaking, the relationship of the *Bromeliaceae* to the *Enantioblastae, Juncales,* and *Scitamineae* is more prominent than to the *Liliiflorae* (*Velloziaceae* excepted), but they do not fit well into either evolutionary line.

Having now eliminated the families with an embryo laterally attached to the endosperm, there remains a complex of families that uniformly possesses vessels throughout the vegetative body, anemophilous flowers, mostly ulcerate or ulcoidate pollen grains frequently coherent to form a tetrad and, as far as the families embrace more than one genus, with at least one instance of a rosette tree. This group comprises the *Cyperaceae, Juncaceae, Sparganiaceae, Thurniaceae*, and *Typhaceae*, most of which have been attributed by various authorities and to some degree, to the palmaceous monocotyledons—correctly, as I should say, at least as far as the *Cyperales* (with one family only) and *Typhales* (two families) are concerned, although arguments for this attribution admittedly consist rather in absence of serious obstacles than in positive evidence.

As to the *Juncales* (*Juncaceae* and *Thurniaceae*), I am less confident to take their palmaceous affinity for granted. Apart from helobial formation of the endosperm, this order rather rigidly maintains pluriovulate ovaries, which is an unexpected feature in an natural unit which should be long adapted to anemophily. The only alternative place for the *Juncales*, if not in the palmaceous group, would be in such a superficially dissimilar assemblage as the *Haemodorales* and *Philydrales*, provided the two eventually prove to be related with each other, which is still uncertain.

To conclude this paper, I shall put forward a few ideas as to the delimitation of dicotyledons from the monocotyledons and the arrangement of the orders and superorders within the latter. The differentiation of the ranalean complex in its broadest sense (i. e., including the monocotyledons) into the ranalean and the monocotyledonous wing has not yet reached such a degree of perfection that intermediates have been eliminated. Characteristic specializations in the central groups, both of the ranalean dicotyledons and the monocotyledons, frequently allow attribution of intergrades to either extreme, e.g. the *Annonaceae* and *Aristolochiaceae* to the ranalean wing; the *Dioscoreales* and *Taccales* to the monocotyledonous wing; the *Piperales* and *Nymphaeales* are best placed halfway in between the two extremes. It would not appear logical to consider the break between the *Piperales* and *Nymphaeales* on the one side and the *Arales, Dioscoreales, Helobiae*, and *Scitamineae* on the other, deeper or more indicative as to isolation than discrepancies between the *Piperales* and *Nymphaeales*, or between the *Arales, Dioscoreales*, and *Scitamineae*.

Provided systematics maintains its claim to express natural and not administrative units, it is suggested that monocotyledons should not be considered a class of their own, but rather a bundle of loosely coherent evolutionary lines, some of them more isolated than others and almost deserving the rank of a subclass, others intricately reticulate and, in this case, only with hesitation being ranked as superorders.

Table 1. Conspectus of the monocotyledonous superorders. (It should be held in mind that the *Juncales* and *Velloziales* can not be attributed definitely to any superorder)

Dicotyledonous features prevailing	Intermediate evolutionary state	Dicotyledonous features rare or absent
a) isolated evolutionary lines:		
Piperiflorae (one order: *Piperales*)	**Ariflorae** (one order: *Arales*)	**Scitamineae** (one order: *Zingiberales*)
Nymphaeiflorae (one order: *Nymphaeales*)	**Helobiae** or **Alismatiflorae** (2–4 orders)	
b) reticulate evolutionary lines:		
	Liliiflorae (probably 5 orders: *Dioscoreales, Stemonales, Asparagales, Liliales, Orchidales*)	**Haemodoriflorae** (one family: *Haemodoraceae*)
	Tacciflorae (one family: *Taccaceae*)	**Pontederiiflorae** (2 orders: *Pontederiales* and *Philydrales*)
	Palmiflorae (with 5 or 6 orders if construed broadly: *Arecales, Cyclanthales, Pandanales, Cyperales, Typhales,* and doubtfully *Juncales*)	**Bromeliiflorae** (possibly 2 orders: apart from *Bromeliales* also *Velloziales*)
		Enantioblastae or **Commeliniflorae** (superorder embracing all families with mealy endosperm and lateral embryo except *Bromeliaceae*; these families have been arranged in one [THORNE 1968] to 4 [DAHLGREN 1976] orders)

The angiosperms, in my opinion, represent one single and in spite of their diversification remarkably homogeneous class which hardly exhibits discontinuities deep enough to justify the recognition of subclasses. I therefore tend to arrange the angiosperm superorders in an informal way, uniting the ranalean and monocotyledonous superorders so as to contrast them against with the non-ranalean dicotyledons, and within this enlarged ranalean complex, the already mentioned two extreme wings become obvious, with a few ambiguities in between. For didactic purposes, I would propose an arrangement of the monocotyledonous superorders as indicated in Table 1. All of these monocotyledonous superorders, the bromeliads and *Enantioblastae* perhaps excepted, exhibit a certain proportion of primitive, that is to say dicotyledonous characters, although there is a tendency for them to be lost as a higher evolutionary level is attained.

With respect to their origin, an arrangement according to their dicotyledonous similarities would appear feasible but as such a procedure would lead towards a linear sequence this would necessarily obscure the reticulation of affinities which is so prominent in this assemblage. A clearer method of illustrating relationships seems to consist in plotting essential character constellations (Fig. 1) instead of presumptive natural units as is usually done. Thus at least partly the procedure on which systematic decisions are based, can be visualized, what may well contribute towards a more objective system of classification.

References

BEHNKE, H.-D., 1975: The bases of angiosperm phylogeny: Ultrastructure. Annals Missouri Bot. Gard. **62**, 647—663.

CARLQUIST, S., 1975: Ecological strategies of xylem evolution. Berkeley-Los Angeles-London: University of California Press.

CHEADLE, V. I., 1953: Independent origin of vessels in the monocotyledons and dicotyledons. Phytomorphology **3**, 23—44.

—— and TUCKER, I. M., 1961: Vessels and phylogeny of monocotyledons. In: Recent Advances in Botany (IX. International Bot. Congress 1959) **1**, 161—165. Toronto.

CORNER, E. J. H., 1976: The Seeds of the Dicotyledons. Vol. **1, 2**. Cambridge: University Press.

DAHLGREN, R., 1974—1976: Angiospermernes taxonomi, bind **1—4**. København: Akademisk Forlag.

DAVIS, G. L., 1966: Systematic Embryology of the Angiosperms. New York-London-Sidney: John Wiley & Sons.

EMBERGER, L., 1960: Les végétaux vasculaires. In: Traité de botanique (Systématique), (M. CHADEFAUD et L. EMBERGER), tome **2**. Paris: Masson & Cie.

ENGLER, A., 1964: Syllabus der Pflanzenfamilien, 12. Aufl. (MELCHIOR, H., Hrsg.). Berlin: Borntraeger.

ERDTMAN, G., 1952: Pollen Morphology and Plant Taxonomy: Angiosperms. Stockholm: Almquist & Wiksell.

FRIES, R. E., 1959: *Annonaceae*. In: Die natürlichen Pflanzenfamilien, 2. Aufl., Band **17 a II**. Berlin: Duncker & Humblot.

HAINES, R. W., and LYE, K. A., 1975: Seedlings of *Nymphaeaceae*. Bot. Journ. Linn. Soc. **70**, 255—265.

HAMANN, U., 1961: Merkmalsbestand und Verwandtschaftsbeziehungen der *Farinosae*. Willdenowia **2**, 639—768.

—— 1962: Weiteres über Merkmalsbestand und Verwandtschaftsbeziehungen der "*Farinosae*". Willdenowia **3**, 169—207.

HÄNSEL, R., LEUSCHKE, A., and GOMEZ-POMPA, A., 1975: Aporphine-type alkaloids from *Piper auritum*. Lloydia **38**, 529—530.

HEGNAUER, R., 1963: Chemotaxonomie der Pflanzen, Band **2**: Monocotyledoneae. Basel und Stuttgart: Birkhäuser.

HUBER, H., 1969: Die Samenmerkmale und Verwandtschaftsverhältnisse der Liliifloren. Mitt. Bot. Staatss. München **8**, 219—538.

Meeuse, A. D. J., 1975: Aspects of the evolution of the monocotyledons. Acta Bot. Neerl. **24**, 421—436.

Moore, H. E., Jr., and Uhl, N. W., 1973: Palms and the origin and evolution of the monocotyledons. Quart. Rev. Biol. **48**, 414—436.

Netolitzky, F., 1926: Anatomie der Angiospermen-Samen. Handbuch der Pflanzenanatomie, Band **10**. Berlin: Borntraeger.

Porsch, O., 1914: Die Abstammung der Monokotylen und die Blüten-nektarien. Ber. dtsch. bot. Ges. **31**, 580—590.

Ramstad, E., 1953: Über das Vorkommen und die Verbreitung von Cheli-donsäure in einigen Pflanzenfamilien. Pharm. Act. Helv. **28**, 45—57.

Stebbins, G. L., 1974: Flowering Plants. Evolution Above the Species Level. London: E. Arnold.

— and Khush, G. S., 1961: Variation in the organisation of the stomatal complex in the leaf epidermis of monocotyledons and its bearing on their phylogeny. Am. Journ. Bot. **48**, 51—59.

Stone, B. C., 1972: A reconsideration of the evolutionary status of the family *Pandanaceae* and its significance in monocotyledon phylogeny. Quart. Rev. Biol. **47**, 34—45.

Suessenguth, K., 1920: Beiträge zur Frage des systematischen Anschlusses der Monocotylen. Beih. Bot. Zbl. **38** (II), 1—79.

Thorne, R. T., 1968: Synopsis of a putatively phylogenetic classification of the flowering plants. Aliso **6**, 57—66.

Address of the author: Prof. Dr. Herbert Huber, Institut für Allgemeine Botanik und Botanischer Garten der Universität Hamburg, Jungiusstr. 6—8, D-2000 Hamburg 36, Federal Republic of Germany.

Plant Syst. Evol., Suppl. 1, 299—319 (1977)

Rancho Santa Ana Botanic Garden, Claremont, California, U.S.A.

Some Realignments in the *Angiospermae*

By

Robert F. Thorne, Claremont, Calif.

Abstract: The system of classification that I have developed since 1950 to indicate the phylogenetic relationships of the higher taxa of the flowering plants has for the most part been published only in synoptical form. Since some of my more iconoclastic alignments have not been fully explained in print, this paper attempts to explain my reasons for some of the more innovative of these positionings.

My classification of the *Angiospermae* deviates considerably from others now widely accepted in several major ways. Because I attempt to stress relationships more than differences, my taxa tend to be more inclusive, while the differences within the major categories are recognized through the use of subcategories like suborders and subfamilies. Family names are in accord with the International Code of Botanical Nomenclature though the nine "exceptional" names are dropped as obsolete. Also for the sake of uniformity in phylogeny I have extended the principle of priority to the names of orders and other higher categories up to the class, anticipating a future rule or recommendation of the Code.

Taxa that are treated in some detail with explanation of the alignments are: the *Aquifoliaceae*, *Sarraceniaceae*, and *Nepenthaceae* of the *Theales* and *Plumbaginaceae* of the *Primulales* in the *Theiflorae*; *Fouquieriaceae* with *Polemoniineae* and *Solanineae* of the *Solanales* in the *Malviflorae*; *Gyrostemonaceae*, *Stylobasium* DESF., and *Emblingia* F. MUELL. of *Sapindineae* and *Batis* L. of *Batineae*, *Rutales*, in the *Rutiflorae*; *Crossosomataceae* of *Rosineae*, *Rosales*, and *Buxaceae* (excluding *Simmondsia* NUTT.), *Buxineae*, and *Balanopaceae*, *Daphniphyllineae*, in the *Pittosporales* of the *Rosiflorae*; *Asteraceae* of the *Asteriflorae* near the *Corniflorae* and *Lamiiflorae* and all three less closely with *Saxifragineae* of the *Rosiflorae*; *Alismatiflorae* as less primitive in the *Monocotyledoneae* than the *Liliiflorae*; and *Juncineae* and *Poineae* in the *Commeliniflorae*.

Introduction

Since 1950 I have been developing a system of classification of the *Angiospermae* that presents as accurately as possible the relationships of the higher taxa in that class. As a teacher of taxonomy I learned very early how resentful students can be when expected to learn a system of classification that they and their instructor realize is highly artificial,

grouping together taxa of dubious or distant relationship. In the early
1950's there were available no classifications that could be accepted
as even approximately phylogenetic. Therefore, I started to construct
my own system on a tentative and synthetic basis, stripping orders of
obviously or probably unrelated families and transferring them to other
orders where they did seem to belong or to a temporary fluctuating pool
of taxa incertae sedis.

Elsewhere I have published (THORNE 1958, 1963, 1974 c, 1976) the
principles which I have followed, my modus operandi, in this attempt
to devise a phylogenetic classification. I mention here only that basically
I have tried to find and use information that I regard as pertinent from
all parts of the flowering plants at all stages of their development. I
have never hesitated to use information made available by morphologists,
cytologists, biochemists, physiologists, geologists, herbivores, predators,
parasites, or any other group deeply interested in the flowering plants
as objects of study, food, shelter, or hosts. It is, of course, presumptious
of me and of my fellow phylogenists, to attempt to gather, evaluate, and
apply this mass of data, but someone has to do it. There are always
available some presumptious fools who will march in where angels are
too prudent to tread.

Fortunately today there are a number, some would say a plethora, of
classifications that purport to be phylogenetic, and that are indeed far more
natural than the classifications earlier in the century. None, including my
own, is yet truly phylogenetic, nor probably will any one of them attain
perfection for botanical generations to come, if ever. What bothers students
of the angiosperms about these various systems is that they differ both in
minor details and sometimes in major groupings. This is to be expected,
considering the different emphasis that each phylogenist places upon the
data available to him. None of us is completely right nor completely wrong,
and each makes important suggestions that specialists can examine and
accept, or discard if found inadequate. Surely all good phylogenists examine
carefully the systems and the justifications given by their fellow botanists.
Where these phylogenists continue to differ, each regards his own interpretation of the data as more realistic than that of his rivals.

My system of classification differs considerably from other systems
now widely accepted, and much of it I have presented only in synoptical
form without thorough discussion of my reasons for various realignments. Because my system attempts to stress relationships more than
differences, my taxa tend to be more inclusive than those of my fellows
with more disintegrative tendencies. As a teacher I consider it pedagog-
ically unsound to expand the number of taxa unnecessarily through
the narrow definition of categories. I realize that in this period of taxonomic and economic inflation I am fighting a rearguard action, but there
is a quixotic excitement about tilting with authoritative windmills and

dodging juggernaut bandwagons. I think the students, and therefore time, are on my side in this fight against debasement of categories.

In another departure from current practice I have extended the principle of priority to orders and other higher categories up to the class. Because LINDLEY first consistently applied the ending "-ales" to generic roots, I have used his "Nixus Plantarum" (1833) as the starting work for ordinal names. Family names are in accord with the "International Code of Botanical Nomenclature" (STAFLEU 1972) though the nine exceptional names are here disregarded as obsolete. In these matters I think I am merely anticipating ultimate decisions for the "International Code".

The divergences, however, of my system from other recent systems that probably interest the reader most are those rather startling realignments that are new or at least different from standard treatments. I shall try to explain my reasons for accepting some of these more iconoclastic alignments, selecting primarily those groupings that I have not discussed at some length elsewhere (THORNE 1973, 1974 a, 1974 b, 1974 c, 1975, 1976).

In a long paper (THORNE 1976: Fig. 2) I have attempted a phylogenetic shrub that presents visually but somewhat imperfectly how I relate the major superorders, orders, and suborders to one another in the *Angiospermae*. Placed toward the center are the superorders that I think retain the most primitive original features of the protoangiosperms, as the *Annoniflorae*, *Hamamelidiflorae*, and *Rosiflorae*. Radiating from the center are the other taxa that have achieved higher evolutionary plateaux in their characteristics, and farthest from the center are the taxa that deviate most from their presumed protoangiospermous ancestry. I do accept both the class and the two subclasses as monophyletic taxa. In reference to this phyletic shrub I shall here work outward from the center in discussing those realignments that may seem most novel to the reader.

Theiflorae

My treatment of the *Annoniflorae* (THORNE 1974) has been available for some time. A similar treatment of the *Theiflorae*, a related but somewhat more specialized superorder, is not yet completed. I shall discuss, therefore, some of the members of the group that are not generally placed in or near the *Theales* by other taxonomists, particularly the *Aquifoliaceae*, *Sarraceniaceae*, *Nepenthaceae*, and *Plumbaginaceae*.

Aquifoliaceae. A prime target for phyletic demolition is the extraordinary melange of largely unrelated families with small haplostemonous flowers, with or without glandular disks, that many taxonomists continue to recognize as the *Celastrales*. The *Aquifoliaceae* seemed to me especially out of place near the *Celastraceae*. The unspecialized stem anatomy, theaceous appearance of the foliage, and generally circumtropical distribution of *Ilex* L. early suggested the *Theaceae* as probable relatives. To students of tropical botany this should not be a startling suggestion. Sterile specimens of *Ilex* are commonly confused with sterile specimens of *Eurya* THUNB., *Adinandra* JACK., *Cleyera* THUNB., *Ternstroemia* MUTIS ex L. f., and other theaceous genera. Indeed, *Ilex* species approach members of the *Theaceae* so closely in stem, leaf, and flower characteristics, that I regard them as haplostemonous, functionally dioecious, small-flowered cousins of the camellias.

Among the features of the more specialized members of the *Theaceae* shared by species of *Ilex* are: basically tropical distribution; ancient fossil record; similar growth habit and foliage; sexuality bisexual to polygamous or functionally dioecious; sepals and petals imbricate; stamens reduced to 5 or 4 and often attached basally to the separate or basally connate petals; syncarpous ovary with mostly fewer than 10 locules; ovules 2 or mostly 1 on each of the axile placentae; single style with lobed or toothed stigma; and fruit a baccate drupe. In xylem anatomy the *Aquifoliaceae* are about as primitive as the *Theoideae* and *Ternstroemioideae* but considerably less specialized than members of the other theaceous subfamilies.

Presumably the close relationship between the two families, which must have stemmed from common Cretaceous ancestry, has been obscured for most temperate taxonomists by their incomplete knowledge of the *Theaceae*, usually restricted to the commonly cultivated species with showy, many-stamened, bisexual flowers and capsular fruits. Actually definitive differences between members of the two families are relatively few, most evident in ovule and seed characteristics. The unitegmic ovules of the *Aquifoliaceae*, for example, develop into rather unspecialized seeds with a rudimentary embryo; whereas, the bitegmic ovules of the *Theaceae* develop into rather specialized seeds with well developed embryo and little or no endosperm. Examined species of *Ilex* lack the sclerenchymatous idioblasts so characteristic of theaceous species though species of both families have mucilaginous epidermal cells. Most hollies have leaves supplied with minute, mostly caducous stipules, thus differing from the exstipulate leaves of the *Theaceae*. Most *Theaceae* have bisexual flowers; most *Ilex* species are functionally dioecious or polygamo-dioecious. Most *Theaceae* have numerous stamens, fascicled or basally monadelphous; most hollies have stamens equal in number to the sepals and separate from one another. Hollies have ovules reduced to 1, less commonly 2, in each carpel; theaceous species mostly have more than 2 in each carpel. The two families thus are now quite distinct from each other but surely closely related within the same suborder. BAAS (1975), however, interprets the relationships of the *Aquifoliaceae* quite differently.

Sarraceniaceae. In view of DeBuhr's recent paper (1975) on the phylogenetic relationships of the *Sarraceniaceae*, no detailed discussion is needed here. I think he has rather definitively placed the *Sarraceniaceae* in its own suborder *Sarraceniineae* adjacent to the *Theineae* of the *Theales*. Although the evidence he presented for this relationship is rather overwhelming, I might add to it that there seems a well-marked tendency within the *Theales* toward leaf dimorphism and ascidiform structures, as displayed also in the saccate floral bracts of the *Marcgraviaceae* and well-known pitchers on many leaves of *Nepenthes* L., discussed below. It may be worth mentioning here too that the *Marcgraviaceae* and *Theaceae*, especially members of the *Bonnetioideae*, *Tetrameristoideae*, and *Kielmeyeroideae* (MAGUIRE 1972, MAGUIRE et al. 1972) are well represented on the Guayana Highlands with *Heliamphora* BENTH. of the *Sarraceniaceae*. It seems quite possible that this ancient area is the original cradle of the *Sarraceniaceae* and many, if not most, of the families of the closely related *Theineae*.

Nepenthaceae. Although I do not regard the *Sarraceniaceae* and the *Nepenthaceae*, the paleotropical analogs of the New World pitcher-plants, as families with direct close relationship, I do now include both in the *Theales* (THORNE 1976), the former as the *Sarraceniineae* adjacent to the *Theineae* (now including the *Clethrineae*) and the latter as the *Nepenthineae* between the *Scytopetalineae* and *Hypericineae*. *Nepenthes* species have some striking stimilarities to the *Dioncophyllaceae*, three monospecific genera of West African lianas, and somewhat fewer resemblances to the *Ancistrocladaceae* and *Hypericaceae* (including the often segregated *Clusiaceae*).

Among the likenesses between *Nepenthes* and the *Dioncophyllaceae* (AIRY SHAW 1952, METCALFE 1952, SCHMID 1964) are: the leaf-climbing lianous habit; alternate, exstipulate, entire, petiolate, often dimorphic leaves, some at least with midrib excurrent into two recurved hooks (*Dioncophyllaceae*) or prolonged into an ascidium-bearing "tendril" (*Nepenthes*); abundant development of glands; stratified phloem; peripheral rings of vascular bundles embedded in cortical fibers; fiber-tracheids; numerous ovules maturing into endospermous (but very different) seeds borne in loculicidally dehiscent capsules. Although some of these similarities may well represent convergences, there appear to be enough basic resemblances to indicate distant common origin despite the quite divergent flowers, pollen, placentation, and seeds.

Ancistrocladus WALL., sole genus of the *Ancistrocladaceae*, shows fewer similarities to *Nepenthes* but more with the *Dioncophyllaceae* (VAN STEENIS 1948, METCALFE 1952, ERDTMAN 1958, KENG 1967, 1970, GOTTWALD & PARAMESWARAN 1968). Aside from the overlapping ranges of the two groups in tropical West Africa, they retain in common: woody, climbing habit by non-homologous hooks; largely exstipulate, alternate, simple, petiolate, evergreen leaves with actinocytic stomata and sunken peltate glands; fiber-tracheids and other similar stem anatomical characteristics, as vascular strands embedded in cortical fibers; contorted petals; stamens mostly 10 in 2 whorls; similar pollen grains; and erect, turbinate embryo with thin leafy cotyledons embedded in starchy endosperm. The differences in calyx, gynoecium, placentation, and fruit indicate rather distant common ancestry, however, within the *Scytopetalineae* of the *Theales*. The suggested relationship of *Ancistrocladus* to the *Hugonieae* of the *Linaceae*, which I had previously accepted (THORNE 1968) seems less likely and probably the result of convergence.

Plumbaginaceae. A review of the similarities between the *Plumbaginaceae* and *Primulaceae* has convinced me that HUTCHINSON's treatment (1973) of the two families in one order, *Primulales*, most nearly fits the phylogenetic facts of any recent treatments. Hence, I have reduced the *Plumbaginales* to subordinal rank, *Plumbaginineae*, in my revised synopsis (THORNE 1976). Among the more convincing characteristics possessed in common by the two families are: similar habit, mostly perennial herbs or subshrubs with leaves mostly exstipulate, simple, and often rosulate; flowers actinomorphic, bisexual, pentamerous, bracteate, and commonly heterostylous; calyx gamosepalous and commonly persistent; corolla sympetalous, though rarely nearly or quite polypetalous, imbricate or convolute in vernation; stamens in a single whorl, mostly epipetalous and oppositipetalous, with the outer, oppositisepalous, whorl obsolete or reduced to staminodia; gynoecium largely superior, unilocular with free-central or free-basal placentation and ovules many to five (*Primulaceae*) or only one (*Plumbaginaceae*); ovules bitegmic, anatropous or semianatropous developing into seeds with straight embryo in copious endosperm; fruits various though sometimes a circumscissily dehiscent capsule (pyxidium) in both families; and similar flavonoids (HARBORNE 1967) and amyloid hemicelluloses in the seeds (BENTVELZEN 1962).

The differences between the two families are suffcient for the recognition of two suborders. The *Primulineae*, with the closely related *Myrsinaceae*, *Theophrastaceae*, and *Primulaceae*, are distinguished by their single style and one or usually more, mostly tenuinucellate ovules on a free-central or basal placenta. The monofamilial *Plumbaginineae* differ by their membranous calyx; wholly or partly distinct styles; solitary, crassinucellate ovule on a basal placenta; large embryo; embryological features; and distinctive pollen grains.

The recent attempts to link the *Primulaceae*, *Plumbaginaceae*, or both to the *Polygonaceae* and *Caryophyllaceae* of the *Chenopodiflorae* over-emphasize the convergences due to common possession of a unilocular ovary with free-central to basal placentation. The anatropous rather than campylotropous ovules, endosperm rather than perisperm, central rather than peripheral embryo, and non-centrospermous pollen grains, seeds, and chemistry of the *Primulales* negate any close common ancestry between the *Primulales* and *Chenopodiales*. The position of the *Polygonales* anent both the *Primulales* and *Chenopodiales* needs further, thorough study.

The relatively close common ancestry of the *Primulales* with the *Ebenales*, *Ericales*, and more primitive *Theales* within the superorder *Theiflorae* would seem to be too obvious to dwell upon here.

Malviflorae

There is much difference of opinion among phylogenists as to whether the *Sympetalae*, as traditionally treated by the Englerians, are a monophyletic group of related orders arising from immediate common ancestors or are a polyphyletic group of non- or only distantly-related orders that have converged in achieving an evolutionary plateau represented by a successful syndrome of evolutionary characteristics. I believe the *Sympetalae*, like the *Polypetalae*, *Apetalae*, *Amentiferae*, *Herbaceae*, and similar artificial groups, are polyphyletic. This is evident already from my treatment in the *Theiflorae* of the *Ericales*, *Ebenales*, and *Primulales*. Another group that I think has little to do with the *Gentianiflorae*, *Lamiiflorae*, and *Asteriflorae*, all indeed more or less related to one another through the more primitive *Rosiflorae*, is the order *Solanales*, which I place in the *Malviflorae* with the *Malvales*, *Urticales*, *Rhamnales*, *Euphorbiales*, and *Campanulales*. Although I have discussed this group briefly elsewhere (THORNE 1974 a), I should like here to consider the North American *Fouquieriineae* of the *Solanales*.

Fouquieriaceae. The largely Mexican *Fouquieriineae*, with the sole family *Fouquieriaceae* and sole genus *Fouquieria* KUNTH, have been ably monographed by HENRICKSON (1967, 1969, 1972, 1973). Because of its distinctive features and rather isolated phyletic position I have treated the family (THORNE 1968, 1976) in its own suborder, equivalent to the *Solanineae* and *Polemonineae*, in the *Solanales*. Despite its specialized habit, it retains from the common protosolanalean ancestors rather more primitive features than members of the other two suborders, among them: relatively unspecialized xylem characteristics; corolla lobes imbricate in bud; separate stamens, non-epipetalous (but filaments connected slightly to petal bases and falling with spent corolla), basically ten in one whorl (but up to 23 in some species); placentation basally axile but imperfectly so above with three septiform placentae nearly meeting but not united by their margins; and bitegmic ovules.

The closeness of relationship of the *Fouquieriineae* to the three families of the other two suborders varies from feature to feature, as should be expected from three distinct evolutionary lines diverging from common ancestors. In general, however, the *Fouquieriaceae* seem to retain more characteristics in common with the more primitive genera, like *Cantua* JUSS. ex LAM., *Cobaea* CAV., and *Huthia* BRAND., of the *Polemoniaceae* (GRANT 1959) than with members of the *Solanaceae* or *Convolvulaceae*. Among these more prominent features are: the shrubby or small tree habit; mostly alternate, simple, exstipulate, deciduous leaves with anomocytic stomata; showy, bisexual, actinomorphic, bracteose flowers with 5 imbricate, persistent sepals or calyx-lobes and 5-lobed, tubular corolla; 3-carpellate and basically 3-loculate gynoecium with 3-branched, single style; fruit a 3-valved, loculicidal capsule; and seeds flat, broadly winged, with firm-fleshy endosperm surrounding a well-developed embryo with spatulate or elliptic cotyledons and a small radicle. An undescribed shrubby genus of *Polemoniaceae* from Baja California under study by REID MORAN (personal communication) has somewhat the habit of *Fouquieria* with substantial woody stem, primary leaves persisting and becoming spinescent, and fascicled secondary leaves arising in the axils of the spines.

My colleague R. SCOGIN has made a thorough study of the comparative biochemistry of the *Fouquieriaceae* as compared with such putative relatives as the *Polemoniaceae*, *Solanaceae*, *Convolvulaceae*, *Tamaricaceae*, *Frankeniaceae*, and *Ericaceae*. His results (personal communication) show a striking correlation chemically between *Fouquieria* and the first three families but little correlation with members of the *Tamaricales* or *Ericales*. Among the chemical characteristics that he reports linking *Fouquieria* closely with other families in the *Solanales* are: presence of coumarins, as scopoletin and scopolin, a pelargonidin glucoside, an acylated anthocyanin (hyacinthin), and triterpene and steroidal saponins; and absence of proanthocyanidins, negatively charged flavonoids, the cyclical polyalcohol pinitol, and tannins. The primary chemical differences between *Fouquieria* and other members of the *Solanales*, according to SCOGIN, are the presence of ellagitannins and iridoids in and absence of alkaloids and methyl ethers of flavonols from *Fouquieria*. DAHLGREN et al. (1976), emphasizing the iridoids in *Fouquieria*, interpret the chemistry of the genus quite differently. It is noteworthy that all families of the *Solanales* are heavily developed in Mexico or the southwestern United States or both, especially *Fouquieria*, the arborescent species of *Ipomoea* L., and many *Polemoniaceae* and *Solanaceae*. *Ericaceae* usually are not found in association with *Fouquieria*.

I am somewhat puzzled by the supposed relationship claimed by some phylogenists between *Fouquieria* and the *Tamaricales* because I can see few resemblances between the *Fouquieriaceae* and the *Tamaricaceae* and *Frankeniaceae*. The *Tamaricaceae* are indigenous only to the Old World and the *Frankeniaceae* are rare within the New World range of *Fouquieria*. I presume those botanists seeing mutual relationships there must have been influenced by the incompletely axile placentation of the upper portion of the fouquieriaceous ovary, often described as parietal, a very widespread placental condition. SCOGIN's study, while dismissing the liklihood of relationship of *Fouquieria* with the *Tamaricales*, does indicate considerable relationship between the *Tamaricaceae* and *Frankeniaceae*. CAMPBELL & THOMSON (1976 and personal communication) have noted the remarkably similar structure of the salt glands of *Frankenia* L. and *Tamarix* L., whereas WALIA & KAPIL (1965) found the embryology of *Frankenia* to be quite different from that of members of the *Tamaricaceae*.

Rutiflorae

Although I have discussed the superorder *Rutiflorae* or some of its component taxa elsewhere (THORNE 1974 a, 1974 b, 1976), my recent transfer of the *Gyrostemonaceae* from the *Chenopodiflorae* to the *Rutiflorae* warrants some discussion here. The *Rutiflorae* are a distinct evolutionary line parallel to but not closely related to the *Theiflorae* and perhaps with closer common ancestry with the *Annoniflorae* than that between the *Theiflorae* and *Annoniflorae*. This is evidenced especially by the striking chemical similarities between the rather primitive *Rutaceae* and the *Annonales* and *Berberidales*. The essentially tropical, large order *Rutales*, including the *Sapindineae*, is especially noteworthy because of its inclusion of the anemophilous *Juglandineae* and probable close common origin with the also anemophilous *Myricales* and *Leitneriales*. The very strong tendency within this superorder toward anemophily has largely been overlooked by many phylogenists, who prefer to trace most of the old amentiferous families to the *Hamamelidiflorae*, which do indeed also display strong anemophilous tendencies.

Gyrostemonaceae. The recent purification of the *Chenopodiales* (*Centrospermae*) by the removal of extraneous unrelated taxa formerly forced unnaturally into dubious alliance with the centrosperms has led to some unexpected alignments. *Theligonum* L. has found a good home in the *Rubiaceae* (WUNDERLICH 1971); the *Primulaceae* and *Plumbaginaceae* are discussed above; *Batis* L. is discussed below; and *Simmondsia* NUTT. still defies all efforts to place it properly. A small Australian family, *Gyrostemonaceae*, earlier removed from the *Phytolaccaceae*, has lately been excluded, quite correctly, from the *Chenopodiales* (GOLDBLATT et al. 1976). Since we have had considerable field experience in Australia with this family and have rather good representation of the five genera in our herbarium, L. DeBUHR, recently returned from Western

Australia, and I have studied the group to see if we could place it with any of the other Australian families. We found that the *Gyrostemonaceae* possess many features in common with the largely Australian *Dodonaeeae* of the *Sapindaceae* and the West Australian *Stylobasium* DESF. (PRANCE 1965) and some features with the West Australian *Emblingia* F. MUELL. and the widely disjunct *Batis* L. We suspect that this group of taxa, mostly anemophilous shrubs or subshrubs of the more arid areas of Australia, form a related complex within or near the *Sapindaceae*. I have elsewhere (THORNE 1976) treated the *Gyrostemonaceae* as a family immediately adjacent to the *Sapindaceae*, including the *Dodonaeoideae* and *Stylobasioideae*, and suggested that LEINS (ERDTMAN et al. 1969) was probably correct in assigning *Emblingia* to this grouping. Once again anemophily in the *Rutiflorae* seems to be far more prevalent than previously realized.

Most of the highly distinctive combination of features that characterize the *Gyrostemonaceae* can be found among the *Sapindaceae*. In addition to the overlapping Australian ranges are: the primarily shrubby or subshrubby habit and stems with normal growth and moderately specialized stem anatomy; alternate, simple, entire, spatulate to commonly linear leaves with very small or no stipules and anomocytic stomata; apetalous, unisexual flowers (species monoecious or dioecious) with connate calyx 8–4–2-lobed or entire; stamens 6 to numerous on the rim or top of a discoid or convex receptacle, producing mostly isopolar, tricolpate, prolate-spheroidal to prolate pollen grains of medium size (PRIJANTO 1970 a); gynoecium of 1 or 2 to numerous (not more than 6 in *Sapindaceae*), connate carpels with more or less free styles; locules of ovary each with 1 axile, campylotropous, bitegmic ovule; fruit dry, an indehiscent nutlet or a schizocarp in which each membranous carpel dehisces and separates from the others and the discoid axis; and a curved seed with small, basal arillode. This distinctive combination of features, not duplicated in any one group of the *Sapindaceae*, plus the often numerous carpels, rather peculiar central, disk-like axis, and central, fleshy or oily endosperm does clearly separate the *Gyrostemonaceae* from the apparently closely related but exendospermous *Sapindaceae*.

Bataceae. This monogeneric family consists of two species, the dioecious American species *Batis maritima* L. and the Australian-Papuan monoecious species *B. argillicola* VAN ROYEN. Because of its usual association in coastal habitats with members of the *Chenopodiaceae* and because of its obvious succulence, *Batis* was long assumed to be centrospermous despite the absence of the usual centrospermous characteristics. The strikingly similar pollen grains of *Batis* and the *Gyrostemonaceae* have been noted by several palynologists (ERDTMAN 1952,

PRIJANTO 1970 a, 1970 b, NOWICKE in GOLDBLATT et al. 1976). Because of these palynological similarities and the reported presence of glucosinolates, DAHLGREN (1975) and GOLDBLATT et al. (1976) have placed the *Bataceae* and *Gyrostemonaceae* in the *Capparales*. Although I can see no more natural relationship of these families to the *Capparales* than to the *Chenopodiales*, there does seem to be great merit in the suggested relationship of the two families to each other.

Batis species resemble plants of the *Gyrostemonaceae* in: their unisexuality and presumed anemophily; shrubby habit with normal, relatively specialized stem anatomy; linear leaves with minute stipules; apetalous, unisexual flowers, some (*B. argillicola*) borne singly in the leaf axils; isopolar, tricolporoidate, subprolate to prolate, pertectate, non-baculate pollen grains of medium size (PRIJANTO 1970 b); pistil with one ovule in each locule; fruit a "septicidal berry" (fleshy capsule?) in *B. argillicola*. It is significant that the Australian-Papuan species (VAN ROYEN 1956, 1957) is considerably less specialized than the American species in its less tidal habitat, less sprawling habit of growth, monoecism, mostly solitary flowers, and free, septicidally-dehiscent fruits. The ultimate antecedents of *Batis maritima*, therefore, like those of the often associated *Dodonaea viscosa* L. (*Sapindaceae*), *Scaevola plumieri* (L.) VAHL (*Goodeniaceae*), and *Cassytha filiformis* L. (*Lauraceae*) of maritime tropical America, may well be Australasian.

Batis is separated from the *Gyrostemonaceae* by a number of distinctive features, such as: leaves with paracytic stomata; membranous spathella (or calyx) and spatulate, valvate tepals (or staminodia); 4 or 5 alternitepalous stamens; pollen grains with compound apertures without opercules; bicarpellate but tetralocular pistil; and anatropous, basal ovules ripening into exalbuminous seeds with a large, straight embryo with spatulate cotyledons. I now prefer tentatively to treat *Batis* as a separate family *Bataceae* and suborder *Batineae* following the *Gyrostemonaceae* and *Sapindaceae* in the *Sapindineae* of the *Rutales*. The absence of glucosinolates from most of the *Sapindineae* might be bothersome to some botanists, but myrosin cells, often associated with these sulphur compounds, have been reported in the sapindaceous *Bretschneidera* HEMSL. (SCHOLZ 1964). If one regards the differences between the *Bataceae* and *Sapindineae* as of the same approximate magnitude as those differences that separate the *Juglandineae* and *Myricales* or the *Anacardiaceae* and *Leitneriales*, the *Bataceae* then should be treated as the *Batales* following the *Rutales* with the *Myricales* and *Leitneriales* in the *Rutiflorae*. Certainly, this whole Australasian anemophilous complex (*Gyrostemonaceae, Stylobasium, Emblingia*, and *Batis*) should be examined carefully by experts to see if the apparent relationship with the *Sapindaceae* is truly the result of common ancestry.

Rosiflorae

The *Rosiflorae* are, like the *Annoniflorae* and *Hamamelidiflorae*, a phyletically important superorder which retains a large number of primitive features. It is central to the understanding of the relationships and origins of many other superorders of flowering plants. The *Rosiflorae* appear to be more closely related to the *Hamamelidiflorae* than to the *Annoniflorae*, but surely all three had at least distant common ancestors in early Cretaceous time. Space does not permit extended discussion of the *Rosiflorae*, but I wish to consider here one representative family, the *Crossosomataceae*, which I include in the *Rosineae* of the *Rosales* within this superorder, along with the *Pittosporales* and *Proteales*.

Crossomataceae. This small family of southwestern North America, containing probably only three species in two genera, has been variously placed by different botanists, usually in the *Rosales* near the *Rosaceae*, *Chrysobalanaceae*, *Connaraceae*, and *Fabaceae* or in the *Dilleniales* (*Dilliniineae* of my *Theales*) with the *Dilleniaceae* and *Paeoniaceae*. To resolve the question of relationship, TATSUNO (1976) has made a thorough study of the *Crossosomataceae* and its reputed relatives, emphasizing biochemistry. Her research did reach two important conclusions: a) that the new Arizona genus *Apacheria* MASON fits well chemically with *Crossosoma* NUTT.; and b) that stamen initiation and maturation in *Crossosoma* is centripetal, not centrifugal as reported previously. My conclusion from TATSUNO's evidence is that the *Crossosomataceae* must be retained in the *Rosineae* since they show overwhelming morphological and other similarities to the other families of that suborder. Among the more striking, non-chemical similarities found in the *Crossosomataceae* and various members of the *Rosaceae*, *Fabaceae*, *Connaraceae*, and *Chrysobalanaceae* are: woody habit, from small tree to low shrub, with relatively specialized wood (such as vessel elements with simple perforation plates and alternate intervascular pitting); leaves alternate, simple, and exstipulate with anomocytic stomata; flowers bisexual, perigynous, and basically pentamerous; androecium with many stamens (15–50), centripetally initiated; gynoecium of 5–3 distinct, stipitate carpels; ovules campylotropous, bitegmic, and crassinucellate with a multicellular archesporium; and fruits follicular producing black, reniform seeds with a fimbriate aril and a large, curved embryo in abundant endosperm. The relatively specialized stem anatomy, absence of fasciculate grouping of stamen traces, and the centripetal development of the stamens, among many other features, surely rule out close relationship with members of the *Dillenineae* of the *Theales*. Although one must view popular names with considerable skepticism, the only bona fide common name I have heard for a member of the *Crossosomataceae* is "wild-apple" for *Crossosoma californicum* NUTT. on Santa Catalina Island, where it is a frequent shrub and indigenous.

Buxaceae. I have long been fascinated by the disjunct, relictual, world distribution and primitive morphology of the *Buxaceae* and much puzzled by its probable relationships. It surely is an ancient family without close relatives. Overly impressed by their reported crotonoid pollen grains, I first placed the *Buxaceae* and the *Thymelaeaceae* (THORNE 1968) in the *Euphorbiales*. I have more recently (THORNE 1976) removed both families from the *Euphorbiales*, the *Thymelaeaceae* to the *Myrtales* and the *Buxaceae* to the *Pittosporales*. The former is still under study, but I wish to discuss the *Buxaceae* here.

The extremely primitive xylem anatomy of the *Buxaceae* combined with the somewhat specialized features of the flower is reminiscent of the rather similar combination in some other pittosporalean refugees from the *Euphorbiaceae*, as *Daphniphyllum* BLUME and *Balanops* BAILL. In comparing the *Buxaceae*, excluding the highly aberrant *Simmondsia* NUTT., with more primitive members of my *Pittosporales*, I found the following features to be retained in common: woody habit, small trees or shrubs, to perennial herbs; wood primitive with very small, medium to very long vessel elements, exclusively solitary and with 15–20-barred scalariform perforation plates, and very small, scalariform to opposite or transitional intervascular pitting; xylem parenchyma diffuse and apotracheal; rays markedly heterocellular, to 2 or 4 cells wide with numerous uniseriates; fiber-tracheids with large to small bordered pits and very short to moderately long; leaves evergreen, exstipulate, simple, mostly glabrous and entire; flowers mostly unisexual (species then usually monoecious), apetalous and sometimes naked, in tight bracteate racemes or spikes; sepals imbricate, usually 4, or absent; stamens 6–4 (rarely numerous) with large anther, dehiscing longitudinally; gynoecium syncarpous (usually abortive in or absent from male flowers) with 3 or 2 carpels and locules each with 2 anatropous ovules, bitegmic and crassinucellar, pendulous from the apex of each locule; fruit a loculicidally dehiscent capsule or a drupe; and seeds with straight, linear embryo in copious fleshy endosperm. Although tempted to combine the *Buxineae* and *Daphniphyllineae* into one suborder, I have tentatively retained the two as distinct because of the rather notable buxaceous features such as stomata surrounded by a rosette of subsidiary cells, predominant monoecism, axile placentation, and relatively well-developed embryo in the seed.

Simmondsia chinensis (LINK) SCHNEID., the popular jojoba of southwestern North America, has recently been studied by BROWN (1976), to try to ascertain biochemically and otherwise what its relationships to the

Buxaceae or other putative relatives might be. Although her biochemical studies strongly support the removal of *Simmondsia* from the *Buxaceae*, they did not place the genus definitively near any other putative relatives. Among the more striking features in which *Simmondsia* differs from the *Buxaceae* are: the relatively specialized vessel elements with simple perforation plates and alternate intervascular pitting; tracheids with large, distinctly bordered pits; anomalous growth from successive cambia; anomocytic stomata on both surfaces of the leaf; petiole with arcshaped vascular strand; anemophily, with no trace of nectaries nor odor; dioecism; tricolpate, oblate-spheroidal pollen grains; deciduous styles; absence of obturator and other embryological differences; large seeds with investing embryo, cotyledons with abundant wax, and little or no endosperm; and numerous biochemical differences as presence of leucocyanidin, leucodelphinidin, and tannins, and lack of alkaloids, and sinapic and ferulic acids. I have accordingly placed *Simmondsia* in my list of taxa incertae sedis.

Balanopaceae. *Balanops* has commonly been placed near the *Fagales* presumably due to the superficial resemblance of the drupe subtended by an involucre to the fagaceous acorn in an acorn-cup. Some time ago I removed *Balanops* from near the *Fagales* to my taxa incertae sedis (THORNE 1974 a). Later in examining our herbarium material in an effort to place the genus, I was impressed by its resemblance in clustered, pseudoverticillate, leathery leaves to species of *Daphniphyllum* (HUANG 1965). In drawing up a chart of likenesses between the two genera, I was astonished by the remarkable "fit" between them: Dioecious trees or shrubs; primitive xylem anatomy with vessel elements solitary, with many-barred (more than 20) scalariform perforation plates and rare scalariform intervascular pitting; narrow rays with numerous uniseriates composed of upright cells; cork sub-epidermal; leaves simple, exstipulate, alternate to subverticillate, mostly entire, with stomata only on the lower surface; flowers unisexual, apetalous or naked; staminate flowers with 6–3 or no sepals, stamens 14–2, usually 5 or 6, on short filaments, sometimes with apiculate connectives and sometimes with rudimentary ovary; pollen grains commonly tricolpate, suboblate, the colpi with granular membranes; female flowers with 6–3 or no sepals, staminodia small or absent, ovary imperfectly 4–2-loculate (placentation thus parietal) with 2 anatropous ovules in each locule; and fruit a drupe with 3–1 pyrenes. Surely the two genera must be treated in one order or preferably, I think, in one suborder, *Daphniphyllineae* of the *Pittosporales*, with rather close common origin with the *Buxineae* discussed above and less close common origin with the *Pittosporineae*. With the *Pittosporaceae* there is close resemblance often in habit, foliage, and parietal type of placentation. I do think *Balanops* and *Daphniphyllum* deserve their traditional recognition as monogeneric families because of notable differences in absence of sepals and staminodia from *Balanops*, deeply bipartite styles in *Balanops*, ovules (unitegmic and erect in *Balanops* but bitegmic and pendulous in *Daphniphyllum*), and seeds (rudimentary apical embryo in thick, fleshy endosperm in *Daphniphyllum* but large, straight embryo in sparse endosperm in *Balanops*). These are differences unexpected in a single family.

Daphniphyllum overlaps in range with members of the *Buxaceae* and *Pittosporaceae* and *Balanops* with the *Pittosporaceae*, but the Asiatic-Papuan *Daphniphyllum* is apparently separated from the Australasian-Fijian *Balanops* by at least the Torres Straight. One must admire the early phyletic perceptivity of BENTHAM & HOOKER, who apparently considered the *Buxaceae* and *Daphniphyllum* (their Family 153) and *Balanops* (Family 154) as related taxa (also treated by METCALFE & CHALK (1950) as adjacent families 244–246).

Asteriflorae

Because of our dissatisfaction with the failure of synantherological specialists to come up with a realistic classification of the *Asteraceae*, my colleague S. CARLQUIST and I have proceeded to publish (CARLQUIST 1976, THORNE 1976) a revised classification of the family with a redistribution of the tribes, placing the *Mutisieae*, *Vernonieae*, *Eupatorieae*, *Cardueae*, and *Arctoteae* into the *Cichorioideae* with the *Cichorieae*. The remaining tribes, clustered about the *Heliantheae*, are retained in the *Asteroideae*. I shall leave the justification for this reclassification to CARLQUIST (1976). In a recent paper WAGENITZ (1976) has independently divided the composite tribes into two groups closely resembling ours, though he places the *Eupatorieae* with the *Asteroideae*.

I wish here to give some of my reasons for insisting that the relationships of the *Asteraceae* are not be sought in the *Campanulaceae* nor *Rubiaceae*, as suggested by some syngenesiologists, but in the *Cornales* (see also WAGENITZ 1976), *Dipsacales*, and *Lamiales*, and ultimately for all these in the rosalean *Saxifragineae*. With the *Cornales*, *Dipsacales*, and *Lamiales* the *Asterales* share: primitively woody habit; exstipulate leaves; largely bisexual, tetracyclic flowers; locules mostly reduced to two or one with each commonly bearing a single, anatropous, unitegmic ovule; and marked tendencies toward the herbaceous habit, much dissected leaves, flowers crowded into bracteate capitula, floral dimorphy and zygomorphy, and maturation of indehiscent, dry, one-seeded fruits. With the *Cornales* and *Dipsacales* the *Asterales* share a floral tube adnate to the inferior ovary, calyx lobes greatly reduced or modified, and strong tendencies, as in some *Saxifragineae*, toward the neutralization, enlargement, or other modification of lateral flowers of the crowded inflorescence into a pseudanthium. The *Asteraceae* have advanced considerably over the *Cornales* in their sympetalous corolla, greater dimorphy and zygomorphy of the flowers, syngenesious stamens, and one-seeded cypselar fruits. They differ from the *Lamiales* in their consistent epigyny, cypselas, stamen isomery, and less consistent floral zygomorphy. The basal attachment of the single ovule in the ovary readily distinguishes the *Asteraceae* from members of the *Cornales* and *Dipsacales*.

Liliiflorae vs. Alismatiflorae

By selecting the generally accepted primitive features retained among the least specialized members of all the extant monocot super-orders one can derive a probable description of the common protomonocot ancestor of these groups: terrestrial, rhizomatous perennial with cambium lacking from stem; vessel elements, tracheid-like with scalariform end-plates, evolved from tracheids only in the roots; leaves alternate, simple, entire, broad-lanceolate, parallel-veined, with broad, sheathing base; flowers bisexual, actinomorphic, hypogynous, spirocyclic or pentacyclic, and entomophilous with parts separate and trimerous; perianth with two series of little differentiated, petaloid tepals; stamens six in two whorls, producing anasulcate pollen grains; nectaries probably lacking and flowers pollinated by beetles or other insects with chewing mouth-parts; three separate carpels with unsealed or lightly sealed stigmatic margins bearing bitegmic, crassinucellate, anatropous ovules; and fruits a whorl of three follicles maturing seeds with a rudimentary embryo in abundant non-starchy endosperm. Such a protomonocot would presumably be growing under warm, moist, shaded, forest condi-tions, possibly in riparian swamps but surely not in open water. This description fits more closely some extant forest liliads, aroids, or arecads, rather than *Sagittaria* L. (*Alismataceae*), *Aponogeton* L. f. (*Aponogeto-naceae*), or *Potamogeton* L. (*Potamogetonaceae*).

Some members of the *Alismatiflorae*, as *Butomus* L. (*Butomaceae*), *Aponogeton*, *Stratiotes* L. (*Hydrocharitaceae*), and the *Alismataceae*, retain a variable number of primitive features from their common ancestors. Among these retained characteristics are bisexual, hypogynous, trimerous flowers with separate parts; monosulcate pollen grains; carpels with many anatropous ovules on submarginal placentae; and follicular fruit (*Butomus*). Yet these same plants combine these primi-tive with rather many specialized features: palustrine or aquatic habitats and habits of growth; relatively specialized vessels, often with porous end-plates, in the roots; separate tepals well-differentiated into three outer greenish sepals and three inner white or colored petals; increased number of stamens and carpels, often mono-ovulate, on an elongated receptacle; pollen grains trinucleate at release; gynoecium paracarpous, epigynous, and with laminar placentation; ovules orthotropous or campylotropous; fruit usually one-seeded achenes or many-seeded berries; and mature seeds with enlarged, curved embryo and no endosperm. These are hardly characteristics of a "primitive" monocot. Their more completely aquatic relatives, like *Halophila* THOU. and *Thalassia* BANKS (*Hydrocharitaceae*), *Halodule* ENDL. and *Zanichellia* L. (*Zanichelliaceae*), *Ruppia* L. (*Potamogetonaceae*), and *Najas* L. (*Najadaceae*), are among the most specialized of all monocots, often devoid of vascular tissue, cuticle, and stomata; with flowers naked, unisexual, and reduced sometimes to a single stamen or pseudomonomerous pistil; and pollen grains nonaperturate and convervoid and pollination and dispersal hydrophilous. The mycophytic *Triuridales*, often considered related to the *Alismatales*, may have closer common ancestry with the *Liliiflorae* than with the *Alismatiflorae*.

Because of the apocarpy and other primitive features in the few alismatiflorean taxa mentioned above, many botanists have regarded this relatively specialized superorder as the most primitive group of Monocotyledoneae. Based upon this assumption, the *Alismatiflorae* have often been suggested as transitional between the monocots and the apocarpous *Nymphaeales* or *Berberidales*, possibly because they often share an aquatic or near-aquatic habitat. The similar modifications for successful adaptation to aquatic niches (laminar placentation or increased number of mono-ovulate carpels among them) and possibly the retention of some primitive features (including vessels restricted to the roots and apocarpy) are expected evolutionary convergences in these quite unrelated groups. To obviate any more attempts to use these aquatics as links between the dicots and monocots, I have in my recent synopsis (THORNE 1976) delayed consideration of the *Alismatiflorae* by placing in the basal position in the *Monocotyledoneae* (*Liliidae*) the *Liliiflorae* with the single order *Liliales* and the three suborders *Liliineae*, *Iridineae*, and *Orchidineae*. Some of the apocarpous liliads have many more primitive features than even the least specialized helobiads listed above. Furthermore, the liliads appear to represent the "main-stream" of monocot evolution; whereas, the alismatads (helobiads) are surely a palustrine-aquatic offshoot from the basic, terrestrial, protomonocot stock. Placing the helobiads first in the *Monocotyledoneae* has been as misleading as the Englerian placement of the specialized anemophilous orders in lead-off position for the *Dicotyledoneae*.

Commeliniflorae

In this context it probably is unnecessary to stress the commelinalean affinities of the *Juncaceae*, *Cyperaceae*, and *Poaceae*. In recent years, however, I have several times heard knowledgable botanists, who should have known better, refer to the lilialean ancestry of each of these families. The most conservative members of the *Commelinales* do share some of the characteristics listed above for the protomonocot, but even the least specialized are somewhat more advanced in the di- to hexaperigenous stomata (Fryns-Claessens and van Cotthem 1973); biseriate, well-differentiated perianth; triloculate gynoecium with axile placentation; mostly orthotropous ovules; and seeds with a mostly basal or lateral embryo in starchy, often mealy, endosperm or perisperm.

Juncaceae and *Cyperaceae*. Some of the primitive members of the *Juncaceae* (like *Thurnia* Hook. f.) and *Cyperaceae*, the two families constituting the *Juncineae*, like the primitive genera of the closely related *Bromeliineae*, *Pontederiaceae*, *Commeliniineae*, and *Eriocaulineae*, appear to be centered in tropical South America with special concentrations on the Guayana Highlands and Brazilian Planalto, usually most abundant in moist, open habitats. The *Juncaceae* and *Cyperaceae* are closely linked by their chromosomes with diffuse centromeres; pollen grains dispersed as tetrads or pseudomonads; typical commelinalean seeds with basal, lentiform or derived capitate embryo in abundant starchy endosperm; and overall specialization for anemophily.

Poaceae. Like the *Juncineae* and heavily Australasian *Flagellariineae*, the monofamilial *Poineae* are anemophilous. Which features are part of the anemophilous syndrome and which are retained from protocommelinalean ancestors are not always easy to determine. Presumably characteristics such as the growth habit, stomatal apparatus, primitive grass flower, pollen grains, seeds, and other embryological features, which are markedly commelinalean, are retained characteristics little tainted by the prevailing anemophily. The poaceous seed with marginal embryo lateral to the abundant horny and mealy, starchy endosperm in itself epitomizes the position of the *Poaceae* both as a member of the *Commelinales* and of the independent suborder *Poineae*.

I regret that space does not permit the presentation of arguments for the alignment of the other three hundred or so angiosperm families. This small sampling, however, may give some indications of the kinds of data and the philosophy upon which I have based my alignment of the superorders and their constituent orders, suborders, families, and subfamilies.

A number of my associates at the Rancho Santa Ana Botanic Garden have supplied data, specimens, other materials, and suggestions. I am especially grateful to Ron Scogin, Sherwin Carlquist, Larry DeBuhr, James Henrickson, Alicia Tatsuno, and Shannon Brown. Botanists at other institutions, too numerous to mention most of them, have over the years generously supplied information, separates, specimens, slides, preserved materials, and guidance in the field. Particularly helpful in this project have been REINO ALAVA, University of Turku; B. L. BURTT, Royal Botanic Garden, Edinburgh; ROLF DAHLGREN, Botanical Museum, University of Copenhagen; TH. ECKARDT, Berlin-Dahlem; ARTHUR GIBSON, University of Arizona; B. P. M. HYLAND, Atherton, Australia; HSUAN KENG, University of Singapore; JOHN PARHAM, Queensland State Herbarium, Brisbane; REID MORAN, San Diego Museum of Natural History; G. R. PROCTOR, Institute of Jamaica, Kingston; and RUDOLF SCHMID, University of California, Berkeley.

References

AIRY SHAW, H. K., 1952: On the *Dioncophyllaceae*, a remarkable new family of flowering plants. Kew Bull. **1951**, 327—347.

BAAS, P., 1975: Vegetative anatomy and the affinities of *Aquifoliaceae*, *Sphenostemon*, *Phelline*, and *Oncotheca*. Blumea **22**, 311—407.

BENTVELZEN, P. A. J., 1962: *Primulaceae*. Flora Males., ser. I, 6 (2), 173—192.

BROWN, S. C., 1976: Biochemistry of *Simmondsia chinensis*. M. A. Thesis, Claremont Grad. School, Claremont, Calif. 72 p.

CAMPBELL, N., and THOMSON, W. W., 1976: The ultrastructure of *Frankenia* salt glands. Ann. Bot. **40**, 681—686.

CARLQUIST, S., 1976: Tribal interrelationships and phylogeny of the *Asteraceae*. Aliso **8**, 465—492.

DAHLGREN, R., 1975: A system of classification of the angiosperms to be used to demonstrate the distribution of characters. Bot. Notiser **128**, 119—147.

— JENSEN, S. R., and NIELSEN, B. J., 1976: Iridoid compounds in *Fouquieriaceae* and notes on its possible affinities. Bot. Notiser **129**, 207—212.

DEBUHR, L. E., 1975: Phylogenetic relationships of the *Sarraceniaceae*. Taxon **24**, 297—306.

ERDTMAN, G., 1952: Pollen Morphology and Plant Taxonomy. An Introduction to Palynology. I. Angiosperms. Stockholm: Almqvist & WIKSELL.

— 1958: A note on the pollen morphology in the *Ancistrocladaceae* and *Dioncophyllaceae*. Veröffentlich. geobot. Inst. Rübel Zürich **33**, 47—49.

— LEINS, P., MELVILLE, R., and METCALFE, C. R., 1969: On the relationships of *Emblingia*. Bot. J. Linn. Soc. **62**, 169—186.

FRYNS-CLAESSENS, E., and VAN COTTHEM, W., 1973: A new classification of the ontogenetic types of stomata. Bot. Rev. **39**, 71—138.

GOLDBLATT, P., NOWICKE, J. W., MABRY, T. J., and BEHNKE, H.-D., 1976: *Gyrostemonaceae*: status and affinity. Bot. Notiser **129**, 201—206.

GOTTWALD, H., und PARAMESWARAN, N., 1968: Das sekundäre Xylem und die systematische Stellung der *Ancistrocladaceae* und *Dioncophyllaceae*. Bot. Jb. **88**, 49—69.

GRANT, V., 1959: Natural History of the *Phlox* Family. I. Systematic Botany. The Hague: Martinus Nijhoff.

HARBORNE, J. B., 1967: Comparative biochemistry of the flavonoids—IV. Correlations between chemistry, pollen morphology and systematics in the family *Plumbaginaceae*. Phytochem. 6, 1415—1428.

HENRICKSON, J. S., 1967: Pollen morphology of the *Fouquieriaceae*. Aliso **6**, 137—160.

— 1969: The succulent Fouquierias. Cactus and Succulent J. **41**, 178—184.

— 1972: A taxonomic revision of the *Fouquieriaceae*. Aliso **7**, 439—537.

— 1973: *Fouquieriaceae* DC. World Pollen and Spore Flora **1**, 1—12.

Huang, T. C., 1965: Monograph of *Daphniphyllum* (1). Taiwania **11**, 57—98.

Keng, H., 1967: Observations on *Ancistrocladus tectorius*. Gard. Bull. Singapore 22, 113—121.

— 1970: Further observations on *Ancistrocladus tectorius* (*Ancistrocladaceae*). Gard. Bull. Singapore **25**, 235—237.

Lindley, J., 1833: Nixus Plantarum. London.

Maguire, B., 1972: The botany of the Guayana Highland—Part IX. *Bonnetiaceae*. Mem. N. Y. Bot. Gard. **23**, 131—165.

— De Zeeuw, C., Huang, Y.-C., and Clare, C. C., jr., 1972: The botany of the Guayana Highland—Part IX. *Tetrameristaceae*. Mem. N. Y. Bot. Gard. **23**, 165—192.

Melchior, H. (Ed.), 1964: A. Engler's Syllabus der Pflanzenfamilien, 12. Aufl., Bd. II. Angiospermen. Berlin: Gebr. Borntraeger.

Metcalfe, C. R., 1952: The anatomical structure of the *Dioncophyllaceae* in relation to the taxonomic affinities of the family. Kew Bull. **1951**, 351—368.

— and Chalk, L., 1950: Anatomy of the Dicotyledons: Leaves, Stem, and Wood in Relation to Taxonomy, with Notes on Economic Uses. 2 vols. Oxford: Clarendon Press.

Prance, G. T., 1965: The systematic position of *Stylobasium* Desf. Bull. Jardin Bot. État Bruxelles **35**, 435—448.

Prijanto, B., 1970 a: *Gyrostemonaceae*. In: World Pollen Flora (Erdtman, G., Ed.), **2**, 5—13. Copenhagen: Munksgaard.

— 1970 b: *Batidaceae*. In: World Pollen Flora (Erdtman, G., Ed.), **3**, 5—11. Copenhagen: Munksgaard.

Royen, P. van, 1956: A new *Batidacea, Batis argillicola*. Nova Guinea, n. ser. **7**, 186—196.

— 1957: *Batidaceae*. Fl. Males., ser. I, **5**, 414—415.

Schmid, R., 1964: Die systematische Stellung der *Dioncophyllaceen*. Bot. Jb. **83**, 1—56.

Scholz, H., 1964: *Sapindales*. In: A. Engler's Syllabus der Pflanzenfamilien, 12. Aufl., Bd. II. Angiospermen (Melchior, H., Ed.), 277—288. Berlin: Gebr. Borntraeger.

Stafleu, F. A. (Chairman, Ed. Comm.), 1972: International Code of Botanical Nomenclature. Reg. Veget. **82**, 1—426.

Steenis, C. G. G. J. van, 1948: *Ancistrocladaceae*. Fl. Males., ser. I, **4** (1), 8—10.

Tatsuno, A. M., 1976: A biochemical profile of *Crossosomataceae*. M. A. Thesis, Claremont Grad. School, Claremont, Calif. 125 p.

Thorne, R. F., 1958: Some guiding principles of angiosperm phylogeny. Brittonia **10**, 72—77.

— 1963: Some problems and guiding principles of angiosperm phylogeny. Amer. Natural. **97**, 287—305.

— 1968: Synopsis of a putatively phylogenetic classification of the flowering plants. Aliso **6** (4), 57—66.

— 1973: Inclusion of the *Apiaceae* (*Umbelliferae*) in the *Araliaceae*. Notes Roy. Bot. Gard. Edinburgh **32**, 161—165.

— 1974 a: The "*Amentiferae*" or *Hamamelidae* as an artificial group: a summary statement. Brittonia **25**, 395—405 (for Oct.—Dec., 1973).

— 1974 b: *Sapindales*. Encycl. Brit., Ed. 15, **16**, 239—244.

— 1974 c: A phylogenetic classification of the *Annoniflorae*. Aliso **8**, 147—209.

THORNE, R. F., 1975: Angiosperm phylogeny and geography. Ann. Missouri Bot. Gard. **62**, 362—367.

— 1976: A phylogenetic classification of the *Angiospermae*. Evol. Biol. **9**, 35—106.

WAGENITZ, G., 1976: Systematics and phylogeny of the *Compositae (Asteraceae)*. Plant Syst. Evol. **125**, 29—46.

WALIA, K., and KAPIL, R. N., 1965: Embryology of *Frankenia* LINN. with some comments on the systematic position of the *Frankeniaceae*. Bot. Notiser **118**, 412—429.

WUNDERLICH, R., 1971: Die systematische Stellung von *Theligonum*. (Zugleich eine kritische Zusammenstellung einiger embryologischer, anatomischer und morphologischer Merkmale der *Rubiaceae*.) Österr. Bot. Z. **119**, 329—394.

Address of the author: Dr. ROBERT F. THORNE, Rancho Santa Ana Botanic Garden, Claremont, CA 91711, U.S.A.

Plant Syst. Evol. Suppl. 1, 321—347 (1977)
© by Springer-Verlag 1977

Mitteilungen aus dem Botanischen Museum der Universität Zürich Nr. 288

Evolutionary Trends in the *Hamamelidales-Fagales*-Group *

By

Peter K. Endress, Zürich

Abstract: Within the *Hamamelidales-Fagales*-complex putative evolutionary trends are outlined. Structure of inflorescences, flowers, pollen, fruits, and seeds, and peculiarities of the life history, particularly the general delay of ovule development and syngamy in relation to anthesis are demonstrated in the context of pollination and dispersal biology. The most outstanding trend with many correlated characteristics is exhibited by the adaptation to wind pollination in some groups of the *Hamamelidales* and especially in almost the whole of the *Fagales*. This seems to be one of the reasons for the dominant position of the *Fagales* in the vegetation of temperate regions.

Introduction

Ten years ago I have demonstrated similarities between the *Hamamelidales* and *Fagales* (ENDRESS 1967). They have turned out to be a surprisingly systematically homogeneous group. At the same time, it was a contribution to the complex problem of a monophyletic or pleiophyletic origin of the angiosperms, which had been discussed earlier with controversial interpretations. In the meantime, the view of the homogeneity of the *Hamamelidales-Fagales*-complex together with the idea of a monophyletic origin of the angiosperms has widely been accepted. The last decade is marked out by an intensive stimulation and increase of knowledge in the field of origin and macrosystematics of the angiosperms, mainly initiated by the works of CRONQUIST (1968) and TAKHTAJAN (1969). The increasing interest has led to the organization of several mutually stimulating symposia devoted to these problems, and also to further surveys and classification proposals by various authors.

The present paper gives a brief review of the actual knowledge of the group with main emphasis on putative evolutionary trends. It should be understood as a sketch with many hypothetical interpretations, which

---

* Dedicated to Professor Dr. FRIEDRICH MARKGRAF, in occasion of his 80th birthday.

may be corrected in the future by further investigations. Besides the already published data some unpublished results of the present author are included.

The group is conceived as containing the following families and subfamilies: *Hamamelidales*: *Hamamelidaceae* (*Disanthoideae, Hamamelidoideae, Exbucklandioideae, Liquidambaroideae, Rhodoleioideae*), *Platanaceae, Myrothamnaceae*; *Fagales*: *Betulaceae* (*Coryloideae, Betuloideae*), *Fagaceae* (*Fagoideae, Castaneoideae, Quercoideae*). In general, only literature of the last decade is cited. The older literature has been reviewed in the following publications: *Hamamelidaceae* (ENDRESS 1967), *Platanaceae* (HUTCHINSON 1967), *Myrothamnaceae* (JÄGER-ZÜRN 1966), *Betulaceae* (ENDRESS 1967, ABBE 1974), *Fagaceae* (ELIAS 1971, SOEPADMO 1972, ABBE 1974). A short review of the *Hamamelidales* is given by ENDRESS (1974 b), of the *Fagales* by ABBE (1974 b) and HJELMQVIST (1974).

General Similarities Between *Hamamelidales* and *Fagales*

Some of the most outstanding similarities will be listed (Fig. 1). The group consists of woody plants with relatively primitive secondary xylem (diffuse porous, vessel elements often with scalariform perforations, vessels thin-walled, angular and narrow), which on the whole is somewhat more advanced in the *Fagales* than in the *Hamamelidales* (see further SHIMAJI 1962, BAAS 1969, MOSELEY 1974, CARLQUIST 1976). The phyllotaxis is often distichous. The leaves are stipulate. In the deciduous taxa the venation is mostly craspedodromous (see further ENDRESS 1968, SKVORTSOVA 1975). In bud the position of the leaf blade with plicate vernation and the arrangement of the stipules may be identical (Fig. 1). The midrib may have the same configuration of vasculature (Fig. 1). Massive, multicellular glandular hairs may be present on the petiole (Fig. 1). Expanding buds often show epinastic movements. Sieve-tube plastids are uniformly of the S-type (BEHNKE 1973). Chemically, the common accumulation of polyphenols (quercetin,

Fig. 1. Some of the most striking structural similarities between *Hamamelidales* and *Fagales*. Stipulate leaf with craspedodromous venation. Vasculature pattern in the midrib. Glandular hairs on the petiole. Folding of leaf blade in bud. Diagram of bud with distichous phyllotaxis, orientation of leaf blade and of adaxial and abaxial stipules. Secondary xylem with scalariform perforation plates in vessels and alternate intervascular pitting. Transverse section of young ovary with identical position of the ovules. Inflorescences: epinastic spikes, terminating a vegetative shoot (from ENDRESS 1967, except SEM photographs, leaf of *Corylus ferox*, and shoot diagram, which are original)

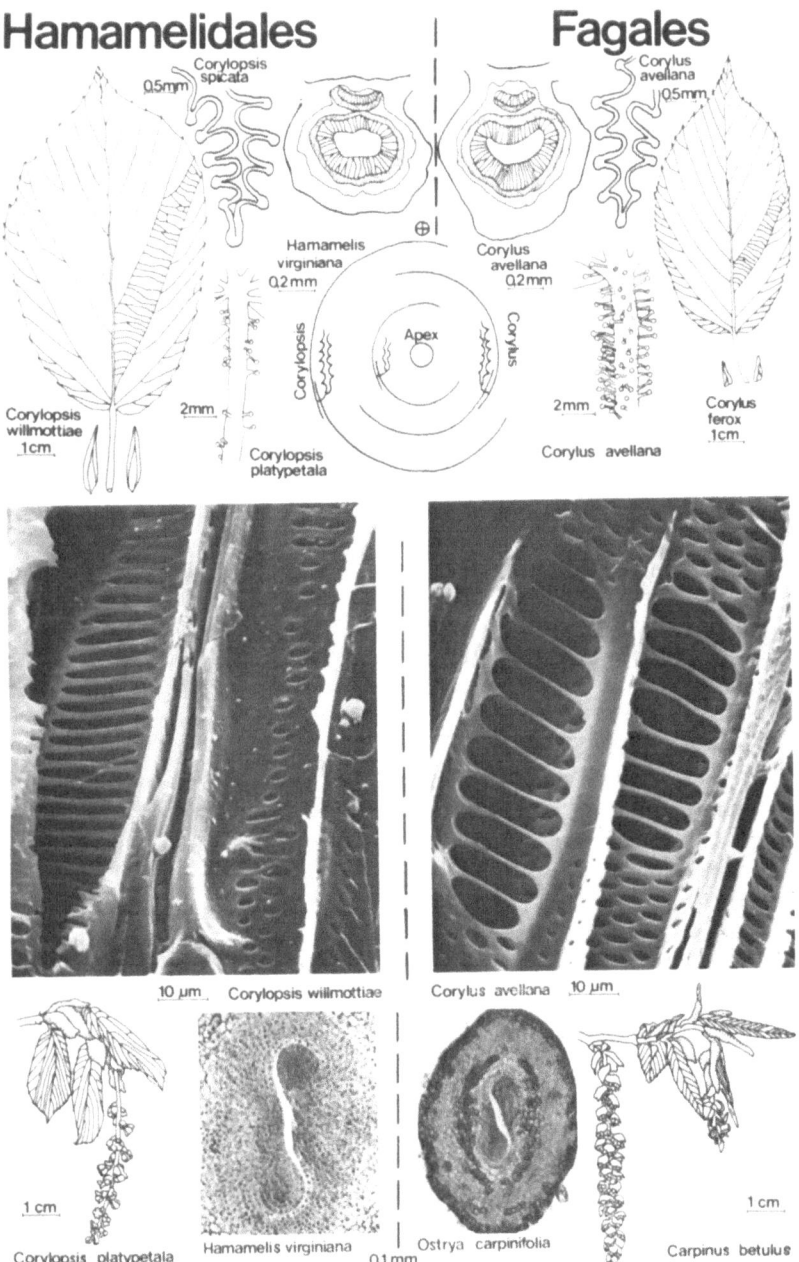

Fig. 1

	Rhodo-leioideae	Disan-thoideae	Hamamelidoideae				
Pollinators	Birds	Flies ———————→	Various insects (few specialized)				Wind or insects
Pollination unit	Rhodoleia inflorescence	Disanthus flower ———→	Hamamelis	Corylopsis inflor. flower	Fothergilla	Parrotiopsis inflorescence	Parrotia/Sycopsis
Sex distribution	☿ ———→						☿ + ♂
Visual attractants	Petals (red) ————————————→		(yellow-red) +shiny sepals	(yellow) +yellow bracts	Filaments (thickened, white)	Bracts (enlarged, white) +yellow stamens	Stamens (orange-red)
Kind of food as attractant	Nectar (at stamen base)	(at petal base)	(on stami-nodes)	(on scales between stam.+ovary)	Pollen ——————————————————→		
Stigma	small ——→						large ——
Anthers	large	small —————————————————————————————————————→					large ——
Stamens/flower	± 10	5	4	5	12-32	15-20	9-14/6-10
Flowers/inflor	5-10	2	3	3-30	20-30	10-20	4-8

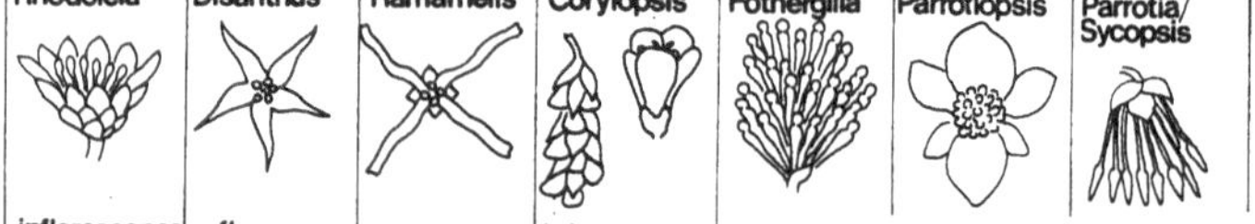

Fig. 2. Structural base of divergence

myricetin, ellagic acid, leucoanthocyans) (Hegnauer, cited in Endress 1967, Jay 1968, Mears 1973, Lebreton 1976) is noteworthy. The inflorescences are basically racemous, often spikes or pendulous catkins, which may be enriched by further dichasial ramification (also in *Hamamelidaceae*: Endress 1967). The ontogeny of unisexual inflorescences may be markedly different for males and females (Fig. 10). The floral structure, especially the gynoecium, shows an astonishing congruence in representatives of both orders, when the whole ontogeny from floral initiation up to fruiting is considered (Figs. 8, 9). Further similarities and various evolutionary trends of the reproductive organs are presented in the following sections.

Pollination Biology — Inflorescence, Flower and Pollen Structure

In contrast to the relatively uniform ground plan (Figs. 3, 4) the external appearance of the inflorescences and flowers varies considerably within the group, especially in the *Hamamelidales* (Endress 1967, 1970, 1974 a, Bogle 1970). Here, obviously, an adaptive radiation has led to mainly bee, fly, bird, and wind pollinated flowers.

The basic inflorescence type in *Hamamelidaceae* is a spike or a system of several spikes. Only in *Distylium*, *Distyliopsis*, and *Matudaea* this system of spikes is reduced to a system of single flowers, arranged in a terminated raceme or a panicle. The possibility of such an evo-

	+Myrotham-naceae	Liquidam-baroideae + Platanaceae	Betulaceae	Fagoideae		Quercoideae	Casta-neoideae
Wind						►	Insects
	Sinowilsonia	Distylium Myrothamnus Liquidambar Platanus		Nothofagus Fagus			
	♂	♂	♂	♂	♂	♂	♂
			►	flower	inflorescence		►
	♂ ♀	♀♂ / ♂♀	♂♀				►
						Mass of filaments	
						Nectar (Castanea)	
							small
		►	medium	large	►	medium	small
5	1-8/3-8	6-10/3-5	1-4	8-90	±6	►	±12
±60	1-30/10-30	±20/±100	♂>100	♂1-3	♂2-20	♂20->100	♂>100

in pollination biology (Original)

lutionary direction has largely been neglected so far in the literature (ENDRESS 1969, 1970, 1971).

Largely unspecialized insect (often bee) pollinated flowers as in *Corylopsis* are medium sized, bisexual, pentamerous, diplo- or haplostemonous, nectariferous (on scales surrounding the ovary base), and bear a double perianth with petals of a conventional shape. The general appearance resembles flowers of for example, the genus *Saxifraga*.

From such forms other types can be derived. *Hamamelis* and *Disanthus* are adapted to fly pollination. The flowers are relatively small. The long, ribbon-like petals are greenish, yellow or red. Shiny surfaces have evolved: in *Hamamelis* on the exposed inside of the sepals and nectariferous staminodia, in *Disanthus* on the nectariferous base of the petals. An unpleasant floral smell is common in both genera. The flowers of other genera closely related to *Hamamelis*, such as *Loropetalum* and *Trichocladus* are apparently not adapted to fly pollination.

A completely other type of insect (mainly bee and bumble-bee) pollinated flowers has evolved in the genus *Fothergilla* with its bottle-brush-like inflorescences. The insects feed on pollen. No nectar is produced. The corolla is totally lacking. The calyx is minute. At the same time, the stamen filaments are thickened beneath the anthers, they are white and showy. Furthermore, the stamen number is increased (Fig. 5). Thereby, two different processes are apparently involved: an increase of the floral sectors to mostly 7, and an increase of the

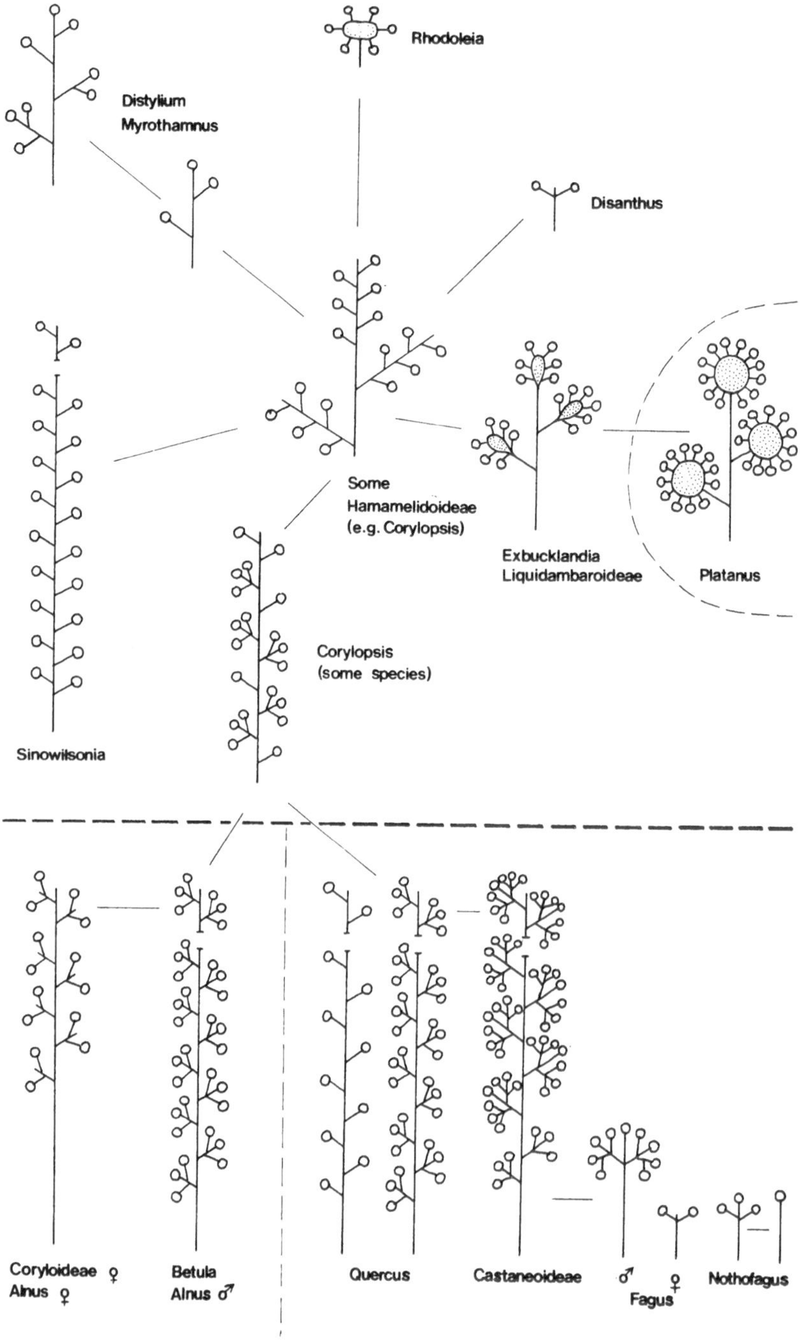

P. K. Endress:

Rhodoleia

Distylium
Myrothamnus

Disanthus

Some
Hamamelidoideae
(e.g. Corylopsis)

Exbucklandia
Liquidambaroideae   Platanus

Corylopsis
(some species)

Sinowilsonia

Coryloideae ♀   Betula   Quercus   Castaneoideae   ♂   ♀   Nothofagus
Alnus ♀   Alnus ♂   Fagus

Fig. 3. Inflorescences (Original, partly after Endress 1967, 1970)

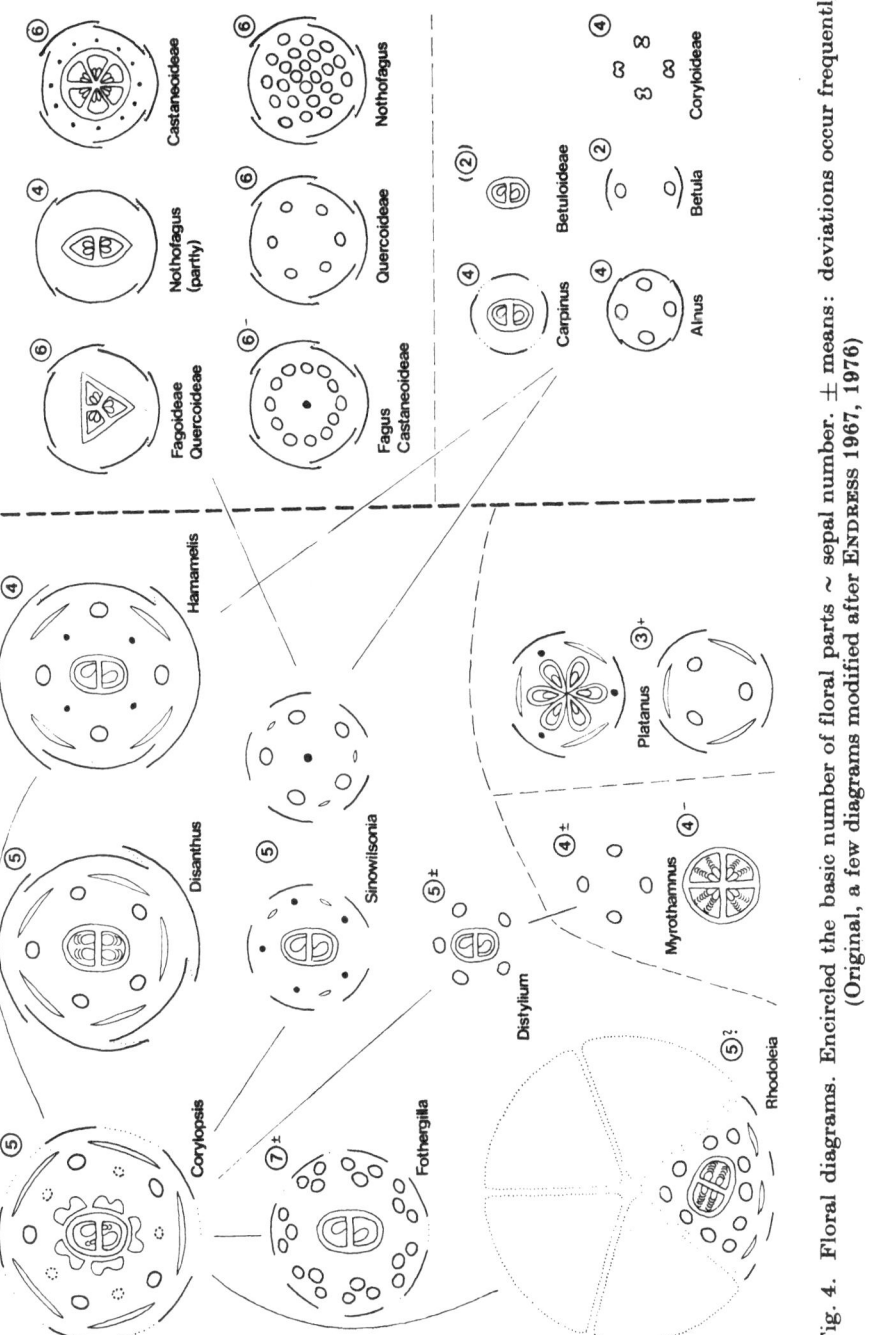

Fig. 4. Floral diagrams. Encircled the basic number of floral parts ~ sepal number. ± means: deviations occur frequently (Original, a few diagrams modified after ENDRESS 1967, 1976)

stamens per sector to 3 or 4. The following considerations favour the hypothesis of a secondary polyandry (ENDRESS 1976): 1. The onto-genetical sequence of the stamen initiation is centrifugal. 2. The initiation pattern in the other polyandrous genus of the family, *Matudaea*, is completely different. Therefore, no "common polyandrous ground plan" of the hamamelidaceous flower has to be supposed. 3. The stamen groups are opposite to the sepals, which suggest a loss of a whorl of petals. 4. A secondary increase in stamen number favoured the evolu-

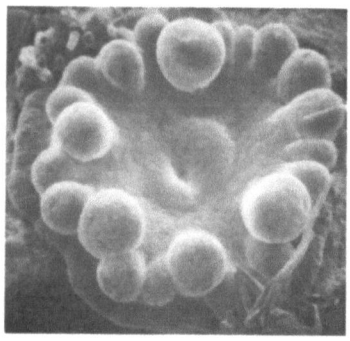
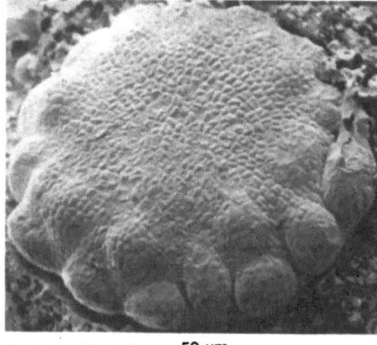

Fothergilla major   100 μm   Matudaea trinervia   50 μm

Fig. 5. Divergent androecium initiation patterns in polyandrous *Hamame-lidaceae* (from ENDRESS 1976)

tion of the special pollination syndrome by an improvement of the visual attraction apparatus and an increased food supply for pollinators.

Parrotiopsis, closely related to *Fothergilla*, has less showy stamens, somewhat fewer in number, without thickened filaments. Visual attractants are, in addition, enlarged white bracts subtending the yellow inflorescences and thus forming a pseudanthium.

Another type of pseudanthium, still more elaborated, has evolved in *Rhodoleia*. Here, the flowers are arranged in heads. Petals are only formed at the periphery of the inflorescence. This is the only case of an extreme floral zygomorphy in the *Hamamelidales-Fagales*-complex. The inflorescence is subtended by brown scales. It resembles a large red flower and is comparable to the capitulum of *Compositae*. Nectar is produced abundantly. Birds have been observed as visitors of these flowers.

Many genera are adapted to wind pollination. As a rule, the flowers are unisexual, apetalous, inconspicuous. The stigmatic surface is large (Fig. 8). The pollen production is high, effected by an enlargement of the anthers (Fig. 6), by an increase in the stamen number per flower (e.g. *Parrotia*, *Matudaea*) or by an increase in the flower number per inflorescence (e.g. *Sinowilsonia*). As already discussed, in the highly polyandrous genus *Matudaea* the stamen initiation pattern is completely different from *Fothergilla*. In *Parrotia*, *Sycopsis*, *Distylium*, and *Matudaea* with incomplete sexual separation, pollination may be effected by both wind or insects. In *Distylium* and *Distyliopsis* the flowers are completely devoid of a perianth, but they are clearly definable (ENDRESS 1970, in contrast to BOGLE 1970, MEEUSE 1975). *Parrotia* and *Sycopsis* have pendent flowering heads with long, but only slightly flexible filaments. *Sinowilsonia* shows unisexual pendulous catkins, which are only slightly flexible. The closely related genus *Fortunearia* is somewhat less specialized, the sexual separation less marked, the inflorescences erect, the flowers with rudimentary petals, inconspicuous, and at least partly wind pollinated.

In the *Liquidambaroideae* the flowers are densely arranged in unisexual, globose heads. A perianth is almost totally lacking.

The inflorescences of the *Platanaceae* are very similar. However, the flowers show a double, although inconspicuous perianth, and 3 to 9 free instead of 2 united carpels (see following section).

The *Myrothamnaceae* closely resembles the genus *Distylium* in general floral appearance. The flowers are without a perianth, the number of floral parts is highly variable, even the inflorescence structure is similar (JÄGER-ZÜRN 1966).

The *Fagales* exhibit a simplification of flowers and specialization of inflorescences as general trends connected with their extreme adaptation to wind pollination. In the *Betulaceae* (ENDRESS 1967, MACDONALD 1971, SATTLER 1973, ABBE 1974 a, KORCHAGINA 1974) the male inflorescences are lax, pendulous, many-flowered catkins, the female ones pendent or erect small spikes. The flowers are minute, but increased in number, especially in the male inflorescences, firstly, by the occurrence of dichasia of threes or twos instead of single flowers, and secondly, by an increased number of primary flowers as well. It is interesting to note that also *Corylopsis* of the *Hamamelidaceae* sometimes shows trimerous dichasia, moreover with the same position of the 2-merous gynoecium in respect to the bract as in *Carpinus* (ENDRESS 1967). *Corylus* exhibits the most extreme specialization and divergent evolution of male and female inflorescences. At anthesis, the female ones do not differ superficially from purely vegetative buds, with the exception of the exserted stigmas of the 6 to 10 flowers, whereas the male catkins in *Corylus avellana* bear more than 100 flowers. In the betulaceous flower, which is always unisexual, the simple perianth is 4- or 2-merous or even lacking. The tepals correspond to the sepals of the *Hamame-*

lidaceae. The androecium contains 4 to 1 stamens, opposite the sepals. The gynoecium is always 2-merous.

In the *Fagaceae* (SOEPADMO 1972, ABBE 1974 a), the *Fagoideae* and *Quercoideae* are generally wind pollinated, the *Castaneoideae* at least partly insect pollinated. In the wind pollinated genera, the male inflorescences, catkins or heads (*Fagus*) or at least the filaments (*Nothofagus*) are lax and pendulous; the anthers and the stigma tend to be large. In the insect pollinated taxa, the inflorescences are stiff and erect; nectaries may be present (*Castanea*); the anthers tend to be very small, dorsifixed and versatile, the stigma seems to be extremely small. The flowers are basically 6-merous, not 3-merous as is usually believed, because, firstly, male flowers mostly have 12 stamens in a diplostemonous arrangement, secondly, the perianth is uniform, all parts behave ontogenetically like the sepals in *Hamamelidaceae*, thirdly, also in a 5-merous or 4-merous flower the sepals usually are not arranged in one whorl, but in a spiral or they are decussate (see e.g. *Disanthus*, *Hamamelis*, Fig. 4). I suppose that the 6-merous condition is derived from a 5-merous one. The increase in stamen number by this process (see also *Fothergilla* of the *Hamamelidaceae*) may again be favourable in pollination biology in different ways. In *Nothofagus* with only few-flowered inflorescences the stamen number may be increased up to 90 per flower in exceptional cases. In contrary, the catkins of the *Quercoideae* have only 6-staminate flowers, but instead of single flowers there are often many dichasial clusters composing the catkin, which increase the stamen number considerably, up to 70 per cluster.

Unexpectedly, in the insect pollinated *Castaneoideae* the stamen number per cluster may even be higher. But this seems to be compatible with the absence of petals. Again, bottlebrush-like inflorescences have evolved, in which the showyness is effected by the dense mass of stamen filaments in many hundreds of flowers. In compensation, the anthers are extremely small. These extreme features: the many-flowered dichasial clusters, the minute anthers, the unisexual flowers (present in all *Fagales*) and the pluriovulate but one-seeded fruits (see following section) could favour the hypothesis of a secondary reversal to entomophily in the *Castaneoideae*.

Bracts, sepals and cupule. In many *Hamamelidaceae* and *Fagaceae* the young flowers are protected until anthesis by their own sepals, in the beginning also by their bracts. In other *Hamamelidaceae* (*Corylopsis, Fothergilla, Parrotiopsis, Parrotia, Sycopsis, Distylium*, etc.), in the *Myrothamnaceae*, and in most *Betulaceae* the sepals are much reduced or even lacking. Here, the floral bracts ("Tragblätter") have taken over the protective function up to the last stages before anthesis (ENDRESS 1975). In some *Betulaceae* (*Alnus, Betula, Corylus*) they are highly specialized to "hypopeltate" organs. All these genera are characterized by dense inflorescences, often catkins. In *Betulaceae*, these peltate scales have taken over an additional function as an arresting mechanism for the released pollen grains (FAEGRI and VAN DER PIJL 1971).

Another specialization is found in the *Liquidambaroideae* and *Platanaceae*. Here, both sepals and bracts, of the single flowers are reduced. But the sterile apices of the anthers are broad and massive. They are contiguous and provide some kind of protection in the last developmental stages before anthesis for the fertile, lower parts of the anthers in the male inflorescences (ENDRESS 1975).

In the female flowers of the *Fagaceae*, the cupule is an additional protective organ, which may also have certain functions in fruit dispersal. The diversity in its evolution has rendered the cupule an important organ in the classification of the family (see e.g. FORMAN 1966).

Petals. In the *Hamamelidaceae* various stages of reduction can be observed. *Corylopsis* and *Rhodoleia* show petals of a normal shape. In *Disanthus, Hamamelis, Loropetalum*, etc., they are narrow and ribbon-like. *Fortunearia* and *Sinowilsonia* have tiny, filamentous petals, often lacking in *Sinowilsonia*. Many other *Hamamelidaceae*, the *Myrothamnaceae*, and the *Fagales* have apparently lost their petals.

Stamens. The *Hamamelidales* uniformly show an apical connective protrusion of various length (Fig. 6) and mostly basifixed anthers. The same features are realized in *Nothofagus*. In the insect pollinated *Hamamelidaceae* and in the *Platanaceae* the anthers open by means of valves—a very unusual character in angiosperms. Most of the wind pollinated *Hamamelidaceae*, the *Myrothamnaceae*, and the *Fagales* have simple slits. In the *Coryloideae, Nothofagus* species, and certain *Hamamelidaceae* with a reduced calyx (*Matudaea, Distylium*), the anthers are hairy. Generally, large anthers have evolved in wind-pollinated genera, whereas small anthers are present in insect pollinated groups.

Pollen grains exhibit a wide range of exine structure and aperture patterns, especially in the *Hamamelidaceae* (CHANG 1964, WALKER and DOYLE 1976, preliminary personal SEM observations, see Fig. 7). Surveys on the *Fagales* are given by TAKEOKA and STIX 1963, KUPRIANOVA 1965, BURRICHTER et al. 1968, STRAKA 1975, SUROVA 1975, HANKS and FAIRBROTHERS 1976). The distribution of the types shows correlations with the general degree of advancement ot the genera, in particular with the pollination type (general discussion in WHITEHEAD 1969).

In the *Hamamelidaceae* the insect pollinated groups have tricolpate pollen with a reticulate surface pattern and with long and often granulate colpi. The meshes of the reticulum are especially large in *Disanthus, Exbucklandia*, and *Dicoryphe*. In the wind pollinated genera, there

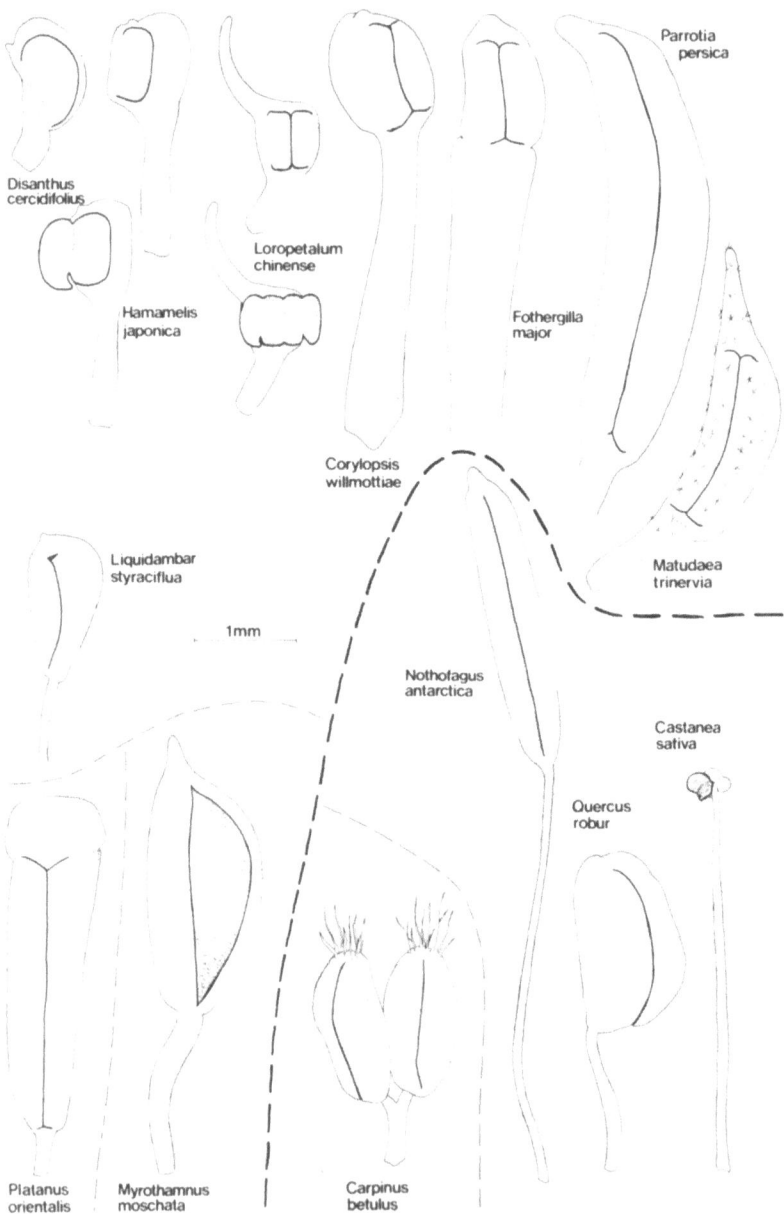

Fig. 6. Stamens. Form, size, and dehiscence patterns. Dehiscing lines marked with thick lines (Original, *Myrothamnus* after JÄGER-ZÜRN 1966)

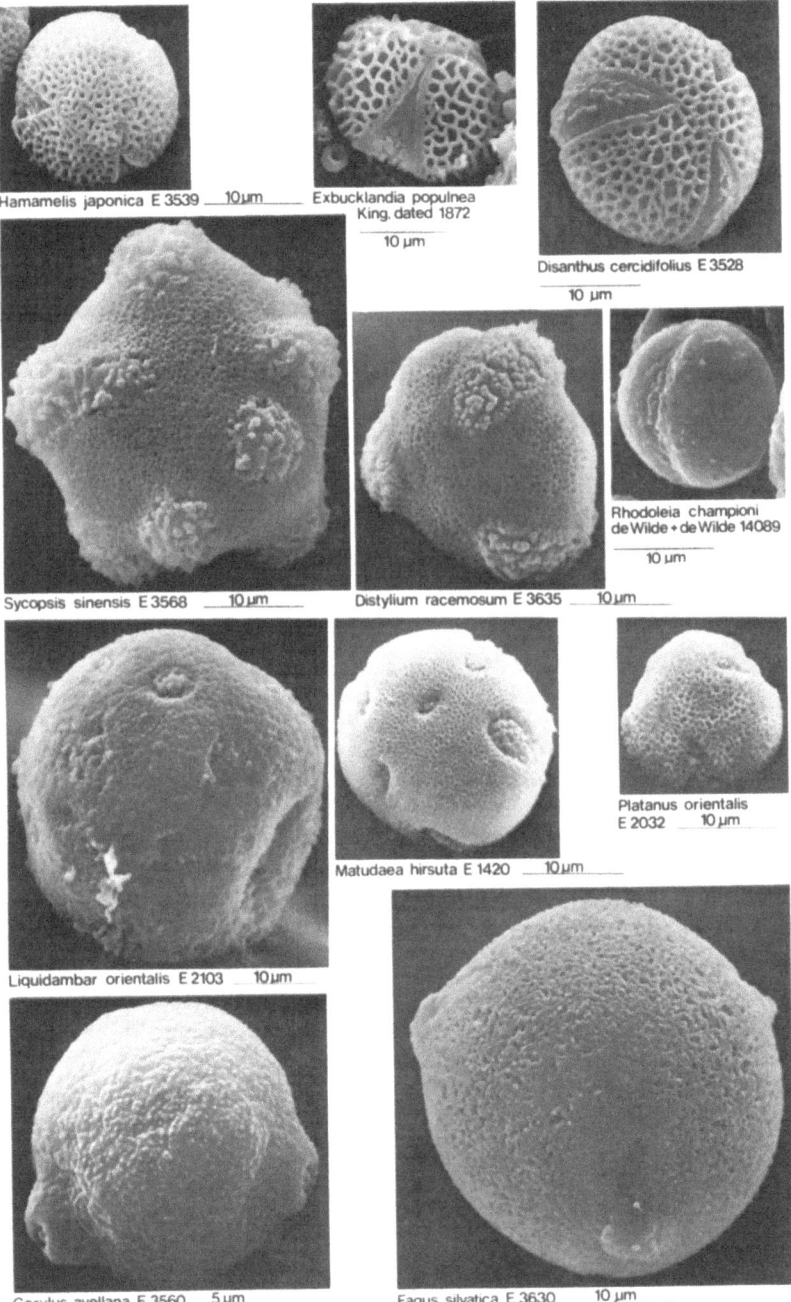

Hamamelis japonica E 3539 ___10μm___ Exbucklandia populnea King, dated 1872
10 μm

Disanthus cercidifolius E 3528
10 μm

Rhodoleia championi de Wilde + de Wilde 14089
10 μm

Sycopsis sinensis E 3568 ___10μm___ Distylium racemosum E 3635 ___10μm___

Platanus orientalis E 2032 ___10μm___

Matudaea hirsuta E 1420 ___10μm___

Liquidambar orientalis E 2103 ___10μm___

Corylus avellana E 3560 5μm   Fagus silvatica E 3630   10 μm

Fig. 7. Pollen (Original). Voucher numbers indicated (E = ENDRESS)

is a clear tendency towards a decrease in the diameter of the surface meshes (*Distylium, Sycopsis, Matudaea*, etc.) or even to a nearly imperforate tectum forming a smooth surface (*Liquidambar*), which is also realized predominantly in the *Fagales*. A second trend is shown in the shortening of the colpi (e.g. *Distylium*) and at the same time in an increase in number of the apertures, as e.g. in *Sycopsis, Matudaea*, and *Liquidambar* with polyrugate or polyporate pollen grains. *Rhodoleia* superficially resembles the 3-colporate and smooth-surfaced *Fagaceae*. The *Platanaceae* are similar to the tricolpate and reticulate *Hamamelidaceae*. The *Myrothamnaceae* differ by having tetrad pollen. The strongest adaptation to wind pollination has apparently occurred in the pollen grains of the *Betulaceae*. They are 3- (or more) -pororate and vestibulate (anguloaperturate) which makes them more buoyant. (Also some *Hamamelidaceae* show a tendency towards anguloaperturate grains, e.g. *Distylium* and *Sycopsis*.) The psilate, smooth surface in *Betulaceae* is accompanied by a granulate, and not columellate, infratectal exine structure, a very rare and scattered feature in angiosperms (VAN CAMPO and LUGARDON 1973, DOYLE, VAN CAMPO and LUGARDON 1975, WALKER 1976).

Dispersal Biology — Gynoecium and Fruit Structure

In the whole group the gynoecium structure is rather uniform (Figs. 1, 8, 9). In the *Hamamelidaceae* the gynoecium is 2-merous, the ovary syncarpous and 2-locular, the styles and stigmas free. The position of the ovary varies between superior and inferior. Each carpel is markedly ascidiform. In contrast to certain other groups of the angiosperms, the ascidiform part begins to form only after the ovule initiation (ENDRESS 1967, 1974 a). It widens as the ovule(s) enlarge(s). It produces a downward prolongation of the ovarial cavity giving space to the developing pendulous ovule(s). It is relatively less marked in the many-ovulate gynoecia. The ovules, if one per carpel, are pendulous, diagonally positioned in the two carpels (Figs. 1, 9), emerging immediately above the upper end of the ascidiform region. The placentation is very uniform. However, there is some confusion between axile and parietal (ENDRESS 1967, WEAVER 1969, BOGLE 1970). In some genera the ovary development is markedly delayed at anthesis (see following section), and therefore the ventral slit of the carpels still open. But up to the stage of syngamy the ventral slit has generally closed, and hence, the placentation is clearly axile.

Variations in gynoecium structure are exhibited by the stigmatic surface area (see third section). It is restricted to the stylar tip in entomophilous genera, but covers the whole ventral side of the styles in anemophilous taxa (Fig. 8). Another kind of variation is represented by the ovular number. An intermediate number is shown by *Disanthus*, *Chunia*, and *Exbucklandia* (6), *Neostrearia* (3), and *Corylopsis* (3, 2 of them reduced). *Rhodoleia* and the *Liquidambaroideae* develop many ovules per carpel in an acropetal succession. In most or all other genera, there is only one ovule in a carpel.

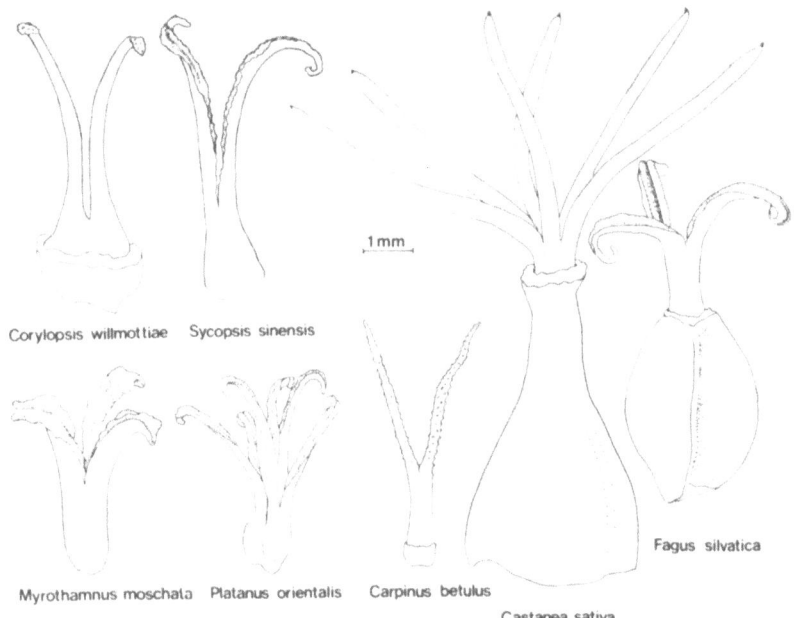

Fig. 8. Gynoecium structure. Stigmas stippled (Original, *Myrothamnus* after Jäger-Zürn 1966)

Also in the multiovulate genera only one or few ovules ripen in each carpel. But the seeds are relatively small, the testa thin (Melikian 1973, Mohana Rao 1974), and often more or less winged. Here, wind dispersal may play a role. In the genera with uniovulate carpels a uniformly, very elaborated and peculiar ejection mechanism for a short range dispersal has evolved. The ripe thick-coated, smooth-surfaced, and very hard seeds are shot out of the loculicidal and ventricidal capsules for a distance of several meters. As many representatives of the family occur on river banks, flowing water is presumably an additional dispersal agent for longer distances. The seed surface is water repellent and the seeds float freely on the water surface for several

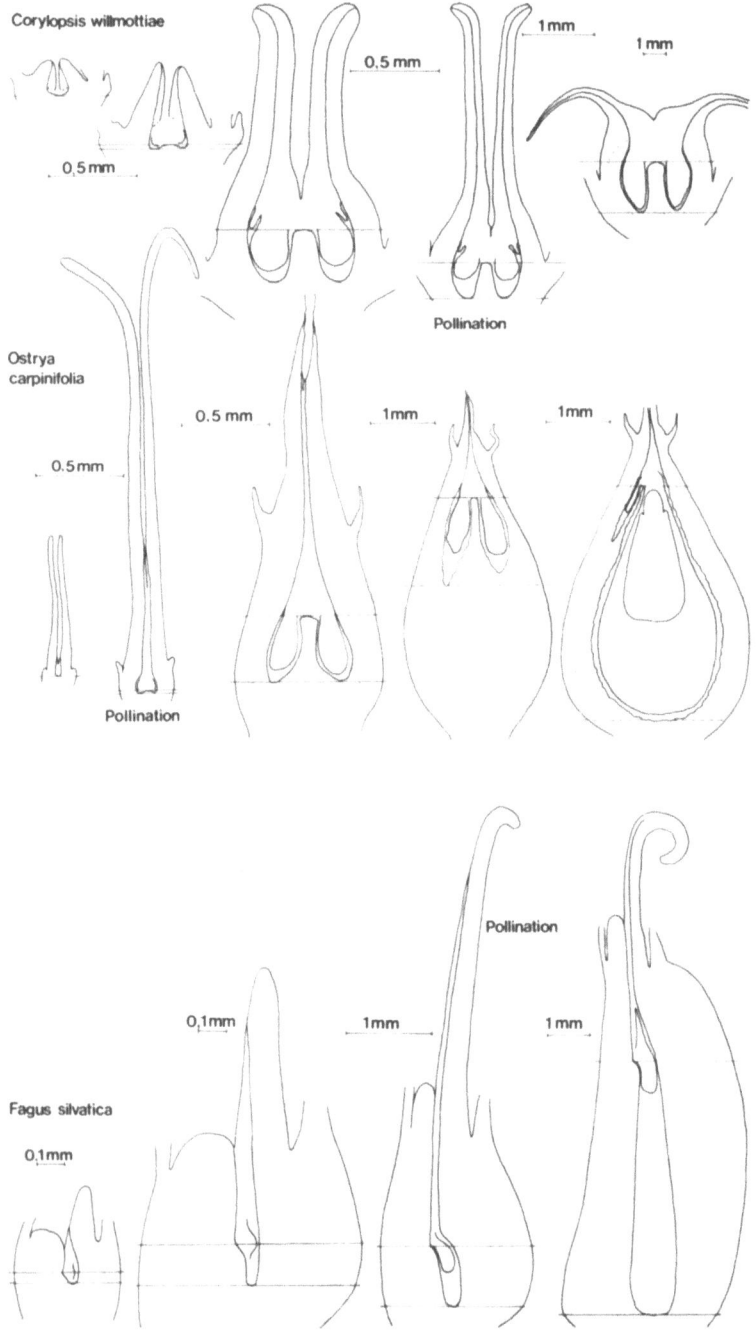

Fig. 9. Gynoecium development. Median longitudinal sections. The ascidi-form part of the ovarial cavity is marked by horizontal lines (amplified after ENDRESS 1967, *Fagus* after FEY)

hours or days. Other additional agents, e.g. animals, have not been reported so far.

In the *Myrothamnaceae* the 3- to 4-merous gynoecium is syncarpous, the carpels ascidiform only at the lowest part of the ovary, the numerous ovules have an axile placentation above the ascidiform portion. The fruit is a ventricidal capsule. Dispersal agents are not known.

The *Platanaceae* have evolved a much more effective dispersal mechanism than the probably closely related *Liquidambaroideae*. It is likely that the most important differences between these two groups are just represented by the features of their two divergent dispersal syndromes. The carpels bear a single orthotropous ovule in their ascidiform region. They are indehiscent and wind dispersed individually by means of a pappus-like hairy tuft at their base. A nearly apocarpous gynoecium with an increased number of carpels per flower (3 up to 9) is correlated with this dispersal syndrome.

In the *Fagales* the fruits are one-seeded and indehiscent throughout as a consequence of their strong adaptation to wind pollination. Therefore, many features of fruit differentiation diverge from the *Hamamelidaceae*. On the other hand, it is most important for the understanding of the uniformity of the group, that the early stages until fertilization are strikingly similar in both orders. This is especially true for the *Betulaceae* when compared with the *Hamamelidaceae* (Figs. 1, 9) (degree of congenital and postgenital "fusion" of the 2 carpels, placentation of the single ovule per carpel).

In the *Fagaceae* the gynoecium is often 3-carpellate and 6-ovulate (constantly 2 ovules per carpel). But in both families only one ovule per gynoecium gives rise to a seed. The other ones degenerate after fertilization. Therefore, the proportions of the ripe ovary differ from the *Hamamelidaceae*. The single seed may become much larger than the two seeds of the *Hamamelidaceae* in relation to the fruit size. The "fertile" one of the two or more locules finally occupies the whole space within the hard outer coat of the fruit wall. This originates in 2 different ways. In many *Fagaceae* (perhaps excluding *Quercus*) the ascidiform part of the carpels becomes extremely elongate (Fig. 9). At least in some genera the cavity is filled with unicellular hairs below or around the ovules. Finally, the hairs are pressed against the walls by the large seed together with the remains of the "sterile" carpels. In *Betulaceae*, in contrast, the massive base below the locules enlarges. In consequence, a large amount of the ovarial tissue around the developing seed degenerates (Fig. 9). *Carpinus* takes an intermediate position. Therefore, in both families, the ripe fruit has functionally only one large ovarial cavity. This has led to misinterpretations of the gynoecium morphology of the *Fagales*.

Inspite of the uniform indehiscence of the fruits in the *Fagales*, the dispersal mechanisms show even more variations than in the *Hamamelidaceae*. In *Betulaceae* and *Fagaceae* smaller, wind dispersed, and larger, animal and water dispersed fruits have evolved. *Alnus, Betula*, and partly *Nothofagus* (WARDLE 1970) show wings on the outer side of the fruit itself as attributes of wind dispersal. In *Carpinus* and *Ostrya* wings are formed by the floral bracts. *Corylus, Ostryopsis* and most *Fagaceae* form larger nuts, which are predominantly dispersed by animals. These nuts have evolved a hard coat, which can, but only with difficulty, be opened by such animals (STEBBINS 1974).

Further Ecological Features—Peculiarities of the Life History

The most striking trend in the life history of the group is the delay of the ovule development in relation to pollination (Figs. 9, 10). Generally, syngamy takes place weeks or even months after pollination. This is possibly true for all *Fagales* and for most of the *Hamamelidales*. Preliminary studies in the *Hamamelidaceae* have shown that from 16 species investigated out of 16 genera only 3 have fully developed embryosacs at anthesis, in 6 cases not even the meiocyte has differentiated. In this respect *Hamamelis virginiana* and *Corylus avellana* are the most extreme species known so far of the two orders with a gap of up to respectively 7 and 4 months between pollination and syngamy. In *Corylus avellana* not even the ovules are initiated at anthesis. It would be interesting to know, if they develop only in pollinated flowers.

At least the extratropical members of both orders are predominantly spring or even late winter flowering. Here, a wide gap has evolved between male and female meiosis. Male meiosis often takes place in autumn, possibly in order to facilitate the disposal of ripe pollen grains in time, whereas female meiosis occurs long after pollination. In *Hamamelis* and in *Corylus avellana* this gap is up to 8 months.

The general adaptation of the life history to spring flowering in the *Hamamelidales*, or ancestors of the group, may have been an important basis (preadaptation) for the evolution of the wind pollinated *Fagales* and for their successful radiation and spreading in temperate regions, in contrast to the less adapted *Hamamelidales*, which, today, are in a stage of isolated survival or even of extinction. They have predominantly a deciduous habit. They grow in communities with low number of tree species, but large number of individuals of the same species, or even in pure stands. And they flower at the end of the deciduous season. Pollination is favoured by the absence of leaves (WHITEHEAD 1969), and at the same time the following development of the fruits is favoured by the beginning of the vegetation season.

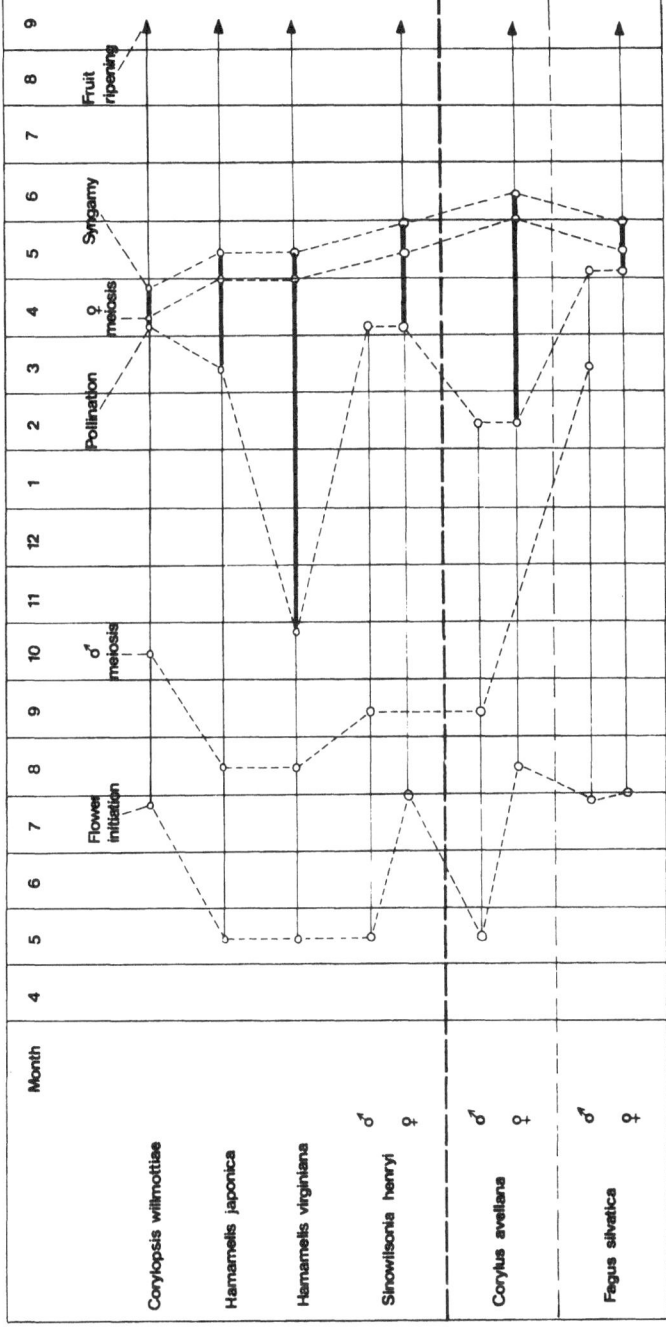

Fig. 10. Timing of major events in flower and fruit development, showing the tendency towards a remarkable delay of ovule development and syngamy in relation to pollination (marked with thick line). Approximate times for wild or cultivated plants in Central Europe or North America (Original, after data from SHOEMAKER 1905, FEY, and personal observations)

As a much less important parallel the fly pollinated syndrome in *Disanthus* and *Hamamelis* (third section) may have evolved with a shift of the flowering time to winter or even autumn. Diptera are the first insects to be active in the cold season during any short period of

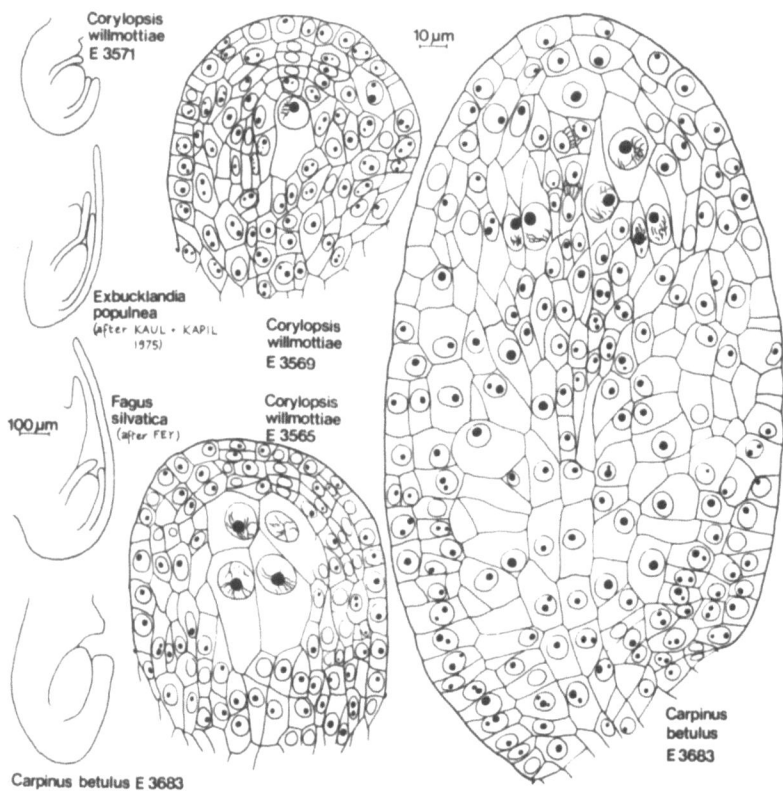

Fig. 11. Ovules. Diverging integument structure (at the tetrad stage). Ovules with a single sporogenous cell and with a massive sporogenous tissue (Original, except indicated figures). Voucher numbers indicated (E = END-RESS)

relatively high temperature. Therefore, they are the most important pollinators among the insects during the winter.

The structure and ontogeny of the ovules and seeds is a particularly interesting aspect of the life history of this group. The ovules are crassi-nucellar and anatropous (orthotropous in *Platanus*). In the *Hama-melidales* and *Fagaceae* (except *Nothofagus*) they are bitegmic, in the *Betulaceae* and in *Nothofagus* unitegmic. A tendency towards an extremely long outer integument is apparent in some groups: *Exbucklandia* of the *Hamamelidaceae* (KAUL and KAPIL 1975), *Myrothamnaceae* (JÄGER-ZÜRN 1966) and *Fagaceae* (Fig. 11). It may partly function as an obturator. In *Exbucklandia* it contributes the wing of the seed. The differentiation of a massive sporogenous tissue in the ovule has been observed in *Corylopsis* species of the *Hamamelidaceae* and in the *Coryloideae* (Fig. 11). In the other genera investigated so far only a single meiocyte has been observed. The frequent occurrence of more than one young embryosac in one ovule is common in *Hamamelidaceae* and *Coryloideae*. The embryosac development conforms to the Polygonum type in all species investigated. *Quercus gambelii* is the only species which has been studied extensively in its ultrastructure (MOGENSEN 1972, 1973, 1975a, b, BROWN and MOGENSEN 1972, SINGH and MOGENSEN 1975). The knowledge of embryo formation is still too scanty for a comparative survey.

Certain inhomogeneous "embryological" characters in the group may be correlated with diversity in seed and embryo size. Firstly, in the medium- and large-seeded *Fagales*, but not in *Hamamelidales* nor in the small-seeded *Fagales (Betuloideae)*, is the outer or the single integument vascularized. Secondly, the endosperm formation is nuclear or cellular (KAPIL and KAUL 1974) in *Hamamelidales*, and extremely nuclear in *Fagales*. In general, small embryos (as in *Hamamelidales*) may possibly rather be accompanied by an endosperm which becomes cellular after only a short nuclear phase or even from initiation; in contrast, large embryos (as in *Fagales*) have a faster growing endosperm with a long nuclear phase.

EHRENDORFER (1973) and STONE (1974) point to further ecological functions of the differing seed size. The small fruits and seeds show an epigeal germination (*Hamamelidales*, *Betulaceae* except *Corylus*, *Fagoideae*), and large fruits are hypogeal (*Corylus*, *Fagaceae* except *Fagoideae*). Large fruits with high storage capacity are further adapted to shady climax forest environments, whereas small diaspores with less nutrient supply and epigeal germination are better adapted to open woodlands.

Cytology

Recent surveys of the chromosome numbers of the whole group are given by EHRENDORFER (1976) and RAVEN (1976), of the *Hamamelidaceae* by GOLDBLATT and ENDRESS (1977). The chromosome numbers are relatively constant for genera or even subfamilies with the exception of polyploid series occurring in several genera. Base numbers seem to be $x = 6$, 7, and 8. The numbers $x = 6$ and $x = 8$ probably have evolved several times from $x = 7$. *Hamamelidoideae, Rhodo-*

leioideae, and *Fagaceae* (except *Nothofagus* and *Trigonobalanus*) show
n = 12 (or higher ploidy levels). *Platanaceae, Betuloideae* and partly
Corylus with at least n = 14 are based on x = 7. *Disanthoideae, Exbucklandioideae, Liquidambaroideae, Carpinus,* and *Ostrya* exhibit
the x = 8 assemblage. *Disanthus* of the *Hamamelidaceae* (GOLDBLATT &
ENDRESS 1977), *Carpinus* and *Ostrya,* all with n = 8, seem to be the
only paleodiploids known so far from the whole complex.

Distribution and Phylogeny

Early fossils are still too scanty for reconstructing the phylogeny
of the group in some detail. However, some extrapolations are possible.
Earliest probable remains of the *Hamamelidales* are known from the
Upper Cretaceous (KNAPPE and RÜFFLE 1975, DOYLE and HICKEY 1976),
of the *Fagales* from the latest Cretaceous (WOLFE 1974, RAVEN and
AXELROD 1974). A time sequence of the pollen types represented in
the group is shown by WOLFE, DOYLE and PAGE (1976): 1. reticulate
tricolpates (as in primitive *Hamamelidaceae*): Lower Cretaceous, 2. tri-
colporates (as in *Fagaceae*): Upper Cretaceous, 3. triporates (as in
Betulaceae): Late Cretaceous.

The main distribution centre of the group is Eastern Asia. Paleobotani-
cally speaking, the distribution is predominantly Laurasian (RAVEN and
AXELROD 1974) except the African respectively Madagascan *Dicoryphe,
Trichocladus* (*Hamamelidaceae*), and *Myrothamnaceae.* The way of migra-
tion of *Nothofagus* to the southern hemisphere and of its radiation is still
controversial (VAN STEENIS 1971, SCHUSTER 1976).

Features of living genera may contribute knowledge to the problem
as well. Some evidence has been discussed in the preceding sections.
In *Hamamelidaceae, Disanthus* is especially interesting with an unusual
combination of primitive features: the only known paleodiploid of
the family (n = 8), ovary superior, medium number of ovules per
carpel, embryo sac ripe at anthesis, insect pollinated, flowers pentamerous, bisexual, perianth double and well developed, calyx imbricate,
pollen reticulate and tricolpate. *Corylopsis,* also with relatively many
primitive characters, is of particular interest as well, because of the
remarkable similarity with certain *Betulaceae.* Both genera are temperate and deciduous. In addition, it is important, that polyandry
in *Hamamelidaceae* has turned out to be advanced. In *Betulaceae,*
primitive and advanced features are more mixed, however, the *Coryloideae,* on the whole, show more resemblance to the *Hamamelidaceae*
than the *Betuloideae.* In the *Fagaceae, Nothofagus* probably shows
the closest affinity to *Hamamelidaceae* and *Betulaceae:* gynoecium

partly 2-merous with a 2-winged fruit, a few species completely lacking a cupule, only a single integument (as *Betulaceae*), anthers long and with connective protrusion (as *Hamamelidales*); some vegetative characters (mostly common with *Fagus* and partly other genera): phyllotaxis distichous, wood diffuse porous, vessels thin and angular, vessel elements with scalariform perforations, ray cells of the type 3 of KRIBS, germination epigeal.

The insect pollinated genera of the *Hamamelidaceae* are partly tropical, partly temperate. The wind pollinated representatives of both orders are predominantly temperate. However, some probably wind pollinated genera have a more tropical or not typically temperate distribution. But, at least, they often occur in open habitats such as river banks (*Distylium*, *Matudaea*) or semideserts (*Myrothamnaceae*). The *Myrothamnaceae* are adapted to dry regions by their extremely poikilohydrous behaviour (VIEWEG & ZIEGLER 1969).

From various grounds, there is some evidence, that the *Fagales*, particularly the *Betulaceae*, may have originated from temperate deciduous *Hamamelidaceae* or *Hamamelidaceae*-like ancestors. On the whole, the elaborate adaptation to wind pollination seems to have been one of the most important prerequisites for the successful adaptive radiation and distribution of the *Fagales* and for the conquest of their dominant position in wide parts of the temperate regions.

I thank Professor Dr. C. D. K. COOK for reading the manuscript and Professor Dr. H. R. HOHL for use of the scanning electron microscope. DORIS METZGER and J. J. PITTET have made the SEM photographs, which is also gratefully acknowledged.

References

ABBE, E. C., 1974 a: Flowers and inflorescences of the "*Amentiferae*". Bot. Rev. **40**, 159—261.
— 1974 b: *Betulales*. Encylopaedia Britannica, Macropaedia (ed. 15), **2**, 872—875. Chicago: Helen Hemingway Benton Publ.
BAAS, P., 1969: Comparative anatomy of *Platanus kerrii* GAGNEP. Bot. J. Linn. Soc. (London) **62**, 413—421.
BEHNKE, H. D., 1973: Sieve-tube plastids of *Hamamelididae*. Electron microscopic investigations with special reference to *Urticales*. Taxon **22**, 205—210.
BOGLE, A. L., 1970: Floral morphology and vascular anatomy of the *Hamamelidaceae*: The apetalous genera of *Hamamelidoideae*. J. Arn. Arb. **51**, 310—366.
BROWN, R. C., and MOGENSEN, H. L., 1972: Late ovule and early embryo development in *Quercus gambelii*. Amer. J. Bot. **59**, 311—316.
BURRICHTER, E., AMELUNXEN, F., VAHL, J., and GIELE, T., 1968: Pollen- und Sporenuntersuchungen mit dem Oberflächen-Rasterelektronen-mikroskop. Z. Pflanzenphysiol. **59**, 226—237.

Carlquist, S., 1976: Wood anatomy of *Myrothamnus flabellifolia* (*Myrothamnaceae*) and the problem of multiperforate perforation plates. J. Arn. Arb. **57**, 119—126.

Chang, K.-T., 1964: Pollen morphology in the families *Hamamelidaceae* and *Altingiaceae* (in russ.). Flora and systematics of the higher plants **13**, 173—232. Moscow-Leningrad: Nauka.

Cronquist, A., 1968: The evolution and classification of flowering plants. London-Edinburgh: Nelson.

Doyle, J. A., and Hickey, L. J., 1976: Pollen and leaves from the Mid-Cretaceous Potomac group and their bearing on early angiosperm evolution. In: Origin and early evolution of angiosperms (Beck, C. B., Ed.), 139—206. New York-London: Columbia Univ. Press.

— Van Campo, Madeleine, and Lugardon, B., 1975: Observations on exine structure of *Eucommiidites* and Lower Cretaceous angiosperm pollen. Pollen Spores **17**, 429—486.

Ehrendorfer, F., 1973: Adaptive significance of major taxonomic characters and morphological trends in angiosperms. In: Taxonomy and ecology (Heywood, V. H., Ed.), 317—327. London-New York: Academic Press.

— 1976: Evolutionary significance of chromosomal differentiation patterns in gymnosperms and primitive angiosperms. In: Origin and early evolution of angiosperms (Beck, C. B., Ed.), 220—240. New York-London: Columbia Univ. Press.

Elias, T. S., 1971: The genera of *Fagaceae* in the southeastern United States. J. Arn. Arb. **52**, 159—195.

Endress, P. K., 1967: Systematische Studie über die verwandtschaftlichen Beziehungen zwischen den Hamamelidaceen und Betulaceen. Bot. Jahrb. Syst. **87**, 431—525.

— 1968: Zur systematischen Stellung von *Parrotia* C. A. Mey. und *Sycopsis* Oliv. und zum neuen Bastard × *Sycoparrotia semidecidua* P. Endress et J. Anliker. Schweiz. Beitr. Dendrol. **16—18**, 5—22.

— 1969: *Molinadendron*, eine neue Hamamelidaceen-Gattung aus Zentralamerika. Bot. Jahrb. Syst. **89**, 353—358.

— 1970: Die Infloreszenzen der apetalen Hamamelidaceen, ihre grundsätzliche morphologische und systematische Bedeutung. Bot. Jahrb. Syst. **90**, 1—54.

— 1971: Blütenstände und morphologische Interpretation der Blüten bei apetalen Hamamelidaceen. Ber. dtsch. bot. Ges. **84**, 183—185.

— 1974a: Unbekannte Blütenpflanzen — Probleme der Großsystematik. Vierteljschr. naturf. Ges. Zürich **119**, 1—21.

— 1974b: *Hamamelidales*. Encylopaedia Britannica, Macropaedia (ed. 15), **8**, 578—580. Chicago: Helen Hemingway Benton Publ.

— 1975: Nachbarliche Formbeziehungen mit Hüllfunktion im Infloreszenz- und Blütenbereich. Bot. Jahrb. Syst. **96**, 1—44.

— 1976: Die Androeciumanlage bei polyandrischen Hamamelidaceen und ihre systematische Bedeutung. Bot. Jahrb. Syst. **97**, 436—457.

Faegri, K., and Van der Pijl, L., 1971: The principles of pollination ecology, 2nd Ed. Oxford-New York-Toronto-Sydney-Braunschweig: Pergamon Press.

Fey, B., In preparation (Doctoral dissertation): University of Zürich, Switzerland.

FORMAN, L. L., 1966: On the evolution of cupules in the *Fagaceae*. Kew Bull. **18**, 385—419.

GOLDBLATT, P., and ENDRESS, P. K., 1977: Cytology and evolution in *Hamamelidaceae*. J. Arn. Arb. **58**, 67—71.

GUSEJNOVA, N. A., 1976: On cytoembryology in *Platanaceae* (russ.). Bjull. Glav. Bot. Sada (Moscow) **102**, 67—71.

HANKS, S. L., and FAIRBROTHERS, D. E., 1976: Palynotaxonomic investigation of *Fagus* L. and *Nothofagus* BL.: light microscopy, scanning electron microscopy, and computer analyses. In: Botanical Systematics **1**, 1—141 + II (HEYWOOD, V. H., Ed.). London-New York-San Francisco: Academic Press.

HJELMVUIST, H., 1974: *Fagales*. Encyclopaedia Britannica, Macropaedia (ed. 15), **7**, 139—142. Chicago: Helen Hemingway Benton Publ.

HUTCHINSON, J., 1967: The genera of flowering plants. Dicotyledones, Vol. 2. Oxford: Clarendon Press.

JÄGER-ZÜRN, IRMGARD, 1966: Infloreszenz- und blütenmorphologische sowie embryologische Untersuchungen an *Myrothamnus* WELW. Beitr. Biol. Pfl. **42**, 241—271.

JAY, M., 1968: Distribution des flavonoides chez les Hamamélidacées et familles affines. Taxon **17**, 136—147.

KAPIL, R. N., and KAUL, USHA, 1974: Embryologically little known taxon—*Parrotiopsis jacquemontiana*. Phytomorphol. **22**, 234—245.

KAUL, USHA, and KAPIL, R. N., 1975: *Exbucklandia populnea*—from flower to fruit. Phytomorphol. **24**; 217—228.

KNAPPE, H., and RÜFFLE, L., 1975: Beiträge zu den Platanaceen-Funden und einigen *Hamamelidales* der Oberkreide. Wiss. Z. Humboldt-Univ. Berlin, math.-nat. Reihe **4**, 487—492.

KORCHAGINA, I. A., 1974: On the nature of the flower of *Betulaceae* (russ.). Trans. Moscow Soc. Naturalists, Biol. Ser., Sect. Bot. **51**, 50—74.

KUPRIANOVA, L. A., 1965: The palynology of the *Amentiferae* (russ.). Moscow-Leningrad: Nauka.

LEBRETON, PH., 1976: Quelques données chimiotaxinomiques relatives aux Fagacées. Bull. Soc. bot. France **123**, 293—298.

MACDONALD, A. D., 1971: Floral development in the "*Amentiferae*". Ph.D. Thesis, McGill University, Montreal, Quebec.

MEARS, J. A., 1974: Chemical constituents and systematics of *Amentiferae*. Brittonia **25**, 385—394.

MEEUSE, A. D. J., 1975: Floral evolution in the *Hamamelididae*. Acta bot. Neerl. **24**, 155—191.

MELIKIAN, A. P., 1973: Seed coat types of *Hamamelidaceae* and allied families in relation to their systematics (russ.). Bot. Zhurn. U.S.S.R. **58**, 350—359.

MOGENSEN, H. L., 1972: Fine structure and composition of the egg apparatus before and after fertilization in *Quercus gambelii*: the functional ovule. Amer. J. Bot. **59**, 931—941.

— 1973: Some histochemical, ultrastructural, and nutritional aspects of the ovule of *Quercus gambelii*. Amer. J. Bot. **60**, 48—54.

— 1975 a: Ovule abortion in *Quercus* (*Fagaceae*). Amer. J. Bot. **62**, 160—165.

— 1975 b: Fine structure of the unfertilized, abortive egg apparatus in *Quercus gambelii*. Phytomorphol. **25**, 19—30.

MOHANA RAO, P. R., 1974: Seed anatomy in some *Hamamelidaceae* and phylogeny. Phytomorphol. **24**, 113—139.

Moseley, M. F., Jr., 1974: Vegetative anatomy and morphology of *Amentiferae*. Brittonia **25**, 356—370.

Raven, P. H., 1976: The bases of angiosperm phylogeny: cytology. Ann. Missouri bot. Gard. **62**, 724—764.

— and Axelrod, D. I., 1974: Angiosperm biogeography and past continental movements. Ann. Missouri bot. Gard. **61**, 539—673.

Sattler, R., 1973: Organogenesis of flowers. A photographic text-atlas. Toronto-Buffalo: Univ. of Toronto Press.

Schuster, R. M., 1976: Plate tectonics and its bearing on the geographical origin and dispersal of angiosperms. In: Origin and early evolution of angiosperms (Beck, C. B., Ed.), 48—138. New York-London: Columbia Univ. Press.

Shimaji, K., 1962: Anatomical studies on the phylogenetic interrelationship of the genera in the *Fagaceae*. Bull. Tokyo Univ. Forests **57**, 1—64.

Shoemaker, D. N., 1905: On the development of *Hamamelis virginiana*. Bot. Gaz. **39**, 248—266.

Singh, A. P., and Mogensen, H. L., 1975: Fine structure of the zygote and early embryo in *Quercus gambelii*. Amer. J. Bot. **62**, 105—115.

Skvortsova, N. T., 1975: Comparative morphological investigations in representatives of the family *Hamamelidaceae* and their phylogenetic relationships (russ.). In: Problems of comparative morphology of the seed plants (Budantsjev, L. J., Ed.), 7—24. Leningrad: Nauka.

Soepadmo, E., 1972: *Fagaceae*. In: Flora Malesiana (Van Steenis, C. G. G. J., Ed.), **7**, 2, 265—403. Groningen: Wolters-Noordhoff Publ.

Stebbins, G. L., 1974: Flowering plants. Evolution above the species level. Cambridge, Mass.: Harvard Univ. Press.

Stone, D. E., 1974: Patterns in the evolution of amentiferous fruits. Brittonia **25**, 371—384.

Straka, H., 1975: Pollen- und Sporenkunde. Eine Einführung in die Palynologie. Stuttgart: Fischer.

Surova, T. G., 1975: Electronmicroscopical studies on pollen and spores of plants (russ.). Moscow: Nauka.

Takeoka, M., and Stix, Erika, 1963: On the fine structure of the pollen walls in some Scandinavian *Betulaceae*. Grana Palynol. **4**, 161—188.

Takhtajan, A., 1969: Flowering plants. Origin and dispersal. Edinburgh: Oliver & Boyd.

Van Campo, Madeleine, and Lugardon, B., 1973: Structure grenue infratectale de l'extexine des pollens de quelques gymnospermes et angiospermes. Pollen Spores **15**, 171—187.

Van Steenis, C. G. G. J., 1971: *Nothofagus*, key genus of plant geography, in time and space, living and fossil, ecology and phylogeny. Blumea **19**, 65—98.

Vieweg, G. H., and Ziegler, H., 1969: Zur Physiologie von *Myrothamnus flabellifolia*. Ber. dtsch. bot. Ges. **82**, 29—36.

Walker, J. W., 1976: Evolutionary significance of the exine in the pollen of primitive angiosperms. In: The evolutionary significance of the exine (Ferguson, I. K., and Muller, J., Eds.), 251—308. Linn. Soc. Symp. Ser. **1**. London: Academic Press.

— and Doyle, J. A., 1976: The bases of angiosperm phylogeny: palynology. Ann. Missouri bot. Gard. **62**, 664—723.

Weaver, R. E., Jr., 1969: Studies in the North American genus *Fothergilla* (*Hamamelidaceae*). J. Arn. Arb. **50**, 599—619.

WARDLE, J., 1970: The ecology of *Nothofagus solandri*. 3. Regeneration. New Zeal. J. Bot. **8**, 571—608.

WHITEHEAD, D. R., 1969: Wind pollination in the angiosperms: evolutionary and environmental considerations. Evolution **23**, 28—35.

WOLFE, J. A., 1974: Fossil forms of *Amentiferae*. Brittonia **25**, 334—355.

— DOYLE, J. A., and PAGE, VIRGINIA M., 1976: The bases of angiosperm phylogeny: paleobotany. Ann. Missouri Bot. Gard. **62**, 801—824.

Address of the author: Prof. Dr. PETER K. ENDRESS, Institut für Systematische Botanik der Universität, Zollikerstr. 107, CH-8008 Zürich, Switzerland.

Plant Syst. Evol., Suppl. 1, 349—374 (1977)

Institute for Systematic Botany, State University Utrecht, The Netherlands

Urticales, Their Differentiation and Systematic Position

By

C. C. Berg, Utrecht

Abstract: The *Urticales* constitute a group of plants with surprising diversity and confusing complexity, especially in the *Moraceae* round which family the present paper is centred. Main trends in the differentiation of the growth habit, flower, fruit, and inflorescence are sketched and as far as possible connected to distribution, pollination, and dispersal. The reduced state of the urticalean flower is a main factor in the differentiation of the reproductive structures. Two main complexes may be distinguished in the variation patterns: one centred round adaptations to wind pollination, the other round adaptations to insect pollination, the latter resulting in several types of pseudanthous or pseudocarpous inflorescences. In both complexes there occur well adapted large genera (*Ficus, Cecropia*) and apparently less well adapted, as well large genera (*Dorstenia, Elatostema*). The *Urticales* appear to belong to a central complex in the Angiosperms. The affinity of the order with the *Malvales* is reasonably clear, while its connections with the *Hamamelidales* and their allies are much more vague.

Introduction

The *Urticales* constitute a distinct order, the circumscription of which has hardly changed since about 1850. It comprises the *Ulmaceae, Moraceae, Urticaceae,* and *Cannabaceae.* A few minor unigeneric families have been inserted in the *Urticales: Theligonaceae, Barbeyaceae,* and *Eucommiaceae,* but either they have been removed subsequently or their inclusion in the *Urticales* is still questionalble. The *Hymenocar-diaceae* yet another unigeneric family, has recently been placed in the *Urticales* (cf. DAHLGREN 1975); according to WEBSTER (1967, 1975) it is euphorbiaceous. Morphological, anatomical, and cytological char-acters place the *Cannabaceae* in a rather isolated position within the *Urticales* and might even be considered to make its position in the order questionable (cf. BECHTEL 1921, TIPPO 1938, RAVEN 1975). This family will not be discussed further in the present paper. Within the *Urticales* the *Ulmaceae* can be clearly delimited. In contrast the limits between the *Moraceae* and the *Urticaceae* are confused, and transferring the

Conocephaloideae from the *Moraceae* to the *Urticaceae* (CORNER 1962) does nothing to clarify the distinction between the two families. It is preferable to regard the *Moraceae* s. str., *Urticaceae* s. str., and *Conocephaloideae* as separate groups, perhaps as subfamilies of a single family (cf. BERG 1973). The *Moraceae* s. str. can be conveniently subdivided into 4 tribes (BERG 1973):

Moreae: with among others *Antiaropsis, Artocarpus, Bagassa, Batocarpus, Bleekrodea, Broussonetia, Cardiogyne, Clarisia, Cudrania, Fatoua, Maclura, Malaisia, Morus, Olmedia, Phyllochlamys, Plecospermum, Sloetiopsis, Sorocea, Sparattosyce, Streblus, Treculia, Trophis.*

Dorstenieae: with among others *Bosqueiopsis, Brosimum, Dorstenia, Helianthostylis, Scyphosyce, Trilepisium* (= *Bosqueia*), *Utsetela.*

Castilleae C. C. BERG (ined.): with *Antiaris, Castilla, Helicostylis, Maquira, Mesogyne, Naucleopsis, Perebea, Pseudolmedia.*

Ficeae: with *Ficus* only.

The *Conocephaloideae* comprise: *Cecropia, Coussapoa, Myrianthus, Musanga, Poikilospermum, Pourouma.*

On the basis of flower (cf. BECHTEL 1921) and wood (cf. TIPPO 1938) characters the *Ulmaceae* are regarded as the most primitive group in the order and the *Urticaceae* as the most advanced. The *Conocephaloideae* seem to be intermediate between the *Moraceae* and *Urticaceae* in this respect. While the *Ulmaceae* and *Urticaceae* are rather uniform, the *Conocephaloideae* show some diversity and the *Moraceae* exhibit an even greater diversity. This diversity is found in the growth habit, flower, fruit, seed and particularly in the inflorescence and infructescence.

Habit

There is a wide variation in habit among the *Urticales*: trees, from huge to small (*Ulmaceae, Moraceae,* and *Conocephaloideae*), and shrubs (mainly *Urticaceae* and *Moraceae*); woody climbers, climbing with the aid of roots (in *Ficus, Urera*) or spines (*Maclura* s.l.), or without special organs |e.g. *Malaisia scandens* (LOUR.) PLANCH., *Myrianthus scandens* LOUIS ex HAUMAN and African species of *Urera*); (hemi-)epiphytes, strangling or not (in *Ficus, Coussapoa* and *Poikilospermum*): undershrubs (mainly *Urticaceae*, but also in some moraceous genera such as *Dorstenia, Fatoua,* and *Perebea*); herbs, perennial (mainly *Urticaceae,* but also *Dorstenia* and *Fatoua*) or annual (*Urticaceae*): succulents (*Dorstenia*), and tuber-geophytes (*Dorstenia*).

Knowledge of the characters differentiating growth habit (life form, tree form, growth-rhythm, differences between juvenile and adult trees, shoot-morphology, deciduous habit, protection of apical meristem, shedding of branches, shedding and abortion of shoot apices, bud characters, and bud formation) in tropical trees is still fragmentary and superficial. This is a field which is still in the early stages of being explored (cf. TOMLINSON and GILL 1973) and is a work for which HALLÉ and OLDEMAN's monograph on the habit and growth of tropical trees (1970) can serve as a basis.

The importance of a thorough knowledge of growth-habit differentiation for tropical forestry and for the cultivation of tropical crops is evident, but its significance for classification, evolution, and plant geography has still to be investigated.

An attempt to arrange and interpret the available data relating to the *Urticales* is presented below. From HALLÉ and OLDEMAN's study, work on moraceous taxa by JARRETT (1959, 1960) and CORNER (1967), and my own observations it can be concluded that in arborescent *Urticales* three, roughly definable growth habits are dominant. These types of growth habits match or comprise categories recognized and described by HALLÉ and OLDEMAN, who called them the "modèle de CORNER", "modèle de RAUH", "modèle de TROLL", etc.

1. Trees with a monopodial trunk, having monopodial subverticillate to more or less diffusely arranged branches. The leaves are borne in spirals or if the branches are thin they tend to be distichous. This growth habit comprises the related models of RAUH and ATTIMS. It is common in *Ficus* and the *Conocephaloideae* and it appears to characterize *Artocarpus* subg. *Artocarpus* (cf. JARRETT 1959, 1960).

2. Trees with a monopodial trunk and a sparse to dense system of more or less vertically growing monopodial branches bearing the leaves in spirals. In the axils of the leaves more or less distinctly phyllomorphic horizontally growing branches with distichous, often more or less asymmetrical leaves. These phyllomorphic branches arise in succession from almost every node. In the *Castilleae* they are shed, a phenomenon which sets this tribe apart. This growth habit, which matches the related models of COOK and ROUX, is characteristic for the *Castilleae* and occurs in the *Ulmaceae* (*Trema*, *Celtis*) and in some *Moreae* (*Chlorophora regia* A. CHEV. and *Artocarpus lakoocha* ROXB. which belongs to the subg. *Pseudojaca*). The genera *Celtis*, *Chlorophora* GAUD., and *Artocarpus* (subg. *Pseudojaca*) also have species which show the next type of growth habit, which matches the model of TROLL. The model of ROUX can be converted into that of TROLL experimentally (ROUX 1968, HALLÉ and OLDEMAN 1970) and this change probably occurs in *C. regia* with age.

3. Trees in which the trunk and main branches are sympodial. The often asymmetrical leaves are borne in two rows, but spirally arranged leaves can be found on initially more or less vertical growing parts of the trunk (of the sapling) or branches. The leaves may already

be distichous in the seedling. This growth habit occurs in many *Ulmaceae* (*Ulmus, Holoptelea, Celtis*), in many *Dorstenieae*, and in many *Moreae* (e.g. in *Chlorophora excelsa* (WELW.) BENTH. and HOOK. and in *Artocarpus nitidus* TRÉC. which is also a member of the subg. *Pseudojaca*).

Besides these three predominant tree forms, there is a less common form found in the *Moraceae*: a monocaul (= unbranched) or sparingly branched (small) tree with relatively short internodes and relatively large, shortly petiolate, entire leaves arranged in spirals. This form is known for several *Ficus* species (cf. CORNER 1967), *Naucleopsis stipularis* DUCKE, and *Dorstenia oligogyna* (PELLEGRIN) C. C. BERG (ined.). In spite of its occurrence in quite different groups of the *Moraceae*, the plants are strikingly similar, even in the leaf shape, the leaf venation, and the presence of large (sub)persistent stipules. CORNER (1967) assigned them to the group of pachycaul trees. In *Ficus* he found that pachycauly is correlated with the presence of primitive characters in the inflorescence and flowers. The same is found in *Naucleopsis stipularis* (cf. BERG 1972) and in *Dorstenia*, in which the West African *D. djettii* GUILLAUMET, a very close relative of *D. oligogyna*, can be regarded as the most primitive species in the genus.

In *Ficus*, which exhibits an extremely wide range of growth habits indeed, relatives of species with the simple tree form show the features of the first type of growth habit described above, the type which is the dominant tree form in the genus.

In *Naucleopsis* there is a gradation in the tree forms observed. The series starts with *N. stipularis* and is followed by a group of species with rather small pachycaul to sub-pachycaul trees having few rather thick and long phyllomorphic self-pruning branches with relatively large leaves [this group matches the features of the model of COOK and includes, e.g. *N. ulei* (WARB.) DUCKE and *N. macrophylla* MIQ.] and goes on to a group of species comprising leptocaul trees, up to 30 m tall, with many relatively slender permanent branches and small phyllomorphic branches with small leaves [this group matches the model of ROUX and includes, e.g. *N. mello-barretoi* (STANDL.) C. C. BERG]. These tree forms can be found in other taxa belonging to the *Castilleae* as well.

The species closest to *Dorstenia oligogyna* and *D. djettii*, like *D. africana* (BAILL.) C. C. BERG (ined.) and *D. kameruniana* ENGL. show the features of the third type of growth habit described above, the type which is dominant in the tribe *Dorstenieae*. On the other hand, *D. oligogyna* and *D. djettii* are related to *D. elliptica* BUR., which mostly forms unbranched or sparingly branched shoots with rhizomes and spirally arranged leaves.

Growth-habit differentiation in *Ficus, Naucleopsis* (and related genera, see BERG 1977), and *Dorstenia* appears to fit the picture of tree-form evolution drawn by CORNER (1949). Although the presence of simply structured trees in these three genera might be the result of convergent tree-form reduction, in agreement with CORNER (1967) I am inclined to regard the simple tree-form (without relating it to the primitive Angiosperm habit or monocaul non-angiospermous seed plants) as representing one of the first stages of tree differentiation. Not only this seems to be the case in the *Moraceae*, but also in other groups, as this tree form is rather widely distributed among tropical arborescent families. On the other hand, the inflorescences of these three moraceous genera belong to the most advanced ones in the family.

Monocauly also occurs in *Cecropia ulei* SNETHLAGE. The genus *Cecropia* and several species of *Artocarpus* (e.g. *A. communis* J. R. and G. FORSTER) have also been counted among pachycaul trees (CORNER 1949). The pachycaul, and sometimes monocaul, trees of these taxa (as well as those of, e.g., the euphorbiaceous genus *Macaranga*) are components of secondary vegetation and have thick, rapidly growing and soft woody stems with long internodes and large and often incised leaves having long petioles. I am not sure whether it is correct to group this type of tree with the pachycaul trees occuring in *Ficus, Naucleopsis*, and *Dorstenia*, which are components of the lower storeys of evergreen tropical forests. Monocauly in *Cecropia* is probably due to reduction.

Most of the species exhibiting the features of the second type of growth habit described above are usually genuine components of evergreen tropical lowland forests with relatively small distribution areas. Many of the species showing the first, but more having features of the third growth habit described above occur in drier tropical habitats like (semi-)deciduous forests and savannas.

Ulmaceae (species of *Ulmus* and *Celtis*) and *Moraceae* (species of *Morus* and *Broussonetia*), which are trees with sympodial growth, even occur in extra-tropical regions. These species show abortion and shedding of shoot apices at the end of the growth season, a phenomenon common in trees of northern temperate region (cf. MILLINGTON and CHANEY 1973) but apparently rare in tropical trees (cf. KWAN KORIBA 1958, HALLÉ and OLDEMAN 1970). This phenomenon also occurs in *Holoptelea grandis* (HUTCH.) MILDBR., which is related to *Ulmus* and occurs in the lowlands of tropical West Africa. Shedding of shoot tips appears to be a distinctive feature of *Morus* and *Broussonetia*. It is also found in *Morus mesozygia* STAPF, occurring in African tropical lowlands, and in the Madagascan *Broussonetia greveana* (BAILL.) C. C. BERG (ined.); in these taxa it is correlated with deciduous habit and the formation of more or less well-developed scaled axillary resting buds. Abortion of shoot tips is also reported for *Maclura pomifera* (SMITH

1963), a species of the third moraceous genus which has members in northern temperate regions. In *Maclura* s.l. which includes *Chlorophora tinctoria* (L.) GAUD. ex BENTH. & HOOK., *Plecospermum*, *Cardiogyne*, and *Cudrania* (cf. CORNER 1962), short-shoots end in spines. Shedding of shoot apices and the formation of spinous shoot tips (because of abortion of the apical meristem) may be regarded as related phenomena (cf. MILLINGTON and CHANEY 1973).

The group of moraceous genera comprising *Morus*, *Broussonetia*, and *Maclura* s.l. shows in its distribution similarities with families and genera like the *Magnoliaceae* and *Hamamelidaceae* and *Berberis*, *Celastrus*, and *Euonymus* (cf. GOOD 1974), viz. centred more or less distinctly in northern or altitudinally subtropical to temperate regions; the areas range in the Old World from Japan through Sri Lanka to Madagascar and East Africa and through the Philippines and Indonesia to New Caledonia and in the New World from S.E. North America through Central America and the Andes southwards. The distribution of this group of moraceous taxa as well as that of some ulmaceous groups suggests that their adaptations to temperate conditions did not arise in (seasonal) tropical lowland conditions. The presence of such species under these conditions appears to be secondary.

Aerial and stilt roots contribute to the diversity of arborescent taxa in the *Urticales*. Such roots are common in *Ficus*, and they also seem to be a characteristic feature of the *Conocephaloideae*, a group in which they are functional not only in epiphytic taxa (*Coussapoa* and *Poikilospermum*) but also in those occurring in inundated places (*Cecropia* species). On the other hand, whether they are functional or not in other species (e.g. of *Pourouma* and *Myrianthus*) still has to be established.

Variation ranging from an almost arborescent habit through a frutescent, suffrutescent, and herbaceous to a succulent habit is found in *Dorstenia*, which, after *Ficus*, is the most diverse and interesting genus of the *Moraceae*. An outline of this variation is shown in Fig. 1. Most non-succulent frutescent to herbaceous African species of *Dorstenia* have branched underground creeping rhizomes, which can be regarded as important for vegetative reproduction (cf. RICHARDS 1952). These species, as well as the only African species with creeping leafy stems (*D. picta* BUR.), are components of the undergrowth of evergreen forest and most of them have rather small areas of distribution. The phanerophytic forms with thin, often more or less woody rhizomes gradually pass via forms with swollen rhizomes (*D. preussii* ENGL.) and those with more or less irregularly shaped discoid tubers (*D. cuspidata* HOCHST.) into truly geophytic forms with regularly shaped placentiform tubers (*D. benguellensis* WELW.). The acaulescent tuberiferous *D. barnimiana* SCHWEINF. is probably derived from a caulescent tuberiferous form. The tuber-forming taxa occur in the

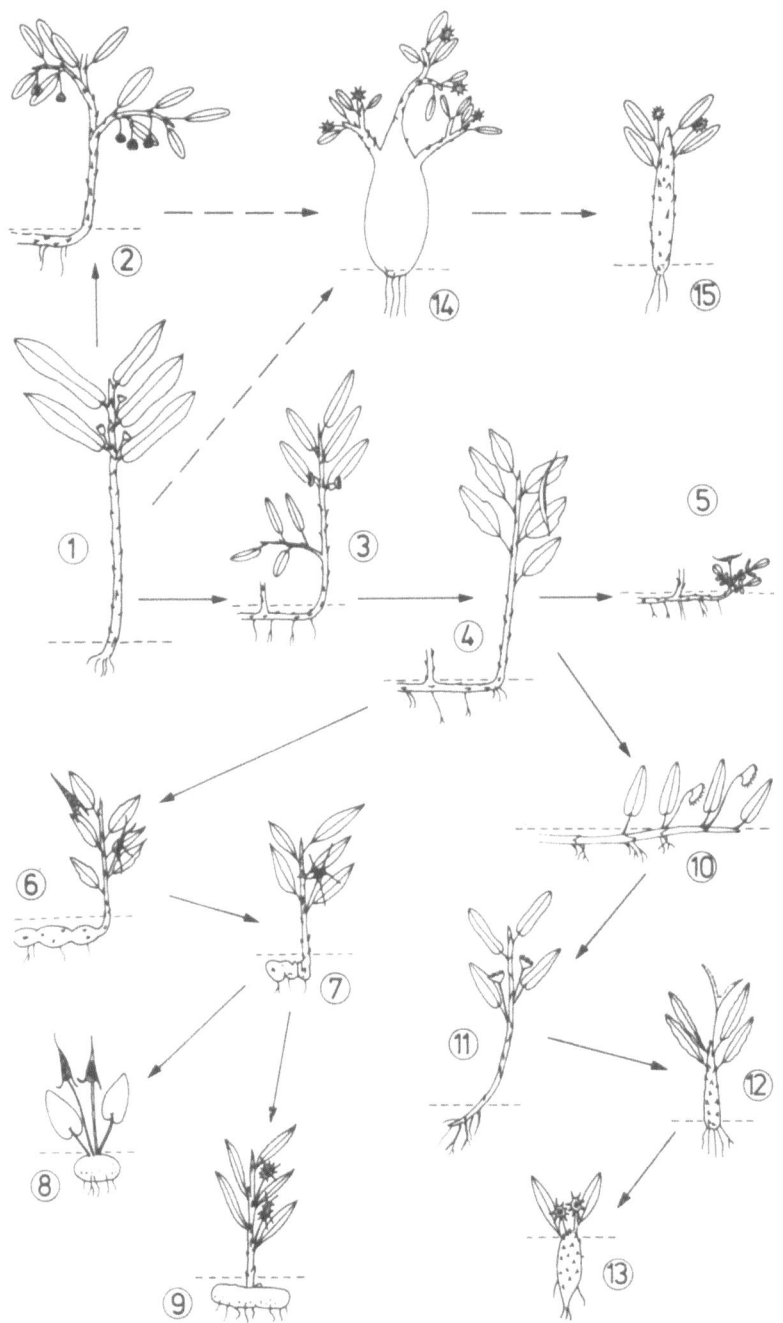

Fig. 1. Schematic drawings of the habits of various *Dorstenia* species, showing the main trends in differentiation. **1** *D. djettii*; **2** *D. africana*; **3** *D. elliptica*; **4** *D. psilurus*; **5** *D. letestui*; **6** *D. preussii*; **7** *D. cuspidata*; **8** *D. barnimiana*; **9** *D. benguellensis*; **10** *D. picta*; **11** *D. elata*; **12** *D. ramosa*; **13** *D. cayapia*; **14** *D. gigas*; **15** *D. crispa*

drier regions of Central and East Africa and have a (rather) wide distribution. The stem succulents like *D. gigas* SCHWEINF., which forms
up to 2 m tall branched shrubs, and *D. crispa* ENGL., which forms
low monocaul stems, cannot be clearly connected with the rhizome-
and tuber-forming group of species.

Dorstenia species from South and Central America can be morphologically connected with *D. picta*. Most of them are components of the
undergrowth of evergreen tropical forests and form short monocaul
stems. In species occurring in relatively dry habitats the stems are
subterranean and somewhat tuberous (cf. CARAUTA, VALENTE and SUCRE
1974). The absence of long creeping rhizomes or tubers derived from
them causes differences in habit and life form between African and
American species, notwithstanding their occurrence in similar habitats.

The above considerations outline the variations in growth-habit
(pachycaul to leptocaul, arborescent to herbaceous) recognizable on
the basis of the available evidence. It is undoubtedly worth-while
investigating this promising field much more closely in dealing with
the distribution, origin, and evolution of plant groups. As is apparent
from the above discussion, growth-habit variation may well be of
importance in taxonomy in spite of statements to the contrary (cf.
WHITMORE 1975).

Flower

The small flowers of the *Urticales* consist basically of 4–5 tepals,
4–5 epitepalous stamens, and a bicarpellate pistil with a single ovule.
There is not the space here to give a detailed survey and the reader is
referred to CORNER's work on the classification of the *Moraceae* (1962),
WEDDELL's monograph of the *Urticaceae* (1856–1857), and PLANCHON's
study on the *Ulmaceae* (1848).

The extremes in the variation of the flower in the *Urticales* are the
bisexual flower with 5 tepals and 10 or sometimes even more stamens, as
in *Ampelocera* (*Ulmaceae*), and the flower consisting of a pistil or a
single stamen only, as occurs in *Treculia* and *Brosimum*, respectively.
Although reduction of the number of floral parts is the main trend,
multiplication sometimes occurs as in the ulmaceous genera *Holoptelea* and *Ulmus* and the moraceous genus *Naucleopsis*.

In the *Ulmaceae* the flowers are bisexual or unisexual by abortion.
Differences between staminate and pistillate flowers are not pronounced
and specialized stamens and strongly reduced flowers do not occur
in this family.

In the other groups of the *Urticales* the flowers are unisexual. Pistillate and staminate flowers can be quite different in size and form of
the perianth. More or less strongly reduced flowers often occur mainly

in groups in which the stamens are straight before anthesis or if inflexed gradually straightening at anthesis, as in *Bosqueiopsis*.

Inflexed stamens which bend outwards suddenly and elastically at anthesis, thus shedding their pollen simultaneously, are distinctive for the *Urticaceae* s. str. and occur in many genera of the *Moreae*. Flowers with such stamens normally have a well-developed perianth with 4–5 tepals, 4–5 stamens, and a more or less conspicuous pistillode. They still show the basic structure of the *Urticales* flower, with the exception of the reduced pistil. This completeness of structure appears to be connected with the presence of the explosive stamens. A more or less strong expression of the bisexual condition is correlated with a less reduced state of the flower, at least as regards the number of tepals and stamens.

In groups of the *Urticales* with straight stamens (*Moraceae* and *Conocephaloideae*) the number of tepals and stamens is often reduced to 3 or 2. Further reduction can result in disappearance of the perianth, leaving a single stamen. *Brosimum lactescens* (S. MOORE) C. C. BERG shows reduction from a 4-merous staminate flower to a 2-merous one and *B. utile* (H. B. K.) PITTIER from a 2-merous one to a perianthless flower with a single stamen (cf. BERG 1972).

Loss of and diminished protection of young stamens by the perianth means that the stamens must be protected in other ways. In *Brosimum* the necessary protection is given by bracts which form a pseudo-perianth, by firm peltate bracts which cover the young flowers, and/or by a foveate receptacle which bears the staminate flowers in pits. In *Castilla* the stamens are protected by the cup-shaped or bivalvate receptacle which is closed until anthesis, in *Pseudolmedia* and *Naucleopsis* by the bracts of the involucre, and in the peculiar inflorescence of *Trilepisium* by an expanded margin of the receptacle. Even if a delicate perianth is present in staminate or pistillate flowers (as in *Treculia* and *Broussonetia*, respectively), bracts appear to have a protective function. In the *Moraceae* and *Dorstenieae* these bracts are usually (hypo)peltate, a form which often arises in connection with the protection of flowers and floral parts (cf. ENDRESS 1975).

In some moraceous genera with straight stamens, as in *Clarisia* and *Antiaris*, the staminate flowers can become more or less disorganized, by losing the distinct relationship between tepals and stamens.

The distribution of straight and inflexed (explosive) stamens suggests that the latter type is derived. The inflexed stamens probably appeared early in the evolution of the *Urticales* and separated its members into two groups. This separation put its stamp on the differentiation of the *Urticales*, especially in relation to pollination. In some genera, such as *Cudrania* and possibly also *Sorocea* and *Antiaropsis*, which taxonomically are more or less closely related to genera with inflexed stamens, the straight stamens are probably derived; this is suggested by their shape and insertion and/or dehiscence and/or the direction of the filament at different flowering stages.

Wind pollination probably occurs in all taxa with inflexed stamens. In the group of *Urticales* without such stamens, more or less distinct adaptations to wind pollination occur. A remarkable case is found in *Cecropia*. In many of its species the anther is detached from the straight filament by abscission, but it remains loosely attached to the flower (by the sticky ends of the appendices of the anther or by a bundle of stretched spiral thickenings of the tracheary elements of the staminal vascular bundle) and is movable; clearly, these species are anemophilous.

In other taxa, e.g. in several representatives of the *Ulmaceae*, but also in *Maquira coriacea* (KARSTEN) C. C. BERG and *Brosimum lactescens* (S. MOORE) C. C. BERG which both belong to groups not generally adapted to wind pollination, adaptation to wind pollination followed a more common way by producing slender filaments and relatively large anthers. In the pendulous catkins one can find another adaptation to wind pollination, which is important especially for taxa with non-specialized stamens (e.g. *Bagassa, Batocarpus, Sorocea*).

The deciduous habit connected with flowering may also be regarded as an adaptation to wind pollination (cf. WHITEHEAD 1969).

An important consequence of wind pollination is that plants depending on it require more or less open habitats, like (semi-)deciduous forests, secondary vegetations, savannas, forest margins, riversides, etc., i.e. places with sufficient air movement. Light demand is correlated with their presence in such habitats. Some arborescent, light-demanding, and more or less distinctly deciduous taxa which depend on wind pollination, like *Chlorophora excelsa* and *Bagassa guianensis* AUBL., can penetrate evergreen forest through accidental clearings. They become tall enough to make wind pollination possible. A few anemophilous taxa, like species of *Elatostema* and *Sloetiopsis usambarensis* ENGL., occur in the undergrowth of evergreen forests and can stand heavy shade, the former as herbs and the latter as small trees. But they chiefly occur near small streams where air movement can be expected. *S. usambarensis* also occurs as shrubs on exposed rocks in river beds and coastal shrub.

Pistillate flowers lacking a perianth or having a reduced one occur in several moraceous genera in which the flowers are enclosed in cavities formed by the receptacle of the infloresecnce (*Ficus*) or by bracts (*Treculia, Broussonetia*). If the pistillate flower is completely immersed in and fused with the receptacle of the inflorescence a perianth can seldom be distinguished. The tepals of the pistillate flower are often fused, forming a tubular perianth which in turn is often fused with the pistil or fruit in megaspermous taxa, and enlarged at fruit. Fusion of the tepals often precedes fusion of the flowers and occurs in many members of the *Castilleae*.

Anatomical studies on the flowers of *Urticales* (BECHTEL 1921) and of *Artocarpus* (SHARMA 1965) indicate that the pistil of the *Urticales* is derived from one with at least 3 carpels forming an ovary which has 3 or more locules with 2 or more axillary ovules per carpel. Ovaries with 2 or 3 locules and/or 3 styles occur rarely in the *Moraceae* and *Ulmaceae*. Further reduction of the gynoecium leads to the pseudo-monomerous pistil with an anatropous basal ovule, as occur in the *Urticaceae* and *Conocephaloideae*. Reduction of one of the two stigmas is observed in several *Moraceae* and can be regarded as a first step in this reduction which has no apparent adaptive significance.

The shape of the stigma varies considerably. In the anemophilous taxa it is usually filiform (with a moderate length and often pilose) or comose.

Fruit

In the *Ulmaceae* the fruits are either samaras (in *Ulmus* and the related genus *Holoptelea*) or drupe(let)s. In the *Urticaceae* the fruits are achenes or drupelets. The achenes can sometimes be enclosed by a winged fruiting perianth (in *Memoralis*) enveloped by a succulent fruiting perianth, forming pseudo-drupelets which are often aggregated (in many taxa of the *Boehmerieae*) or ejected by staminodes (e.g. in *Elatostema*).

The basic type of fruit in the *Moraceae* seems to be a dehiscent drupe and is found in several genera belonging to different tribes: *Antiaropsis, Bleekrodea, Broussonetia, Dorstenia, Fatoua, Ficus* (cf. CORNER 1962), *Scyphosyce, Sparattosyce, Streblus, Sloetiopsis*, and *Utsetela*. This fruit is more or less stipitate and the white fleshy exocarp splits into two valves that squeeze out or eject the endocarp body which is often smooth and blackish in megaspermous taxa, or greyish and tuberculate in microspermous taxa (*Dorstenia* species, *Fatoua*). The endocarp bodies drop near the plants. How they are dispersed subsequently and whether they are eaten by animals is unknown. Common gregarious occurrence of several taxa with these fruits suggests short-distance dispersal. In most cases the fruits or infructescences containg them are inconspicuous (often greenish) and often hidden by the foliage of the plants. In *Antiaropsis* and probably also in *Bleekrodea* the presence of a red fruiting perianth seems to promote dispersal because of the attractive red-black colour contrast.

The stipes and/or the white colour of the indehiscent drupelets such as occur in *Chlorophora excelsa, Bagassa guianensis* AUBL. and several species of *Broussonetia,* the way in which the exocarp is thickened in the indehiscent drupelets of *Morus* and related genera, and the occurrence of a transition from the dehiscent to the indehiscent drupelet in *Broussonetia* suggest that these two types of indehiscent drupes are close to the dehiscent type and that they may well be derived from the latter (see Fig. 2). These indehiscent drupelets are enclosed in succulent and often coloured spicate or globose infructescences which are attractive to animals.

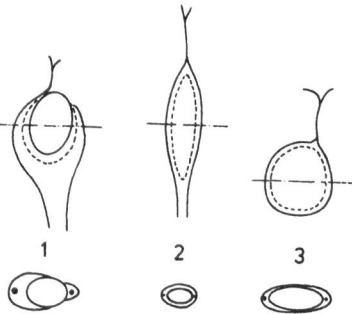

Fig. 2. Three types of moraceous fruits. **1** dehiscent drupelet (*Dorstenia*); **2** stipitate indehiscent drupelet (*Broussonetia*); **3** broadly based indehiscent drupelet (*Morus*)

In many *Moraceae,* especially in megaspermous taxa, the fruit is fused with the tubular enlarged fleshy and coloured perianth or immersed in and fused with an enlarged fleshy and coloured receptacle of the inflorescence, thus forming solitary or aggregated pseudo-drupes or drupaceous infructescences. In most of these cases it is not possible to distinguish a pericarp. The succulent perianths fused with the fruits are often yellow- to red-coloured, but sometimes they are black, e.g. in *Sorocea.* In this genus the pseudo-drupes are borne by contrasting red-coloured (often swollen) pedicels and the rachis of the racemose inflorescence. The enlarged receptacles fused with the fruits are usually yellow- to red-coloured as well, but are black in *Trilepisium.*

If the dehiscent drupe (not being adapted to long-distance dispersal) is the original type of fruit in the *Moraceae,* the coloured perianth in *Antiaropsis,* the formation of indehiscent drupes, enclosure of the fruits in coloured perianths or receptacles, and aggregation into syncarpous infructescences may be regarded as improvements in bringing about dispersal.

In the *Conocephaloideae* there are various kinds of fruit. In *Pourouma* and *Myrianthus* the megaspermous fruits together with the enlarged perianths form pseudo-drupes. The microspermous dry fruits of *Cecropia* and *Musanga* are enclosed by greenish succulent perianths. In the epiphytic genera *Coussapoa* and *Poikilospermum* the pericarp consists of an endocarp with small protuberances and a mucilaginous mesocarp and the endocarp body, surrounded by the mucilaginous layer, is (often) squeezed out by the basally thickened fruiting perianth.

The seeds vary from large, when they are usually without endosperm but with a more or less complicated embryo having thick and often unequal cotyledons and a short radicle, to small, when they are usually with endosperm and a rather simple embryo, rather flat and equal cotyledons, and a long radicle. As pointed out by CORNER (1962) microspermy is connected with a herbaceous habit, occurrence in secondary vegetations, or epiphytism. Megaspermy is more or less clearly linked with an arborescent habit and occurrence in tropical forests. Connections between seed characters and primitiveness appear to be lacking in the *Urticales*. Large seeds occur in many *Moraceae* (*Castilleae*, most *Dorstenieae*, and several *Moreae* like *Artocarpus* and *Treculia*) and in several *Ulmaceae*. Small seeds occur in the *Urticaceae*, in several *Moraceae* (*Ficus, Dorstenia*, and several members of the *Moreae*). *Dorstenia* shows a distinct morphological series in its range of large to small seeds: large seeds occur in the frutescent species, medium-sized ones mainly in the suffrutescent species, and small ones in the herbaceous and succulent species.

The testa is usually thin and its protective function has been transferred to the endocarp, which is usually crustaceous but may even be woody.

Inflorescence

An explanation of the morphological structure of the inflorescence can be given in general terms, although several details are still problematical (cf. GOLENKIN 1894, BERNBECK 1932). The basic structure of the *Urticales* inflorescence appears to be a compound dichasium (or cymose panicle, cf. CORNER 1967). The basal (= central) flowers in the bisexual inflorescence are pistillate but in the *Urticaceae* often staminate (cf. BERNBECK 1932).

In the aggregation of small flowers, the *Urticales* follow a common trend in the Angiosperms.

The *Ulmaceae* mostly have loose or condensed cymose or secondarily racemose inflorescences with bisexual, bisexual and staminate, or unisexual flowers. The number of flowers can be reduced to one in the pistillate inflorescence.

The main variations which are observed in the inflorescences of the *Urticaceae* s.l. are tendencies towards:

1. shortening of the axes;
2. reduction in the number of the axes;
3. fusion of the axes;
4. dorsi-ventral orientation (= adaxial orientation of the flowers);
5. change from dichasial to monochasial growth;
6. reduction in the number of flowers;
7. change from bisexual to unisexual.

Repeatedly branched inflorescences are common in the *Urticaceae* s. str. and the *Conocephaloideae*. In the former group the variation mostly results in paniculate, racemose, spicate, capitate, or sessile glomerate inflorescences, but sometimes in discoid (and involucrate) inflorescences (as in *Elatostema* in which genus they are transformed into a fig-like inflorescence in *Elatostema ficoides* WEDD.). The inflorescences of the *Urticeae* s. str. are either unisexual or bisexual. In most genera of the *Conocephaloideae* the flowers usually become clustered in heads at the ends of the branches of the unisexual inflorescence. Reduction in the ramification of the inflorescence can result in a single head, spike, or umbel in the pistillate inflorescence. In *Cecropia* the flowers are arranged in digitate clusters of spikes which are enveloped by a caducous spathe, thus forming an unique type of inflorescence. In the *Urticeae* s. str. and *Conocephaloideae*, as well as in the *Moraceae*, pistillate and staminate inflorescences are often (quite) different in size and shape; the pistillate inflorescences are usually the most derived ones.

Branched inflorescences are rare in the *Moraceae*. They are found in *Fatoua* and *Bleekrodea* and in both genera are mostly bisexual. The other members of the *Moreae* have mostly unisexual spikes, racemes, heads, or the more derived discoid-involucrate inflorescences

Fig. 3. Schematic drawings of the inflorescences of *Moraceae*, showing some of the main trends in differentiation. **A. *Castilleae*,** pistillate inflorescences. 1–3 hypothetical; **4a** *Perebea, Helicostylis*, etc.; **4b,** *Pseudolmedia, Perebea*, etc.; **5** *Perebea, Castilla*, etc.; **6a** *Castilla*; **6b** *Antiaris, Mesogyne*; **7** *Naucleopsis.* **B. *Castilleae*,** staminate inflorescences. 1–3 hypothetical; **4a** *Perebea, Maquira*, etc.; **4b** *Perebea humilis*; **5** *Helicostylis*; **6** *Naucleopsis*; **7** *Pseudolmedia*; **8** *Castilla*. **C. *Moreae*,** pistillate inflorescences. **1** *Trophis, Sorocea*, etc.; **2** *Trophis, Sorocea*, etc.; **3a** *Trophis involucrata*; **3b** *Olmedia aspera*; **4a** *Sorocea sprucei*; **4b** *Sorocea duckei*; **5a** *Clarisia ilicifolia*; **5b** *Clarisia racemosa*; **6a** *Antiaropsis*; **6b** *Antiaropsis, Phyllochlamys*; **7** *Sparattosyce*

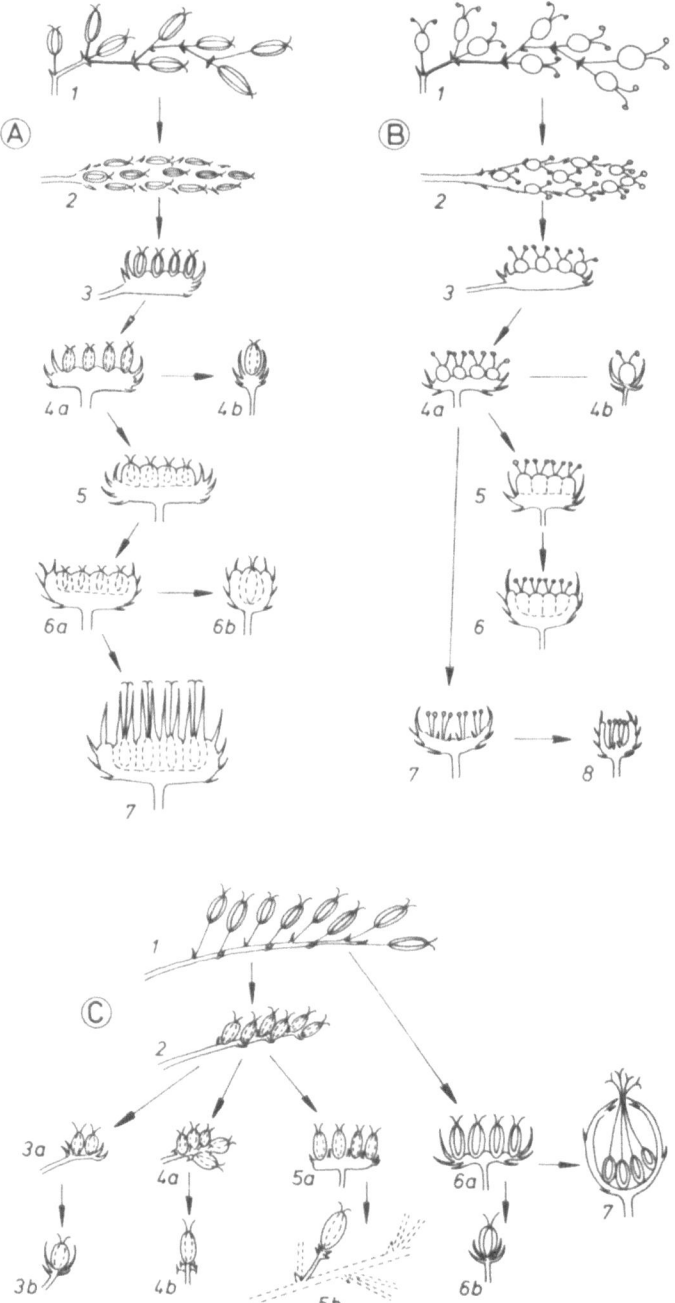

Fig. 3

(as in *Antiaropsis, Olmedia,* and *Phyllochlamys*) or the urceolate·inflorescences (in *Sparattosyce*). Further reduction can result in uniflorous pistillate inflorescences.

The *Castilleae* have unisexual, discoid to globose or to urceolate, involucrate inflorescences, which are often uniflorous in the pistillate ones but rarely so in the staminate ones. The *Dorstenieae* have discoid to globose or to urceolate, basically bisexual inflorescences, which may or may not be more or less distinctly involucrate. They can become unisexual. Pistillate inflorescences, on the other hand, can become uniflorous. The *Ficeae* have bisexual (or functionally unisexual) urceolate inflorescences.

Schematic drawings of the variation in the inflorescences of some groups of the *Moraceae* are given in Figs. 3 and 4.

The more or less distinctly involucrate inflorescences, such as occur in the *Castilleae*, in several *Dorstenieae*, and in a few *Moreae*, are more or less distinctly pseudanthous.

A remarkable aspect in the variation of the more or less pseudanthous inflorescences is that some trends in the differentiation of the angiospermous flower are repeated: many to few parts (= flowers), free to fused parts (= flowers), superior to inferior parts (= flowers), free or adnate to a hypanthium-like of the (cup-shaped) receptacle of the inflorescence. For the genus *Dorstenia* can be added the change from an actinomorphic to a zygomorphic inflorescence.

Distribution of the sexes is very variable. In the predominantly anemophilous families of the *Ulmaceae* and *Urticaceae* bisexual inflorescences are common and dioecism is not a dominant feature. However, dioecism is characteristic for the *Conocephaloideae*. In the *Castilleae*, with their strictly unisexual inflorescences and flowers which seldom show vestiges of the other sex, monoecious taxa are rather common. Furthermore, in some of these taxa (*Antiaris* and *Castilla*) both monoecoius and dioecious specimens can be found. In the *Dorstenieae*, with their basically bisexual inflorescences, staminate inflorescences can occur alongside the bisexual ones (in *Bosquieopsis* and *Helianthostylis*).

The inflorescences are solitary, or in the *Urticaceae* s.l. often in pairs, in the axils of the leaves. Secondary inflorescences can be borne

---

Fig. 4. Schematic drawings of the inflorescences of *Moraceae*, showing some of the main trends in differentiation. **D.** Mostly **Dorstenieae. 1** hypothetical; **2** *Bleekrodea* (*Moreae*); **3a** and **3b**, *Bleekrodea madagascariensis*; **4** *Utsetela*; **5a** and **5b**, *Helianthostylis*; **6** *Trymatococcus*; **7a**, **7b**, and **7c** *Bosquieopsis gilletii*; **8** *Brosimum*; **9a** and **9b**, *Brosimum lactescens*; **10a** and **10b**, *Brosimum utile*; **11** *Dorstenia djetii*; **12** *Dorstenia africana*; **13** *Dorstenia barteri*; **14** *Scyphosyce*; **15** *Trilepisium*; **16** *Ficus*

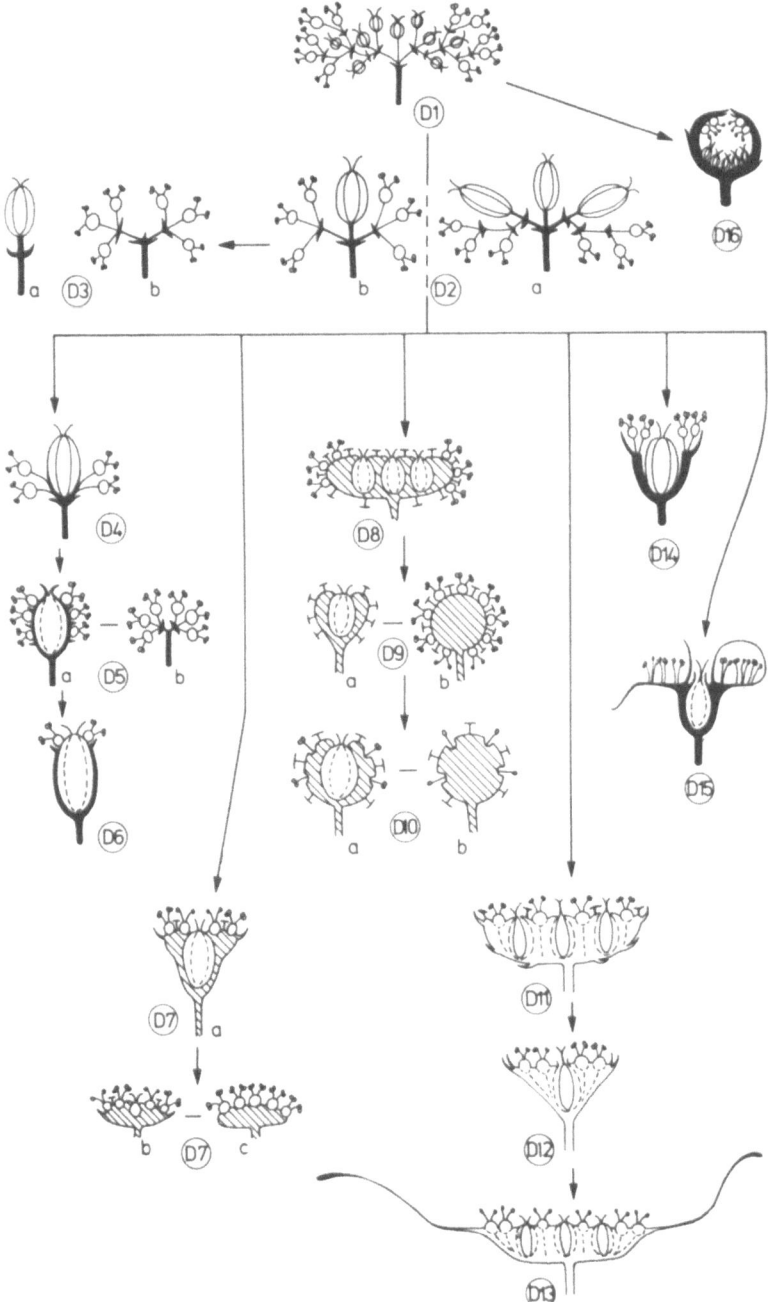

Fig. 4

on more or less condensed leafless shoots as in cauliflorous taxa or in taxa with uniflorous inflorescences like *Clarisia racemosa* R. & P. which has its uniflorous pistillate inflorescences in secondary racemes. In most *Castilleae* the inflorescences are borne in clusters in the leaf axils. In some of them (e.g. *Antiaris, Mesogyne*, and *Castilla*) staminate and pistillate inflorescences can occur in the same cluster; the staminate ones surround the pistillate one, thus simulating to some extent the arrangement of "male" and "female" parts in the bisexual flower.

The variation in the inflorescence of the *Urticales* is related to and partly governed by the structure and variation in the flower and fruit and the dimensions of the seed. Consequently, it is related to pollination and dispersal as well.

The small monochlamydeous flowers of the *Urticales* do not produce nectar. A few taxa are known to produce odour; mostly it is only staminate flowers including those of some anemophilae, which do this. The perianth is uncoloured, at least at anthesis. These disadvantages with regard to insect pollination appear to have been more or less successfully overcome in two ways. First, by adaptation to wind pollination, largely achieved by modification of the stamens (see p. 357–358). In groups having such stamens the shape of the inflorescence does not seem to be so important although arrangement of the staminate flowers in catkins may promote shedding of the pollen. In groups whose stamens are not adapted to wind pollination, pendulous spicate or racemose staminate inflorescences are essential for pollination. The presence of such inflorescences is correlated with an arborescent or frutescent habit and if arborescent often with deciduousness as well. In these groups of the *Urticales* variation in the pistillate inflorescence is chiefly related to dispersal.

The arrangement of flowers (which do not have stamens adapted to wind pollination) in secondary functional units more suited to insect pollination is the other way be which disadvantages due to the reduced state of the flower have been overcome. In many cases these units simulate more or less closely the structure and appearance of flowers. In some *Moraceae* pseudanthy is very pronounced. The staminate inflorescences of most *Naucleopsis* species resemble the flowers of *Tovomita* (*Guttiferae*) and *Ternstroemia* (*Theaceae*) and the bisexual inflorescences of *Trilepisium* resemble flowers of several *Myrtaceae*. The significance of the differences between staminate and pistillate inflorescences in groups which otherwise have more or less distinctly pseudanthous inflorescences is puzzling.

Pseudanthy is connected with occurrence in tropical forests. Unfortunately, data about pollination in groups with pseudanthous inflorescences are (almost) lacking. As a rule, the fructification rate is normal, which, assuming that reproduction is sexual, suggests successful pollination. However, some facts about *Dorstenia* suggest that adaptation to insect pollination is less satisfactory. In many *Dorstenia* species some features (colour, shape of the bract-arms and appendages) of the more or less distinctly pseudanthous bisexual inflorescences with their minute flowers show a tendency towards adaptation to pollination by flies. A study of several populations of *Dorstenia africana* (BAILL.) C. C. BERG (ined.), which has yellow, faintly smelling inflorescences, showed that a very low percentage of the inflorescences set fruit which indicates only occasional pollination. The same is found in *Scyphosyce manniana* BAILL. which in many characters resembles *Dorstenia*. The occurrence of apomixis (cf. GUSTAFSON 1946), as well as the presence of rhizomes to ensure vegetative reproduction (cf. RICHARDS 1952) and possibly the production of a wide range of inflorescence forms, often bizarre and apparently without significance, could well be related to unsuccessful adaptations to insect pollination.

It is remarkable that the urticaceous genus *Elatostema* which resembles *Dorstenia* species in so many features also contains apomictic species (cf. FAGERLIND 1944). As in *Dorstenia*, many species of *Elatostema* have discoid, more or less distinctly involucrate, mostly bisexual inflorescences with short stamens in the staminate flowers. They usually form creeping rhizomes and often occur in the undergrowth of evergreen forests, a habitat which does not promote wind pollination. *Elatostema* even resembles *Dorstenia* in regard to the ejection of the fruits (containing seeds without endosperm in contrast with other *Urticeae* s. str.) and also in its taxonomical complexity.

Successful adaptation to insect pollination is only known for *Ficus*, with its pseudocarpous inflorescences. More or less pseudocarpous (although quite different from *Ficus* in structure and appearance) are the pistillate inflorescences which occur in *Artocarpus* and *Treculia*. In spite of some records on pollination in *Artocarpus* (JARRETT 1959 to 1960, FAEGRI & VAN DER PIJL 1972) how it takes place in these genera is uncertain.

Cauliflory in the *Moraceae* is connected with the presence of these pseudocarpous inflorescences. In *Ficus* one can find almost the whole range of types of cauliflory (cf. CORNER 1933, RICHARDS 1952) and it is probably related to dispersal rather than to pollination (cf. VAN DER PIJL 1972). The same may possibly be true for the other moraceous taxa with pseudocarpous inflorescences.

The more or less pseudanthous and, of course, the pseudocarpous pistillate and bisexual inflorescences form at the fruiting stage one- to many-seeded structures which often resemble true fruits. In *Treculia africana* DECAISNE and several species of *Artocarpus* these structures resemble the fruits of the bombacaceous genus *Durio*. The spinous infructescences of *Treculia acuminata* BAILL., *T. obovoidea* N. E. BROWN, and several species of *Naucleopsis* resemble the fruits of *Oncoba* (*Flacourtiaceae*) and *Nephelium* (*Sapindaceae*). The infructescences formed by several multiflorous *Castilleae* are reminiscent of the compound fruits which occur in the *Annonaceae*. Other inflorescences, like those in species of *Brosimum*, form drupaceous infructescences.

General Remarks

Surveying the differentiation patterns of the *Urticales* I am inclined to regard them for the greater part and especially with reference to reproductive structures as the results of two, not always clearly distinguishable processes. In the first of these processes adaptive aspects are not apparent or may even be lacking, while in the second one developments seem to be adaptive responses to changes caused by the first process. Examples of the first process can be found in the trends towards the reduction of flowers and inflorescences — trends which can also be found in other groups of the Angiosperms. The process with distinct adaptive aspects can be found in developments re-establishing successful pollination and dispersal.

The reduced state of the flower can be regarded as a leading factor concerning the variation in the *Urticales*, especially with regard to their reproductive structures. Two main complexes can be distinguished. In one complex, the adaptations centre round insect pollination; in the other, the adaptations relate to wind pollination. The former complex is clearly linked to the conditions of evergreen (to semi-deciduous) tropical lowland forests. The group of plants involved can be roughly characterized by an arborescent habit (comprising rosette trees), by pseudanthous (to pseudocarpous) inflorescences with further reductions in the flowers and unspecialized stamens, and by megaspermous (aggregated) pseudo-drupes.

The number of *Moraceae* which can be regarded as characteristic representatives of this group of plants is quite small, but the concerned species show considerable diversity, notwithstanding the rather constant conditions under which they probably arose. The large and diverse genera *Ficus* and *Dorstenia* do not fully match the features of this group of *Moraceae*, in view of the occurrence of microspermy in both genera, the herbaceous habit in *Dorstenia*, and the extension of their areas of distribution far outside the rain forest. This last seems in *Dorstenia* to be due mainly to the herbaceous habit and in *Ficus* to the special way of pollination, which makes it independent of its pollinators in growth habit, habitat, and arrangement of the inflorescences (cf. VAN DER PIJL 1972).

The other complex is linked to the occurrence in more or less open (mesic to semi-arid) habitats, chiefly tropical but also subtropical to temperate (see p. 353). This group of plants comprises the majority of the *Urticales* (*Ulmaceae*, *Urticaceae*, and many taxa of the *Moraceae*) and shows a wide variation in habit (arborescent to herbaceous) and inflorescences (showing adaptation to wind pollination or not, seldom pseudanthous), the least reduced staminate flowers and mostly specialized stamens. The fruits are dry or succulent, microspermous or megaspermous. The plants are essentially light-demanders and are often pioneers. In several of its characters, especially in those of the inflorescences (see p. 367), the urticaceous genus *Elatostema* tends towards the group of *Moraceae* belonging to the first complex.

The *Conocephaloideae*, a diverse group of light-demanding plants with aerial or stilt roots, cannot be satisfactorily placed in either of the two complexes. It comprises three groups of morphologically and ecologically vicarious genera: *Coussapoa* and *Poikilospermum*, *Pourouma* and *Myrianthus*, and *Musanga* and *Cecropia*. *Cecropia* alone can be fitted into the second complex, as it exhibits clear adaptations to wind pollination, which are lacking in the other genera.

The four, more or less exceptional genera mentioned above (*Ficus*, *Dorstenia*, *Elatostema*, and *Cecropia*) demonstrate that abundant speciation, occurrence in a wide range of habitats, and considerable morphological diversity do not depend solely on structures well-adapted to pollination and dispersal. These four genera are the largest or second largest ones of the subdivisions to which they belong.

Ficus and *Cecropia* can be regarded as well-adapted to insect and wind pollination, respectively. They are well adapted for dispersal and they occupy a wide or relatively wide range of habitats. Not only successful adaptation, but the fact of having evolved rather complicated and unique structures to which they owe successful pollination and dispersal, seem to promote speciation and diversification. In contrast with *Ficus* and *Cecropia*, developments in *Dorstenia* have not led to successful adaptations either to (insect) pollination or to dispersal. Nevertheless, this genus comprises numerous species, occupies a wide range of habitats, and shows considerable diversity, expecially in the inflorescence and growth habit. The strange patterns in the differentiation of the inflorescence, the switching over to asexual (apomictic and/or vegetative) reproduction, and the considerable cytological differentiation (cf. LE COQ 1963, FEDOROV 1969) can be regarded as responses to adaptive failures in the reproductive structures.

Elatostema is the urticaceous parallel of *Dorstenia* (see p. 367). The combination of the discoid-involucrate inflorescence and reflexed stamens seems to be an important disadvantage with regard to pollination. While the trends in differentiation led *Dorstenia* outside the rain forest, the reverse apparently occurred with *Elatostema*.

For the *Moraceae* STEBBINS (1974) assumed a reversal from wind pollination to insect pollination. This assumption is not supported by the differentiation patterns found in the *Urticales*. STEBBIN's assumption appears to be based on the consideration that (most?) angiospermous groups arose under semi-arid conditions and that their presence in tropical rain forest is secondary.

Interpretation of morphological (and anatomical) differentiations and reading of morphological series depend largely on the nature of the conditions (humid tropical, humid subtropical, montane, semi-arid) which are accepted as being those under which the Angiosperms or angiospermous groups originated. Postulating these conditions seems to be easier for those groups which have retained primitive characters than for old groups with derived characters.

In the *Urticales*, the most derived and advanced forms of flowers and inflorescences and, on the other hand, the presumed primitive and less derived growth habits are bound up with humid tropical low-land conditions.

Position of the *Urticales*

In recent classifications of the Angiosperms the *Urticales* can be found in a position either near the *Hamamelidales* (TAKHTAJAN 1969, CRONQUIST 1968) or near the *Malvales* (THORNE 1968, STEBBINS 1974, DAHLGREN 1975).

The reduced monochlamydeous flower, the common occurrence of anemophily, and the presence of catkins have tied the *Urticales* to the "*Amentiferae*", especially to the *Fagales*. Many of these amentiferous orders have been connected with the *Hamamelidales* and of these relationships, that of the *Fagales* with the *Hamamelidales* is established and generally accepted.

Most arborescent *Urticales* inhabiting subtropical to temperate regions resemble representatives of the *Fagales* in growth habit and morphology of the shoots and leaves. TIPPO (1938) concluded from the results of a comparative anatomical study of the wood from members of the *Urticales* and their presumed allies that the *Urticales* can be considered to be derivatives of the *Hamamelidales*.

On the other hand, differences in the structure of and variation in the inflorescence (ENDRESS 1970), fruit, and seed do not support the presumed relationship between the *Urticales* and *Hamamelidales* and their allies. Furthermore, a 4–(5-)merous flower with a single whorl of small sepaloid perianth parts, bicarpellate pistils, and adaptations to wind pollination appear in reduction series in separate groups of the Angiosperms. Similarities in growth habit and morphology of the shoots and leaves can be ascribed to similarities in environmental conditions. Phytochemical differences between the *Urticales* and *Hamamelidales* shed doubt on the presumed relationship of those groups as well (cf. MEEUSE 1975).

In his monograph of the *Urticaceae* WEDDELL (1856–1857) discussed several characters indicating affinity of this family with the *Malvales*, especially with the *Tiliaceae*. In commenting on the ENGLER system of the Angiosperms HALLIER (1903) found in the characters of the *Urticales* more indications for a connection with the *Malvales* and *Euphorbiaceae* than with the *Hamamelidaceae*, *Fagaceae*, and *Betulaceae*.

In summary, the reasons most of which have already been used by WEDDELL and HALLIER for considering the *Urticales* to be related to the *Malvales* are: the presence of bast fibres and mucilaginous cells and ducts in both orders, the common occurrence of connate stamens (revealed in the *Ulmaceae* by the anatomical study of BECHTEL 1921) and valvate tepals in the *Urticales* and the occurrence of unilocular anthers with two pollen-sacs in *Ficus* subg. *Urostigma* sect. *Malvanthera* (CORNER 1960); similarities between the orders with regard to the variation in growth habit and leaves (simple or, if deeply incised or compound, then palmate), and the patterns in variation of the inflorescence, fruit, and seed of the *Urticales* which matches that of the *Malvales* more clearly than that of the *Hamamelidales* and their allies.

WEDDELL regarded the differences in the androecium of the *Urticaceae* and the *Malvales* and the persistent perianth in the *Urticaceae* as contradicting the affinities of the orders. The latter point does not seem to be very important and the differences in the androecium are less acute if the androecium of the *Ulmaceae* and the results of VAN HEEL's study on the androecium of the *Malvales* (1966) are taken into account. Neither from the anatomical characters of the wood (pers. comm. Dr. A. M. W. MENNEGA) nor from the chromosome numbers arguments against the supposed relationship of the orders can be obtained.

Most facts indicate that the *Urticales* belong to a central complex in the Angiosperms, comprising at least the *Urticales*, *Malvales*, *Euphorbiales*, and the *Violales* sensu CRONQUIST (1968). This is largely in agreement with DAHLGREN's view (1975). The affinities of the *Urticales* with the *Malvales* are closest between the *Ulmaceae* and *Tiliaceae*, the least derived families of the orders. Both orders and the *Euphorbiaceae* provide many tropical pioneer tree species (cf. WHITMORE 1975). This is connected with differentiation in growth habit and other characters and in this respect the *Euphorbiaceae* show similarities to the *Urticales*. The relation between these groups seems to be collateral.

The relationships of the *Urticales* with the *Violales* are rather vague and point to the *Flacourtiaceae*, as do many taxa of the central complex (cf. Takhtajan 1969), and this indicates a key position for that family.

The author is much indebted to Dr. N. G. Bisset for the correction of the English text and for his valuable critical remarks, to Dr. A. M. W. Mennega for her critical reading of the text, to the Netherlands Foundation for Tropical Research (WOTRO) for the grants which enabled the author to carry out field studies on *Moraceae* that form part of the basis of the present paper. Mr. H. Rypkema prepared the drawings.

References

Bechtel, A. R., 1921: The floral anatomy of the *Urticales*. Amer. Jour. Bot. **8**, 386—410.

Berg, C. C., 1972: Flora Neotropica Monograph No. **7**. *Olmedieae* and *Brosimeae* (*Moraceae*). New York: Hafner.

— 1973: Some remarks on the classification and differentiation of *Moraceae*. Meded. Bot. Mus. Herb. Utrecht No. **386**.

— 1977: The *Castilleae*, a tribe of the *Moraceae*, renamed and redefined due to the exclusion of the genus *Olmedia* from the "*Olmedieae*". Acta Bot. Neerl. **26**, 73—82.

Bernbeck, F., 1932: Vergleichende Morphologie der Urticaceen- und Moraceen-Infloreszenzen. Bot. Abhandl. Heft **19**.

Carauta, J. P. P., Valente, M. da C., and Sucre, B. D., 1974: *Dorstenia* L. (*Moraceae*) dos Estados da Guanabara e do Rio de Janeiro. Rodriguésia **27**, 225—278.

Corner, E. J. H., 1933: A revision of the Malayan species of *Ficus*: *Covellia* and *Neomorphe*. Jour. Malay. Br. Asiat. Soc. **11**, 1—65.

— 1949: The durian theory or the origin of the modern tree. Ann. Bot. N. S. **13**, 367—414.

— 1960: Taxonomic notes on *Ficus* Linn., Asia and Australasia. Gard. Bull. Singapore **17**, 368—485.

— 1962: The classification of *Moraceae*. Gard. Bull. Singapore **19**, 187—252.

— 1967: *Ficus* in the Solomon Islands and its bearing on the Post-Jurassic history of Melanesia. Philos. Trans. Roy. Soc. London B. **252**, 23—159.

Cronquist, A., 1968: The evolution and classification of flowering plants. Boston.

Dahlgren, R., 1975: A system of classification of the Angiosperms to be used to demonstrate the distribution of characters. Bot. Notiser. **128**, 119—197.

Endress, P. K., 1970: Die Infloreszenzen der apetalen Hamamelidaceen, ihre grundsätzliche morphologische und systematische Bedeutung. Bot. Jahrb. **90**, 1—54.

— 1975: Nachbarliche Formbeziehungen mit Hüllfunktion im Infloreszenz- und Blütenbereich. Bot. Jahrb. **96**, 1—44.

Faegri, K., and Pijl, L. van der, 1971: The principles of pollination ecology. Oxford etc.: Pergamon Press.

Fagerlind, F., 1944: Die Samenbildung und die Zytology bei agamospermischen und sexuellen Arten von *Elatostema* und einigen nahestehenden Gattungen nebst Beleuchtung einiger damit zusammenhängender Probleme. Kung. Svensk Vet. Akad. Handl. **21**, 1—130.

FEDOROV, A. A., 1969: Chromosome numbers of flowering plants. Leningrad: V. L. Komarov Botan. Institute.

GOLENKIN, M., 1894: Beitrag zur Entwicklungsgeschichte der Inflorescenzen der Urticaceen und Moraceen. Flora **78**, 97—132.

GOOD, R., 1974: The geography of flowering plants. London: Longman.

GUSTAFSON, A., 1946: Apomixis in higher plants. Lunds Univ. Arsskr. N. F. Avd. 2, **42**, 1—66.

HALLÉ, F., and OLDEMAN, R. A. A., 1970: Essai sur l'architecture et la dynamique de croissance des arbres tropicaux. Paris: Masson & Cie.

HALLIER, H., 1903: Über die Verwandtschaftsverhältnisse bei ENGLER's Rosalen, Parietalen, Myrtifloren und in anderen Ordnungen der Dikotylen. Abhandl. Naturw. Hamburg **18**, 3—98.

HEEL, W. A. VAN, 1966: Morphology of the androecium in *Malvales*. Blumea **13**, 174—394.

JARRETT, F. M., 1959—1960: Studies in *Artocarpus* and allied genera. Jour. Arnold Arb. **40**, 1—37, 113—155, 298—368; **41**, 73—140, 320—340.

KWAN KORIBA, 1958: On the periodicity of tree growth in the tropics. Gard. Bull. Singapore **17**, 11—81.

LE COQ, C., 1963: Contributions à l'étude cyto-taxinomique des Moracées et des Urticacées. Revue Gen. Bot. **70**, 385—426.

MEEUSE, A. D. J., 1975: Floral evolution in the *Hamamelididae*. III. *Hamamelidales* and associated groups including *Urticales*, and final conclusions. Acta. Bot. Neerl. **24**, 181—191.

MILLINGTON, W. F., and CHANEY, W. R., 1973: Shedding of shoots and branches. In: Shedding of plant parts (KOZLOWSKI, S. T.), 149—204. New York-London: Academic Press.

PLANCHON, J. E., 1848: Sur les Ulmacées et la famille des Urticées. Ann. Sci. Nat. Paris Bot. III. **10**, 244—341.

PIJL, L. VAN DER, 1972: Principles of dispersal in higher plants. Berlin-Heidelberg-New York: Springer.

RAVEN, P. H., 1975: The bases of Angiosperm phylogeny: Cytology. Ann. Missouri Bot. Gard. **62**, 724—764.

RICHARDS, P. W., 1972: The tropical rain forest. Cambridge: University Press.

ROUX, J., 1968: Sur le comportement des axes aériens chez quelques plantes à rameaux végétatifs polymorphes; le concept de rameaux plagiotrope. Ann. Sci. Nat. Paris. Bot. XII. **9**, 109—256.

SHARMA, M. R., 1965: Morphological and anatomical investigations on *Artocarpus* FORST. III. The flower. Phytomorphology **15**, 185—201.

SMITH, C. A., 1963: Shoot apices in the family *Moraceae* with a seasonal study of *Maclura pomifera* (RAF.) SCHNEID. Bull. Torrey Bot. Club **90**, 237—258.

STEBBINS, G. L., 1974: Flowering plant. Evolution above the species level. Cambridge, Mass.: Belknap Press.

TAKHTAJAN, A., 1969: Flowering plants. Origin and dispersal. Edinburgh: Oliver & Boyd.

THORNE, R. F., 1968: Synopsis of putatively phylogenetic classification of the flowering plants. Aliso **6**, 57—66.

TIPPO, O., 1938: Comparative anatomy of the *Moraceae* and their presumed allies. Bot. Gaz. **100**, 1—99.

TOMLINSON, P. B., and GILL, A. M., 1973: Growth habits of tropical trees: Some guiding principles. In: Tropical forest ecosystems in Africa and South America: A comparative review (MEGGERS, B. J., AYENSU, E. S., and DUCKWORTH, W. D.), 129—143. Washington: Smithsonian Institution Press.

Webster, G. L., 1967: The genera of *Euphorbiaceae* in the Southeastern United States. Jour. Arnold Arb. **48**, 303—430.

— 1975: Conspectus of a new classification of the *Euphorbiaceae*. Taxon **24**, 593—601.

Weddell, H. A., 1856—1857: Monographie de la famille des Urticées. Arch. Mus. Hist. Nat. Paris **19**, 1—591.

Whitehead, R., 1969: Wind pollination in the Angiosperms: Evolutionary and environmental considerations. Evolution **23**, 28—35.

Whitmore, T. C., 1975: Tropical rain forests of the Far East. Oxford: Clarendon Press.

Address of the author: Dr. C. C. Berg, Institute for Systematic Botany, State University Utrecht, Transitorium II, Heidelberglaan 2, Utrecht, The Netherlands.

Plant Syst. Evol., Suppl. 1, 375—395 (1977)
© by Springer-Verlag 1977

Systematisch-Geobotanisches Institut, Universität Göttingen,
Federal Republic of Germany

New Aspects of the Systematics of *Asteridae*

By

Gerhard Wagenitz, Göttingen

Abstract: Five main groups can be distinguished in the *Asteridae*:
1. *Gentianales—Rubiales* (and *Oleales*?); 2. *Polemoniales* (including *Solanaceae*
and *Boraginaceae*); 3. *Scrophulariales—Lamiales*; 4. *Dipsacales*; 5. *Campanulales—Asterales*. The *Loasaceae*, *Fouquieriaceae*, and *Columelliaceae* have
several characteristics of the *Asteridae* but can scarcely find their place in
this group as circumscribed here. Several plants with much reduced flowers
have to be included in the *Asteridae*: *Theligonum* in the *Rubiales*; *Callitriche*,
Hippuris and perhaps also *Hydrostachys* in the *Scrophulariales—Lamiales*
group. For the *Scrophulariales—Lamiales* and the *Dipsacales*, affinities
with the woody *Saxifragales* and *Cornales* are most probable and several
genera have been shifted between these groups. The *Asterales* show remarkable phytochemical agreements with the *Araliales*. Several authors have
proposed a breaking up of the *Asteridae* but the evidence for this does not
seem conclusive.

Introduction

The *Asteridae* comprise the "core" of the old *Sympetalae* or *Gamopetalae*. The name and the status as a subclass *Asteridae* are only 12 years
old (TAKHTAJAN 1964) but the group is not a new one, being equivalent
to a great extent to the "*Sympetalae*, Reihe *Haplostemones*" (EICHLER
1875), "Anisokarpe Gamopetalen" (GOEBEL 1882) and the "*Sympetalae-
Tetracyclicae*" (WARMING & MÖBIUS 1929). The *Asteridae* have been
accepted as a taxonomic and phylogenetic unit in the systems of CRON-
QUIST (1957, without name; 1968), TAKHTAJAN (1959, "Überordnung
Tubiflorae"; 1966—1973), and STEBBINS (1974). In several other recent
systems families of the *Asteridae* may be found dispersed in two or more
places implying a multiple origin of the group (HUTCHINSON 1926–1973,
EMBERGER 1960, SOÓ 1961–1975, THORNE 1968, DAHLGREN 1975).

I shall discuss here the characters of the *Asteridae*, their main groups,
the position of some aberrant families, and lastly the question of affinities with one or more groups outside the *Asteridae*.

Characters and Tendencies Common to the *Asteridae*

According to CRONQUIST (1968) about a third of the species of dicotyledons belong to this subclass. A calculation on the basis of the species numbers given in the "Syllabus der Pflanzen-Familien" (MELCHIOR 1964), with corrections for some families, amounts to about 58,700 species or 34% of the dicotyledons (WAGENITZ 1967). Nevertheless this subclass is more uniform in terms of the floral diagram and embryological characters than the others.

With very few exceptions the flowers are basically pentamerous with a connate or much reduced calyx, a sympetalous corolla and one whorl of stamens alternating with the corolla-lobes and adnate to the tube. The number of stamens is often reduced to 4 or 2 (more rarely 3 or only 1) in zygomorphic flowers. The gynoecium is syncarpous with the notable exception of the *Apocynaceae* and *Asclepiadaceae*. In most cases there are only two carpels, but exceptions to this rule are numerous. Some of the carpels may be sterile, eventually leading to a pseudomonomerous gynoecium. The most constant character is the structure of the ovules which are unitegmic, tenuinucellate and anatropous (or hemianatropous). Very few exceptions are known: campylotropous ovules are said to occur in the *Acanthaceae* and almost orthotropous ones in *Avicennia* of the *Verbenaceae* (DAVIS 1966), while the ovule of *Phryma* reported earlier as orthotropous (COOPER 1941) is hemianatropous (WHIPPLE 1972). The ovules of the *Convolvulaceae* are said to be bitegmic and tenui- or crassinucellate by DAVIS (1966). The structure of the nucellus is apparently variable in this family but the ovules are unitegmic (see SCHNARF 1931, GOVIL 1970), the statement to the contrary in DAVIS must be due to a mistake. Not constant for the group and not restricted to it, but more common here than in any other subclass is the cellular endosperm (in the following "cellular endosperm" is used as an abbreviation for "ab initio cellular endosperm formation").

The morphological trends (progressions) of the flowers in the *Asteridae* are known from other subclasses too. The following are of paramount importance:

1. Flowers (corolla) regular → irregular
2. Reduction in the number of stamens
3. Reduction in the number of ovules
4. Ovary superior → inferior

If flower morphology is surveyed in connection with pollination ecology three diverging tendencies are especially noteworthy. One is found in many members of *Scrophulariaceae*, *Gesneriaceae*, *Acanthaceae*, and *Labiatae* (*Lamiaceae*) exhibiting rather large showy zygomorphic flowers which are adapted to more or less restricted groups of pollinators. The structure of the flower guarantees that the pollen of the four or two stamens is deposited on the insects or birds in such a manner that pollination is effective. Very specialized floral mechanisms have evolved in these families, e.g. in *Salvia* (HILDEBRAND 1866, CORRENS 1891, GRANT & GRANT 1964), *Pedicularis* (SPRAGUE 1962, MACIOR 1968, 1970) or *Calceolaria* and *Diascia* (VOGEL 1974). A second trend leads to the development of long and narrow floral tubes restricting effective visits to animals with long and thin mouth-parts. Good examples may be found in the *Polemoniaceae* (GRANT & GRANT 1965), *Solanaceae*, *Apocynaceae*, and *Rubiaceae* (SILBERBAUER-GOTTSBERGER & GOTTSBERGER 1975), but even in some *Scrophulariaceae* (VOGEL 1954). A third principle has been also very successful: numerous small flowers are aggregated into a dense inflorescence which may finally form a pseudanthium in the sense of TROLL (1928) as in the *Compositae* or *Dipsacaceae*. Besides of these two families a similar type of inflorescence can be found e.g. in the *Polemoniaceae* (*Ipomopsis* sect. *Microgilia*), *Rubiaceae* (*Cephalanthus*), *Valerianaceae*, *Globulariaceae*, *Calyceraceae*, and *Campanulaceae* (*Jasione*, *Phyteuma*). Finally one group should be mentioned which in the dicotyledons is unparalleled in the complication of flower structure: the *Asclepiadaceae* with their elaborate system for the transfer of pollen or pollinia by "translators".

Although I am well aware of all the difficulties involved in the calculation of an "advancement index", I think that in the case of the *Asteridae* it gives a good idea of the relatively advanced position of this group. According to SPORNE (1969, 1974) 38 of the 61 families of dicotyledons with an advancement index of 70% or more belong to the *Asteridae* and the families of this subclass have a total range of this index from 57% (*Apocynaceae*) to 100% (*Callitrichaceae*, *Hippuridaceae*, *Hydrostachyaceae*, *Phrymaceae*), with only 5 families below 70%. The same is shown in the diagram of STEBBINS (1974) where no order of *Asteridae* approaches the center.

The Main Groups of the *Asteridae*

Five main groups (evolutionary lines) seem rather evident in the *Asteridae*, taking into account a combination of morphological, embryological and phytochemical characters (Table 1).

Gentianales — Rubiales (and *Oleales*)

The families of the *Gentianales* (*Contortae*) have already been brought together in one order by BARTLING (1830). The close affinity of the *Rubiaceae*—distinguished mainly by the inferior ovary and the lack of intraxylary phloem—has been noted earlier but was formally accepted

Table 1. Characters distinguishing main groups in the *Asteridae*

	Leaves	Stipules	Inflorescence	Symmetry of Flowers	Stamens	Ovary	Endosperm	Iridoid Compounds
1a. *Gentianales*	opposite	+ → 0	monotelic	*	isomerous	superior	nuclear (cellular)	+
b. *Rubiales*	opposite	+	monotelic	*	isomerous	inferior	nuclear (cellular)	+
c. *Oleales*	opposite	0	monotelic	*	oligomerous	superior	cellular	+
2. *Polemoniales*	alternate (opposite)	0	monotelic (polytelic)	* (↓)	isomerous	superior	nuclear → cellular	0
3. *Scrophulariales-Lamiales*	opposite or alternate	0	polytelic	→	(isomerous) → oligomerous	superior (inferior)	cellular	+
4. *Dipsacales*	opposite	0	monotelic → polytelic	(*) →	isomerous → oligomerous	inferior	cellular	+
5a. *Campanulales*	alternate	0	monotelic → polytelic	*	isomerous (oligomerous)	inferior	cellular	+ 0
b. *Asterales*	opposite or alternate	0	polytelic	*	isomerous	inferior	nuclear or cellular	0

only in recent times (WAGENITZ 1959, CRONQUIST 1968, TAKHTAJAN 1966–1973).

The most important common characters of this group are: leaves opposite, entire, with stipules (sometimes rudimentary, lacking in the *Gentianaceae*), intraxylary phloem present (except in *Rubiaceae*). Flowers actinomorphic with convolute, valvate or rarely imbricate aestivation, stamens isomerous. Endosperm nuclear (cellular in few *Gentianaceae* and *Rubiaceae*). Essential chemical characters are the presence of indole alkaloids (*Loganiaceae*, *Apocynaceae*, *Rubiaceae*) and cardenolides (*Apocynaceae*, especially *Echitoideae*, *Asclepiadaceae*). The indole alkaloids belong to the group of complex seco-iridoids; simple seco-iridoids and a few iridoids also occur (JENSEN et al. 1975).

The *Gentianales-Rubiales* are the only group in the *Asteridae* where true stipules occur (see WEBERLING 1957 on stipule-like organs in *Caprifoliaceae*) and they have the rare combination of nuclear endosperm and presence of iridoids. The position of two families each with only few genera is controversial: the *Menyanthaceae* agree with the *Gentianaceae* (in which they have been included for a long time) in chemical characters (HEGNAUER 1969) but they differ in several anatomical features (LINDSEY 1938) and in having cellular endosperm (VIJAYA-RAGHAVAN & PADMANABAN 1969). CRONQUIST (1968) has consequently transferred this family to the *Polemoniales*, but they have surely no close relative here and I still think it is possible that they are an aberrant member of *Gentianales*. The *Buddlejaceae* (at least the best known genus *Buddleja*) have morphological (lack of true stipules, HASSELBERG 1937), anatomical, embryological and phytochemical characters which differ from those of *Gentianales* and agree with the *Scrophulariales* (WAGENITZ 1959). It must be admitted, however, that the borderline between the *Loganiaceae* and *Buddlejaceae* is not as clearcut as it may appear. This has prompted the monographers of the *Loganiaceae* not to accept the *Buddlejaceae* (LEENHOUTS 1962, LEEUWENBERG 1967), although they do not doubt the affinities of *Buddleja* with the *Scrophulariaceae*, and LEENHOUTS had to admit that for *Buddleja* "palynology reveals a close relationship to the *Scrophulariaceae* and hardly any to the *Loganiaceae*" (PUNT & LEENHOUTS 1967).

The family *Oleaceae* is singular in several respects and this has led to the fact that it is nearly the only one of the "old" members of *Asteridae* whose place in this subclass has recently been debated. TAKHTAJAN (1969, 1973) has placed the *Oleales* near the *Celastrales* (compare MAU-RITZON 1939). But its embryological characters are those of *Asteridae* and so is the floral diagram, although there are a few members with free petals or without petals. *Oleaceae* have often been included in the *Gentianales*, but they differ in the following characters: leaves sometimes

divided (true pinnate leaves are very rare in *Asteridae*! HICKEY &
WOLFE 1975, using leaf characters, include *Oleales* in the ~Rosidae~ near
Rhamnales), without stipules, intraxylary phloem lacking, usually
4 petals and 2 stamens (extremely rare in flowers with a strictly regular
corolla), cellular endosperm.   CRONQUIST (1968) has *Oleaceae* in his
Scrophulariales "for a lack of a better alternative", but they differ e.g.
by the monotelic (determinate) inflorescence (TROLL 1969) and the regular
corolla, and no family which is near to *Oleaceae* can be found in the
Scrophulariales.   *Oleaceae* have quite an array of iridoid compounds,
especially seco-iridoids (JENSEN et al. 1975), with one group (Oleuropein
group) apparently restricted to this family. The spectrum of these
iridoids is more consonant with an affinity to the *Gentianales*. On the
other hand, planteose has recently been found in seeds of *Fraxinus*; it is
an oligosaccharide so far only known from several families of the *Tubiflorae* s.l. (JUKES & LEWIS 1974). On the whole the erection of a separate
order seems fully justified and it is best treated here as an appendix to
the *Gentianales—Rubiales* line.

Polemoniales and *Scrophulariales—Lamiales*

These two groups together correspond to the "*Tubiflorae*" (+ *Plan-
taginales*) in the system of ENGLER (1897; see also MELCHIOR 1964 and
WETTSTEIN 1935). In most other systems the families are grouped into

Table 2. Characters of main groups in the *Tubiflorae*

	Polemoniales	*Scrophulariales-Lamiales*
Flowers	actinomorphic (few exceptions)	± zygomorphic
Stamens	isomerous	mostly oligomerous
Inflorescence	monotelic[1] (polytelic in most *Convolvulaceae*)	polytelic[1]
Endosperm	nuclear or cellular	cellular
Iridoid compounds	0	+ in most families
Alkaloids	+ in several families	0
Stored carbohydrates	often starch	mainly stachyose and other oligosaccharides

[1] The two main-types of inflorescences distinguished by TROLL (1964,
cf. WEBERLING 1965) correspond to a certain degree to the determinate and
indeterminate types of the older morphologists. Data on the occurrence of
these types in the families of the *Tubiflorae* may be found scattered in the
publications of EICHLER 1875, DANERT 1958, TROLL 1964, 1969, WEBERLING
1957b (for additional information I thank Prof. WEBERLING).

two or more different orders but there is no consensus about their delimitation. If morphological and phytochemical data are combined, two main groups emerge (DAHLGREN 1975): *Polemoniales* with *Polemoniaceae, Hydrophyllaceae, Boraginaceae* s.l., *Convolvulaceae, Solanaceae*, and *Nolanaceae* (presumably also the *Lennoaceae*), *Scrophulariales* with the bulk of the *Tubiflorae*, and the allied *Lamiales* with *Verbenaceae* (incl. *Phrymaceae*, WHIPPLE 1972), *Callitrichaceae* and *Labiatae* (*Lamiaceae*). The differences are shown in the following table (Table 2).

The delimitation of the *Polemoniales* adopted here differs from the system of CRONQUIST by the inclusion of the *Boraginaceae*, and from that of TAKHTAJAN by the inclusion of the *Solanaceae*. *Solanaceae* have often been put near the *Scrophulariaceae* and WETTSTEIN (1891) wrote that the border between these families could only be drawn artificially. But in fact differences in the inflorescence-type (DANERT 1958, TROLL 1964: 177–179), the anatomy (lack of intraxylary phloem in *Scrophulariaceae*) and in chemical characters make it easy to distinguish between these families[1]. On the other hand, *Convolvulaceae* and *Solanaceae* both have intraxylary phloem, chemically allied types of alkaloids (HEGNAUER 1973, FROHNE & JENSEN 1973) and they can store starch (sweet potatoes and potatoes are well known exemples). These two families together with the *Nolanaceae* seem to form one group in the *Polemoniales*, *Polemoniaceae* and *Hydrophyllaceae* a second one, while the *Boraginaceae* is rather isolated (this is also in line with anatomical evidence: PATEL & INAMDAR 1971, INAMDAR & PATEL 1973, and corresponds well to the division in small orders by HUTCHINSON 1973). The *Scrophulariales* with a greater number of families are more homogeneous. The close connections between most families of this order have been emphasized by many authors (e.g. TAKHTAJAN 1959, MELCHIOR 1964, CRONQUIST 1968, IHLENFELDT 1967) and cannot be discussed here. In embryogeny *Scrophulariales* and *Lamiales* seem to be very uniform as compared to the *Polemoniales* (YAMAZAKI 1974).

Dipsacales

The *Dipsacales* include the *Caprifoliaceae, Valerianaceae, Morinaceae* (VIJAYARAGHAVAN & SARVESHWARI 1968) and *Dipsacaceae*. This rather homogeneous group was left when the *Rubiaceae* had been removed from the old *Rubiales* (WAGENITZ 1959). Like the *Scrophulariales* they usually have zygomorphic flowers, cellular endosperm and iridoid compounds

---

[1] FAIRBROTHERS et al. (1975) interpreted the serological data of HAWKES & TURNER (1968) as being in better accordance with the inclusion of *Solanaceae* in the *Scrophulariales*. But as only two genera of *Solanaceae* (*Salpiglossis* and *Schizanthus*) were tested and all the reactions were faint, it seems hazardous to use this evidence.

(although mostly of a different type). The ovary is inferior and there is a strong tendency for the reduction of carpels and of stamens although the corolla is never as distinctly zygomorphic as in most *Scrophulariales*. DAHLGREN (1975) has united the *Dipsacales* with the *Oleales, Goodeniales,* and *Gentianales* to form his *Gentiananae*. I should like to stress that in my opinion the *Dipsacales* are nearer to the *Scrophulariales* than to the *Gentianales*.

The position of the genera *Sambucus (Caprifoliaceae* or *Sambucaceae)* and *Adoxa (Adoxaceae)* is doubtful. The isolated position of the genus *Sambucus*, especially as regards morphological and anatomical characters, has been known since long (HÖCK 1892, SCHWERIN 1920, FUKUOKA 1972), it is less evident in chemical characters (HEGNAUER 1964, BOHM & GLENNIE 1971), but this may change if we had more information. In embryology the occurrence of the rare *Adoxa*-type of the development of the embryo sac in *Sambucus* and *Adoxa* is remarkable. DAHLGREN has placed *Sambucus* and *Adoxa* in his *Cornales*, I shall return to this idea later on.

Campanulales—Asterales

Campanulales are a group recognized as early as 1830 (by BARTLING) including *Goodeniaceae* (incl. *Brunoniaceae*, CAROLIN 1959, DUIGAN 1961), *Stylidiaceae* and *Campanulaceae* and *Lobeliaceae*. Only recently this order has been broken up, e.g. by THORNE (1968) and DAHLGREN (1975). Mainly on account of the presence of iridoid compounds DAHLGREN has excluded the *Goodeniaceae* (incl. *Brunoniaceae*) and the *Stylidiaceae*. For the *Goodeniaceae* embryological differences between this family and the *Campanulaceae* are also mentioned (JENSEN et al. 1975), but of these only the absence of endosperm haustoria seems to me important and VIJAYARAGHAVAN & MALIK (1972), while favouring the formation of a separate order *Goodeniales*, would place it "very close to the *Campanulales*". In my opinion the evidence for keeping these families together (morphological characters, presence of inulin) still prevails.

In a recent article (WAGENITZ 1976) I have made some comments on the affinities of the *Campanulales* and *Asterales*. The similar pollen presentation mechanisms and similar morphological tendencies are noteworthy but might be due to parallelism, but they are supported by phytochemical agreements: the common occurrence of inulin is well known, but recently the acetylenes, very characteristic for the *Compositae* (mainly in the *Asteroideae*), have found to be widespread in the *Campanulaceae* (BOHLMANN et al. 1973). The *Calyceraceae* have been of interest in this connection as they approach the *Compositae* quite closely in some characters. But there has always been a debate as to their proper place. TAKHTAJAN (1969) has an order *Calycerales* between *Campanulales* and *Asterales*. With most *Campanulales* they have in common the binucleate pollen (only trinucleate pollen is known from the *Dipsacales*, BREW-BAKER 1967). Quite recently the detection of secologanin, a simple seco-iridoid (JENSEN et al. 1975), has been added to those arguments in favour of a position in or near the *Dipsacales* (CRONQUIST 1968, THORNE 1968, DAHLGREN 1975). But surely this is not a well known family and the same is true for the *Pentaphragmataceae* and *Sphenocleaceae*. According to embryological studies (KAPIL & VIJAYARAGHAVAN 1965, MAHESHWARI & KAPIL 1967) these two monogeneric families are correctly placed in the *Campanulales*, but phytochemical data are totally lacking.

Aberrant Genera Recently Added to the *Asteridae*

As stated above the main groups of *Asteridae* have a good set of common characters and have been put into this subclass (or its older equivalents) by numerous authors. But there are several genera or small families which agree with the *Asteridae* in most of the "key characters" but are quite aberrant in several others. According to floral morphology, they can be put into two different categories.

Families of Plants With Well-developed Perianth

a) *Loasaceae*. The *Loasaceae* have a combination of embryological characters which is rarely (if ever) found outside the *Asteridae*: unitegmic, tenuinucellate ovules, cellular endosperm with the development of terminal haustoria. This makes them highly anomalous in the *Parietales* (or *Cistales*, *Violales*, *Passiflorales*), their traditional place (SCHNARF 1933). In line with the embryological characters is the occurrence of iridoids (KOOIMAN 1974, JENSEN et al. 1975). TAKHTAJAN (1959, 1973) included *Loasaceae* in the *Polemoniales*, but this order has no iridoids according to present knowledge, and of course there are several conspicuous differences in flower morphology. Aberrant in comparison with all groups of the *Asteridae* is the tendency to increase the number of stamens: there are 4–5 stamens, alternating with the free (!) petals (DANDY 1967) only in the small subfamily *Gronovioideae*. The variability of the androecium is astonishing: recently it has been shown that *Mentzelia* differs strikingly in the development of the groups of stamens from three other investigated genera of *Loasaceae* (LEINS & WINHARD 1973). I find it impossible to include this family in any of the existing orders of *Asteridae*, DAHLGREN had apparently the same impression when erecting a separate superorder.

b) *Fouquieriaceae*. Members of this small family with two genera and eleven species have the following characters which might point to an affinity with the *Asteridae*: sympetalous corolla, tenuinucellate ovules (although with two integuments!), cellular endosperm, presence of several iridoid substances (DAHLGREN et al. 1976). In many systems the *Fouquieriaceae* have been placed in the *Parietales* or a smaller group of this affinity (*Violales, Tamaricales*), in modern systems e.g. by TAKHTAJAN (1959–1973), CRONQUIST (1968) and HUTCHINSON (1973), but the characters mentioned above would be very aberrant here. The bitegmic ovule, number of stamens, iridoid substances make the genus equally anomalous in the *Polemoniales* (MELCHIOR 1964 has it near the *Polemoniaceae*, THORNE 1968 in his *Solanales*). DAHLGREN et al. (1976) have thus proposed placing the *Fouquieriaceae* in a monotypic order near the *Ericales* and *Cornales*.

c) *Columelliaceae*. This monogeneric family has been referred to the *Tubiflorae* by many authors since the end of the last century (FRITSCH 1894, WETTSTEIN 1935, MELCHIOR 1964, HUTCHINSON 1926–1973, TAKHTAJAN 1959–1969). THORNE (1968) has included it in his *Saxifragaceae* s.l., CRONQUIST (1968) in the *Rosales*. CRONQUIST was already using information from the work of STERN et al., published 1969, on an exhaustive investigation of the anatomy. No close relation could be detected, but the anatomical evidence was in favour of a connection with the woody *Saxifragales*. No detailed embryological or phytochemical investigations are available, only GIBBS (1974) seems to have made a few simple tests, the most remarkable being a strong positive reaction to the tannin-test (unusual for the *Tubiflorae*).

d) *Hoplestigmataceae*. This family has been placed near the *Boraginaceae* in several recent systems. The basis for this seems to be rather shaky. *Hoplestigma* has many petals and 20–30 stamens free from the corolla to mention only the most conspicuous differences from the *Asteridae*. Other places to which it has been assigned are: *Bixales* (HUTCHINSON 1973), *Violales* (CRONQUIST 1968), *Ebenales* (GILG 1908, WAGENITZ in MELCHIOR 1964). As no recent investigations and no data at all on the embryology and phytochemistry are known to me, a discussion seems rather pointless.

Families of Plants With Reduced Flowers

One of the more conspicuous differences between the modern concept of *Asteridae* and equivalent older groupings is the inclusion of several monogeneric families of plants without a corolla: *Theligonaceae, Hippuridaceae, Hydrostachyaceae*, and *Callitrichaceae*. This has been discussed in some detail elsewhere (WAGENITZ 1975), and it may suffice to repeat here the main arguments for the new systematic position of these four groups in a tabulated form (Table 3).

A leading role in the assignment of these families to the *Asteridae* has been played by embryological characters, especially the unitegmic-

Table 3. Plants with reduced flowers recently added to the *Asteridae*

Family	Old position	New position	Main evidence for new position[1]	Main differences to allied families (in new position)
Theligonaceae	*Caryophyllales* *Myrtales*	*Rubiales*	Endosperm nuclear iridoid substances opposite leaves with stipules raphides present	Unisexual flowers simple perianth male flowers with 6–30 stamens ovary with 1 ovule
Hippuridaceae	*Myrtales* (*Haloragales*)	*Scrophulariales*	Endosperm cellular iridoids: aucubin, catalpol stachyose	Apetalous ovary inferior 1 carpel (?) with one ovule
Hydrostachyaceae	near *Rosales*	*Scrophulariales*	Endosperm cellular micropylar haustorium type of embryo development	Unisexual flowers perianth 0 two styles median stipules (?)
Callitrichaceae	*Euphorbiales* *Myrtales* (*Haloragales*)	*Lamiales*	Endosperm cellular haustoria present iridoids: aucubin, catalpol opposite leaves type of glands	Unisexual flowers perianth 0 carpels transverse two styles

[1] For all groups must be added here: unitegmic-tenuinucellate ovules!

tenuinucellate ovules (PHILIPSON 1974 has *Hippuridaceae* and *Theli-gonaceae* in his group with unitegmic-crassinucellate ovules, but compare JUEL 1911 and WUNDERLICH 1971). Only in the *Theligonaceae* have phytochemical researches, testifying the presence of iridoids (KOOIMAN 1971), given rise to an investigation and evaluation of morphological, anatomical and embryological characters. For *Hydrostachys* phyto-chemical data are still lacking; they would be most welcome to corrob-orate the evidence from the thorough investigations of morphology and embryology by RAUH & JÄGER-ZÜRN (1966).

Affinities of the *Asteridae*

The idea that the *Sympetalae* are not a natural group is an old one and for several decades taxonomists have looked for affinities between the orders of *Sympetalae* and non-sympetalous groups. If only floral morphology is taken into account this is scarcely more than a guess and some of the older suggestions as to the origin of groups of the *Asteridae* are clearly obsolete. I can scarcely see any foundation for the idea that the *Geraniales* (BESSEY 1915, WERNHAM 1913) or more specifically the *Linaceae* (HALLIER 1912) occupy a key position as a starting point for the *Tubiflorae*. For the *Polemoniaceae* recently KAPIL et al. (1969) have again brought forward this proposition, but in fact they advance more embryological differences than morphological similarities! I think we should also dismiss the long cherished idea of a direct connection between the *Umbelliflorae* and *Rubiales*. This seems to have been suggested by a superficial resemblance of the ovary in derived members of these groups: *Umbelliferae* (*Apiaceae*) and the *Galieae* of the *Rubiaceae*.

Recently chemical data together with embryological and morpho-logical characters help to give a better foundation to the ideas on the affinities of some groups of the *Asteridae*. One complex of families gets more and more interest in this respect: the woody *Saxifragales* (*Cuno-niales*, sometimes included in *Rosales*) and *Cornales*. If the broad concept of *Cornales* proposed by HUBER (1963) and DAHLGREN (1975) is accepted, we have to deal with these *Cornales* alone. In these families several characters can be found which are familiar to us as typical for the *Asteridae*: unitegmic-tenuinucellate ovules occur in most *Escalloniaceae* and *Hydrangeaceae*, while unitegmic-crassinucellate ovules are widespread in the *Cornales* s. str. (PHILIPSON 1974), cellular endosperm has been found in the *Hydrangeaceae*, *Davidiaceae*, and *Cornaceae* (and in several herbaceous groups of the *Saxifragales*); finally iridoid compounds are present in the *Escalloniaceae*, *Hydrangeaceae* and many *Cornales*. The flowers are not very specialized; they show a strong tendency towards

Table 4. Groups of uncertain systematic position shifted between the *Saxifragales-Cornales* and *Asteridae*

| | Proposed position in: | |
	Saxifragales or *Cornales*	*Asteridae*
Columellia (*Columelliaceae*)	Saxifragaceae (THORNE 1968) Rosales s. l. (CRONQUIST 1968) Saxifragales (TAKHTAJAN 1973)	Tubiflorae (FRITSCH 1894; MELCHIOR 1964) Scrophulariales (TAKHTAJAN 1966)
Sambucus (*Sambucaceae*)	Cornales (DAHLGREN 1975)	Dipsacales-Caprifoliaceae (most authors)
Adoxa (*Adoxaceae*)	Saxifragaceae (WARMING & MÖBIUS 1929; GUNDERSEN 1950) Saxifragales (HUTCHINSON 1973) Cornales (DAHLGREN 1975)	Rubiales (WETTSTEIN 1935; ENGLER & DIELS 1936) Dipsacales (WAGENITZ 1959; TAKHTAJAN 1966—1973; CRONQUIST 1968; THORNE 1968)
Alseuosmiaceae	Rosales s. l. (CRONQUIST 1968) Saxifragales (TAKHTAJAN 1973) Cornales (DAHLGREN 1975)	Caprifoliaceae (TAKHTAJAN 1969; HUTCHINSON 1973)
Berenice	Saxifragaceae-Escallonioideae (ENGLER 1930)	Campanulaceae (ERDTMAN & METCALFE 1963)
Donatia	Saxifragaceae (ENGLER 1890/91) Saxifragales (HUTCHINSON 1973) Rosales s. l. (THORNE 1968) Cornales (DAHLGREN 1975)	Campanulales-Stylidiaceae (WAGENITZ in MELCHIOR 1964; CRONQUIST 1968) Campanulales-Donatiaceae (TAKHTAJAN 1959—1973)

an inferior ovary; and a sympetalous corolla may occur (e.g. *Polyosma*, more often in the *Crassulaceae* belonging to the herbaceous families of *Saxifragales*). The comparison of *Asteridae* with this complex is rendered difficult by the fact that the *Saxifragales* and *Cornales* are very diverse groups. The "old" *Cornales* have been chosen by GIBBS (1974) as one of his examples for "chaos in taxonomy", and the situation is still more complicated if the broad concept of *Cornales* proposed by DAHLGREN is accepted. It is remarkable that several genera have been shifted between the *Saxifragales—Cornales* and the *Asteridae*. Examples for this have been combined in table 4. The most intimate connection can be seen between *Cornaceae* and *Caprifoliaceae* with *Sambucus* (and *Viburnum?*) as a link, this has been confirmed by serological investigations (HILDE-BRAND & FAIRBROTHERS 1970 a, b). *Hydrostachys* could have been added to the families in table 4. When discussing the systematic position of this genus, RAUH & JÄGER-ZÜRN (1966) had to choose between the *Saxifragales* and *Scrophulariales* (a very valuable conspectus of characters of these groups can be found in this publication).

On a phytochemical basis another connecting line has recently been drawn between the *Araliales* and the *Asterales*, the main point being the common occurrence of polyacetylenes and sesquiterpene lactones (HEGNAUER 1971, 1973, BOHLMANN et al. 1973, WAGENITZ 1976). These substances are rare in the angiosperms, and if confronted with this evidence most systematists with some inclination to chemotaxonomy feel that there "is something in it". In embryology unitegmic-cras-sinucellate ovules occur in the *Araliaceae*, unitegmic-tenuinucellate in *Umbelliferae* (*Apiaceae*).

It should be noted here that the two groups to which these "phyto-chemical connecting lines" go, the *Cornales* and the *Araliales*, have been united in one order in most older systems (and recently re-united in the system of TAKHTAJAN 1973). And RODRÍGUEZ (1971) in a very well documented discussion of the relationships of the *Umbelliflorae* speaks about "a remarkable network of interrelations and similarities across the gap between the two main lines". In this connection RODRÍGUEZ uses the picture of a puzzle which does not fit together and that is exactly the feeling I have.

We shall now ask if the evidence discussed above makes it possible to divide the *Asteridae* into two (or more) distinct phyletic lines with different affinities. The authors advocating such a division have drawn the line in different ways and it will not be possible to discuss here all the proposals which have been made.

One possibility was brought forward by HEGNAUER (1964: 544), FROHNE & JENSEN (1973: 176) and JENSEN et al. (1975: 169). Using mainly the presence of iridoid compounds as a guiding principle JENSEN et al. build up two groups of orders, one (*Gentianales, Rubiales, Oleales, Scrophulariales, Lamiales, Dipsacales*) with iridoids is brought in close connection to the *Cornales* s.l. (and *Ericales*); the other group (*Pole-moniales, Campanulales* s. str., *Asterales*), especially the *Asterales*, show the phytochemical agreements with the *Araliales* mentioned above. If the iridoids are thus taken as a key-character the old *Tubiflorae* have to be put into quite different lines and so have the families of the *Campanulales*. Too much stress is laid now on this very fashionable group of substances. Their presence, which shows such a surprisingly good correlation with cellular endosperm and unitegmic ovules, is surely an important character. But what about their absence? As far as is known the *Asclepiadaceae* differ from all other families of the *Gentianales* by the absence of iridoid compounds, and in the *Scrophulariaceae* there are several tribes without these compounds too (KOOIMAN 1970). In these cases connection with related iridoid-containing groups by common morphological characters has been so strong, that no one has thought of excluding these groups. I am not sure that the differences between the families of *Campanulales* and the orders of the old *Tubiflorae* are so far-reaching as to exclude a common origin. GIBBS (1974) on account of a broad survey of chemical characters thinks the *Tubiflorae* s.l. are "an order of surprising homogeneity". As far as palynology goes it gives apparently more arguments for than against a connection between the different groups of the *Tubiflorae*, and the evidence is at least not incompatible with an origin in the *Rosidae* (WALKER & DOYLE 1975).

On the basis of their studies on leaf-characters (venation, tooth-type, etc.), HICKEY & WOLFE (1975) have quite recently proposed a division of *Asteridae* into two groups:

a) Dilleniid-Leafed Asterids (*Gentianales*, *Rubiales*, *Polemoniales* incl. *Solanaceae*, *Campanulales*, *Asterales*);

b) Rosid-Leafed Asterids (*Dipsacales*, *Scrophulariales*, *Lamiales*).

The *Tubiflorae* are here divided in the same manner as according to the chemical characters, but the *Gentianales*—*Rubiales* are found "on the other side of the line". As expressed by their names, the authors think that the first of these groups is based in the *Dilleniidae*, the other in the *Rosidae*. Taxonomists must surely be thankful to the authors for bringing into the discussion a new set of characters, but I feel that it is going too far to propose such a division on this evidence, especially as it is evident from the description of the leaf characteristics that the sample investigated of the *Asteridae* is still rather small.

In the first attempt to apply numerical methods to the classification of dicotyledons (YOUNG & WATSON 1970) only the larger families are treated. For this and other reasons not too much weight should be attributed to this analysis. But it is interesting to note that in the main group "Tenuinucelli" there are four secondary groups: 1. "Asclepioids" corresponding to the *Gentianales*—*Rubiales* (and *Sapotaceae* and several families of doubtful position), 2. "Acanthoids" which are families of the *Scrophulariales*—*Lamiales*, 3. *Compositae* and 4. *Umbelliferae*. Several other families of the *Asteridae* come out as intermediate between 1 and 2. Unfortunately, the woody families of the *Saxifragales* have not been included in this study.

Conclusion

We cannot prove with certainty that the *Asteridae* are a mono-phyletic group, and doubts about this viewpoint are possible, but to me the evidence seems to be equally inconclusive for a distinction of several phyletic lines of different origin. The pattern of the distribution of characters cannot be explained without parallel evolution in morpho-logical, embryological but surely also phytochemical characters. And we often simply do not know which characters we can rely on as indicating phyletic affinity or only a certain level of evolution. This is a major difficulty, and at the same time we have two other problems: still very inadequate knowledge of the distribution of many characters, and the problem of handling the data. This is too well known to dwell on it any more. I feel strongly after preparing this lecture that the traditional method of looking into the literature and compiling the data is unsatis-factory and that data-banks are necessary. Three titles of articles written by systematists with a wide experience come to my mind. What is Systematic Botany? An unending synthesis (Constance 1964), an unachieved synthesis (Merxmüller 1972) or even the Stone of Sisyphus (Heywood 1974)?

References

Bartling, F. G., 1830: Ordines naturales plantarum eorumque characteres et affinitates. Göttingen: Dieterich.

Bessey, C. E., 1915: The phylogenetic taxonomy of flowering plants. Ann. Missouri Bot. Gard. **2**, 109—164.

Bohlmann, F., Burkhardt, T., and Zdero, C., 1973: Naturally Occurring Acetylenes. London-New York: Academic Press.

Bohm, B. A., and Glennie, C. W., 1971: A chemosystematic study of the *Caprifoliaceae*. Canad. J. Bot. **49**, 1799—1807.

Brewbaker, J. L., 1967: The distribution and phylogenetic significance of binucleate and trinucleate pollen grains in the angiosperms. Amer. J. Bot. **54**, 1069—1083.

Carolin, R. C., 1959: Floral structure and anatomy in the family *Goodenia-ceae* Dumort. Proc. Linn. Soc. New South Wales **84**, 242—255.

Constance, L., 1964: Systematic botany, an unending synthesis. Taxon **13**, 257—273.

Cooper, D. C., 1941: Macrosporogenesis and the development of the seed of *Phryma leptostachya*. Amer. J. Bot. **28**, 755—761.

Correns, C., 1891: Zur Biologie und Anatomie der Salvienblüthe. Jahrb. Wiss. Bot. **22**, 190—240.

Cronquist, A., 1957: Outline of a new system of families and orders of Dicotyledons. Bull. Jard. Bot. Bruxelles **27**, 13—40.

— 1968: The Evolution and Classification of Flowering Plants. London: Nelson.

·DAHLGREN, R., 1975: A system of classification of the angiosperms to be used to demonstrate the distribution of characters. Bot. Not. **128**, 119—147.

—, JENSEN, S. R., and NIELSEN, B. J., 1976: Iridoid compounds in *Fouquieriaceae* and notes on its possible affinities. Bot. Not. **129**, 207—212.

DANDY, J. E., 1967: *Loasaceae*. In: The Genera of Flowering Plants (*Angiospermae*) (HUTCHINSON, J.). Dicotyledones **2**, 353—362. Oxford: Clarendon.

DANERT, S., 1958: Die Verzweigung der Solanaceen im reproduktiven Bereich. Abh. Deutsch. Akad. Wiss. Berlin, Kl. Chem. 1957; Nr. 6, 1—183.

DAVIS, G. L., 1966: Systematic Embryology of the Angiosperms. New York etc.: J. Wiley.

DUIGAN, S. L., 1961: Studies of the pollen grains of plants native to Victoria, Australia. 1: *Goodeniaceae* (including *Brunoniaceae*). Proc. Roy. Soc. Victoria N. S. **74**, 87—109.

EICHLER, A. W., 1875: Blüthendiagramme. Band 1. Leipzig: Engelmann.

EMBERGER, L., 1960: Les végétaux vasculaires. In: Traité de Botanique (Systematique). Tome II. (CHAUDEFAUD, M. et EMBERGER, L.). Paris: Masson.

ENGLER, A., 1890—1891: *Saxifragaceae*. In: Natürl. Pflanzenfam. (ENGLER, A., und PRANTL, K., Hrsg.) **III 2 a**, 41—93. Leipzig: Engelmann.

— 1897: Übersicht über die Unterabteilungen, Klassen, Reihen, Unterreihen und Familien der Embryophyta siphonogama. In: Natürl. Pflanzenfam. (ENGLER, A., und PRANTL, K., Hrsg.), Nachträge zum II.—IV. Teil, 341—357. Leipzig: Engelmann.

— 1930: *Saxifragaceae*. In: Natürl. Pflanzenfam. 2. Aufl. **18a**, (ENGLER, A., und PRANTL, K., Hrsg.), 74—226.

— und DIELS, L., 1936: Syllabus der Pflanzenfamilien. 11. Aufl. Berlin: Borntraeger.

ERDTMAN, G., and METCALFE, C. R., 1963: Affinities of certain genera incertae sedis suggested by pollen morphology and vegetative anatomy. Kew Bull. **17**, 249—256.

FAIRBROTHERS, D. E., MABRY, T. J., SCOGIN, R. L., and TURNER, B. L., 1975: The bases of angiosperm phylogeny: chemotaxonomy. Ann. Missouri Bot. Gard. **62**, 765—800.

FRITSCH, K., 1894: *Columelliaceae*. In: Natürl. Pflanzenfam. **IV. 3b** (ENGLER, A., und PRANTL, K., Hrsg.), 186—188.

FROHNE, D., und JENSEN, U., 1973: Systematik des Pflanzenreichs unter besonderer Berücksichtigung chemischer Merkmale und pflanzlicher Drogen. Stuttgart: G. Fischer.

FUKUOKA, N., 1972: Taxonomic study of the *Caprifoliaceae*. Mem. Fac. Sci. Kyoto Univ. Ser. Biol. **6**, 15—58.

GIBBS, R. DARNLEY, 1974: Chemotaxonomy of Flowering Plants. 4 Vol. Montreal-London: McGill-Queen's Univ. Press.

GILG, E., 1908: Die systematische Stellung der Gattung *Hoplestigma* und einiger anderer zweifelhafter Gattungen. Bot. Jahrb. Syst. **40**, Beibl. 93, 76—84.

GOEBEL, K., 1882: Grundzüge der Systematik und Speciellen Pflanzenmorphologie. Leipzig: Engelmann.

GOVIL, C. M., 1970: *Convolvulaceae*. In: Symposium on Comparative Embryology of Angiosperms. Bull. Indian National Sci. Acad. **41**, 246—249.

GRANT, K. A., and GRANT, V., 1964: Mechanical isolation of *Salvia apiana* and *Salvia mellifera (Labiatae)*. Evolution **18**, 196—212.

GRANT, V., and GRANT, K. A., 1965: Pollination in the Phlox Family. New York-London: Columbia University Press.

GUNDERSEN, A., 1950: Families of Dicotyledons. Waltham, Mass.: Chronica Botanica.

HALLIER, H., 1912: L'origine et le système phylétique des Angiospermes exposés à l'aide de leur arbre généalogique. Arch. Néerl. Sci. Exact. Nat. ser. III. B. **1**, 146—234.

HASSELBERG, G. B. E., 1937: Zur Morphologie des vegetativen Sprosses der Loganiaceen. Symb. Bot. Upsal. **2** (no. 3), 1—170.

HAWKES, J. G., and TUCKER, W. G., 1969: Serological assessment of relationships in a flowering plant family (*Solanaceae*). In: Chemotaxonomy and Serotaxonomy, Syst. Ass. Special Vol. **2** (HAWKES, J. G., Ed.), 77—88. London-New York: Academic Press.

HEGNAUER, R., 1964, 1969, 1973: Chemotaxonomie der Pflanzen. Band **3**, **5**, **6**. Basel-Stuttgart: Birkhäuser.

— 1971: Chemical patterns and relationship of *Umbelliferae*. In: The Biology and Chemistry of the *Umbelliferae* (HEYWOOD, V. H., Ed.), 267—277. London: Academic Press.

HEYWOOD, V. H., 1974: Systematics—the Stone of Sisyphus. Biol. J. Linn. Soc. **6**, 169—178.

HICKEY, L. J., and WOLFE, J. A., 1975: The bases of angiosperm morphology: vegetative morphology. Ann. Missouri Bot. Gard. **62**, 538—589.

HILDEBRAND, F., 1866: Über die Befruchtung der *Salvia*-Arten mit Hülfe von Insekten. Jahrb. Wiss. Bot. **4**, 451—478.

HILLEBRAND, G. R., and FAIRBROTHERS, D. E., 1970a: Serological investigation of the systematic position of the *Caprifoliaceae*, I. Correspondence with selected *Rubiaceae* and *Cornaceae*. Amer. J. Bot. **57**, 810—815.

— — 1970b: Phytoserological systematic survey of the *Caprifoliaceae*. Brittonia **22**, 125—133.

HÖCK, F., 1892: Zur systematischen Stellung von *Sambucus*. Bot. Centralbl. **51**, 233—234.

HUBER, H., 1963: Die Verwandtschaftsverhältnisse der Rosifloren. Mitt. Bot. Staatssamml. München **5**, 1—48.

HUTCHINSON, J., 1926: The Families of Flowering Plants. I. Dicotyledons. London: Macmillan.

— 1973: The Families of Flowering Plants. Ed. 3. London: Oxford University Press.

IHLENFELDT, H.-D., 1967: Über die Abgrenzung und die natürliche Gliederung der *Pedaliaceae* R. BR. Mitt. Staatsinst. Allg. Bot. Hamburg **12**, 43—128.

INAMDAR, J. A., and PATEL, R. C., 1973: Structure, ontogeny, and classification of trichomes in some *Polemoniales*. Feddes Repert. **83**, 473—488.

JENSEN, S. R., NIELSEN, B. J., and DAHLGREN, R., 1975: Iridoid compounds, their occurrence and systematic importance in the angiosperms. Bot. Not. **128**, 148—180.

JUEL, H. O., 1911: Studien über die Entwicklungsgeschichte von *Hippuris vulgaris*. Nova Acta Regiae Soc. Sci. Upsal., ser. IV. **2**, nr. 11.

JUKES, C., and LEWIS, D. H., 1974: Planteose, the major soluble carbohydrate of seeds of *Fraxinus excelsior*. Phytochemistry **13**, 1519—1521.

KAPIL, R. N., RUSTAGI, P. N., and VENKATARAMAN, R., 1969: A contribution to the embryology of *Polemoniaceae*. Phytomorphology **18**, 403—412.

— and VANI, R. S., 1967: *Nyctanthes arbor-tristis* L.: embryology and relationships. Phytomorphology **16**, 553—563.

— and VIJAYARAGHAVAN, M. R., 1965: Embryology of *Pentaphragma horsfieldii* (MIQ.) AIRY SHAW with a discussion on the systematic position of the genus. Phytomorphology **15**, 93—102.

KOOIMAN, P., 1970: The occurrence of iridoid glycosides in the *Scrophulariaceae*. Acta Bot. Neerl. **19**, 329—340.

— 1974: Iridoid glycosides in the *Loasaceae* and the taxonomic position of the family. Acta Bot. Neerl. **23**, 677—679.

LEENHOUTS, P. W., 1962: *Loganiaceae*. In: Flora Malesiana I. **6** (2), 293—387.

LEEUWENBERG, A. J. M., 1967: Notes on American *Loganiaceae* II. Revision of *Peltanthera* BENTH. Acta Bot. Neerl. **16**, 143—146.

LEINS, P., und WINHARD, W., 1973: Entwicklungsgeschichtliche Studien an Loasaceen-Blüten. Österr. Bot. Ztschr. **122**, 145—165.

LINDSEY, A. A., 1938: Anatomical evidence for the *Menyanthaceae*. Amer. J. Bot. **25**, 480—485.

MACIOR, L. W., 1968: Pollination adaptation in *Pedicularis groenlandica*. Amer. J. Bot. **55**, 927—932.

— 1970: The pollination ecology of *Pedicularis* in Colorado. Amer. J. Bot. **57**, 716—728.

MAHESHWARI, P., and KAPIL, R. N., 1967: Some Indian contributions to the embryology of angiosperms. Phytomorphology **16**, 239—291.

MAURITZON, J., 1939: Die Bedeutung der embryologischen Forschung für das natürliche System der Pflanzen. Acta Univ. Lund. N. F. Avd. 2. **35**. Nr. 15, 1—70.

MELCHIOR, H. (Hrsg.), 1964: A. ENGLER's Syllabus der Pflanzenfamilien. 12. Aufl. Band 2. Angiospermen. Berlin-Nikolassee: Gebr. Borntraeger.

MERXMÜLLER, H., 1972: Systematic Botany—an unachieved synthesis. Biol. J. Linn. Soc. **4**, 311—321.

PALIWAL, G. S., and SRIVASTAVA, L. M., 1969: The cambium of *Alseuosmia*. Phytomorphology **19**, 5—8.

PATEL, R. C., and INAMDAR, J. A., 1971: Structure and ontogeny of stomata in some *Polemoniales*. Ann. Bot. (London) **35**, 389—409.

PHILIPSON, W. R., 1974: Ovular morphology and the major classification of the dicotyledons. Bot. J. Linn. Soc. **68**, 89—108.

PUNT, W., and LEENHOUTS, P. W., 1967: Pollen morphology and taxonomy in the *Loganiaceae*. Grana Palyn. **7**, 469—516.

RAUH, W., und JÄGER-ZÜRN, I., 1966: Zur Kenntnis der *Hydrostachyaceae*. 1. Teil, Blütenmorphologische und embryologische Untersuchungen an Hydrostachyaceen unter besonderer Berücksichtigung ihrer systematischen Stellung. Sitzungsber. Heidelb. Akad. Wiss. Math.-Nat. Kl. Jg. 1966, 1. Abh., 1—117.

RODRÍGUEZ, R. L., 1971: The relationships of the *Umbellales*. In: The Biology and Chemistry of the *Umbelliferae* (HEYWOOD, V. H., Ed.), 63—91. London: Academic Press.

SCHNARF, K., 1931: Vergleichende Embryologie der Angiospermen. Berlin: Gebr. Borntraeger.

— 1933: Die Bedeutung der embryologischen Forschung für das natürliche System der Pflanzen. Biol. Gen. **9**, 271—288.

SCHWERIN, F. VON, 1920: Revisio generis *Sambucus*. Mitt. Deutsch. Dendrol. Ges. **29**, 194—231.

SILBERBAUER-GOTTSBERGER, I., und GOTTSBERGER, G., 1975: Über sphingophile Angiospermen Brasiliens. Plant Syst. Evol. **123**, 157—184.

SOÓ, R., 1961: The present aspect of the evolutionary history of *Telomophyta*. Ann. Univ. Sci. Budapest Sect. Biol. **4**, 167—178.

— 1967: Die modernen Systeme der Angiospermen. Acta Bot. Acad. Sci. Hung. **13**, 201—233.

— 1975: A review of the new classification systems of flowering plants (*Angiospermatophyta, Magnoliophytina*). Taxon **24**, 585—592.

SPORNE, K. R., 1969: The ovule as an indicator of evolutionary status in angiosperms. New Phytol. **68**, 555—566.

— 1974: The Morphology of Angiosperms. London: Hutchinson.

SPRAGUE, E. F., 1962: Pollination and evolution in *Pedicularis* (*Scrophulariaceae*). Aliso **5**, 181—209.

SPRAGUE, T. A., 1927: The morphology and taxonomic position of the *Adoxaceae*. J. Linn. Soc. Bot. **47**, 471—487.

STEBBINS, G. L., 1974: Flowering Plants. Evolution above the Species Level. Cambridge, Mass.: Belknap Press.

STERN, W. L., BRIZICKY, G. K., and EYDE, R. H., 1969: Comparative anatomy and relationships of *Columelliaceae*. J. Arnold Arbor. **50**, 36—75.

TAKHTAJAN, A., 1959: Die Evolution der Angiospermen. Jena: VEB G. Fischer.

— 1964: The taxa of the higher plants above the rank of order. Taxon **13**, 160—164.

— 1966: Sistema i filogenija cvetkovych rastenij (Systema et phylogenia Magnoliophytorum). Moskva-Leningrad: Nauka.

— 1969: Flowering Plants. Origin and Dispersal. City of Washington: Smithsonian Inst. Press.

— 1973: Evolution und Ausbreitung der Blütenpflanzen. Stuttgart: G. Fischer.

THORNE, R. F., 1968: Synopsis of a putatively phylogenetic classification of the flowering plants. Aliso **6**, 57—66.

TROLL, W., 1928: Organisation und Gestalt im Bereiche der Blüte. Berlin: J. Springer.

— 1964, 1969: Die Infloreszenzen. Typologie und Stellung im Aufbau des Vegetationskörpers. 1. Band; 2. Band, 1. Teil. Stuttgart: G. Fischer.

VIJAYARAGHAVAN, M. R., and MALIK, U., 1972: Morphology and embryology of *Scaevola frutescens* K. and affinities of the family *Goodeniaceae*. Bot. Not. **125**, 241—254.

— and PADMANABAN, U., 1969: Morphology and embryology of *Centaurium ramosissimum* DRUCE and affinities of the family *Gentianaceae*. Beitr. Biol. Pfl. **46**, 15—37.

— and SARVESHWARI, G. S., 1968: Embryology and systematic position of *Morina longifolia* WALL. Bot. Not. **121**, 383—402.

VOGEL, S., 1954: Blütenbiologische Typen als Elemente der Sippengliederung dargestellt anhand der Flora Südafrikas. Bot. Studien **1**. Jena: VEB G. Fischer.

— 1974: Ölblumen und ölsammelnde Bienen. Tropische und subtropische Pflanzenwelt **7**. Akad. Wiss. Lit. (Mainz), Math.-Naturw. Kl. Wiesbaden: F. Steiner.

WAGENITZ, G., 1959: Die systematische Stellung der *Rubiaceae*. Ein Beitrag zum System der Sympetalen. Bot. Jahrb. Syst. **79**, 17—35.

— 1967: Betrachtungen über die Artenzahlen der Pflanzen und Tiere. Sitzungsber. Ges. Naturf. Freunde Berlin N. F. **7**, 79—93.

— 1975: Blütenreduktion als ein zentrales Problem der Angiospermen-Systematik. Bot. Jahrb. Syst. **96**, 448—470.

— 1976: Systematics and phylogeny of the *Compositae* (*Asteraceae*). Plant Syst. Evol. **125**, 29—46.

WALKER, J. W., and DOYLE, J. A., 1975: The bases of angiosperm phylogeny: palynology. Ann. Missouri Bot. Gard. **62**, 664—723.

WARMING, E., and MÖBIUS, M., 1929: Handbuch der systematischen Botanik. 4. Aufl. Berlin: Gebr. Borntraeger.

WEBERLING, F., 1957a: Morphologische Untersuchungen zur Systematik der Caprifoliaceen. Akad. Wiss. Lit. Mainz, Abh. Math.-Naturw. Kl. Jg. 1957, Nr. 1, 1—50.

— 1957b: Die Infloreszenzen von *Bonplandia* CAV. und *Polemonium micranthum* BENTH. und ihre vermittelnde Sonderstellung unter den Blütenständen der *Polemoniaceae*. Beitr. Biol. Pfl. **34**, 195—211.

— 1965: Typology of inflorescences. J. Linn. Soc. Bot. **59**, 215—221.

WERNHAM, H. F., 1913: Floral evolution: with particular reference to the sympetalous dicotyledons. New Phytol. Reprint no. 5. Cambridge.

WETTSTEIN, R. VON, 1891: *Solanaceae*. In: Natürl. Pflanzenfam. IV, **3b** (ENGLER, A., und PRANTL, K., Hrsg.), 4—38.

— 1935: Handbuch der Systematischen Botanik. 4. Aufl. Leipzig-Wien: F. Deuticke.

WHIPPLE, H. L., 1972: Structure and systematics of *Phryma leptostachya* L. J. Elisha Mitchell Sci. Soc. **88**, 1—17.

WUNDERLICH, R., 1971: Die systematische Stellung von *Theligonum*. Österr. Bot. Ztschr. **119**, 329—394.

YAMAZAKI, T., 1974: A system of *Gamopetalae* based on the embryology. J. Fac. Sci. Sect. 3. Bot. (Tokyo) **11**, 263—281.

YOUNG, D. J., and WATSON, L., 1970: The classification of dicotyledons: a study of the upper levels of the hierarchy. Aust. J. Bot. **18**, 387—433.

Address of the author: Prof. Dr. GERHARD WAGENITZ, Lehrstuhl für Pflanzensystematik, Systematisch-Geobotanisches Institut der Universität, Untere Karspüle 2, D-3400 Göttingen, Federal Republic of Germany.

Plant Syst. Evol., Suppl. 1, 397—405 (1977)
© by Springer-Verlag 1977

Summary Lecture

By

Hermann Merxmüller, Botanische Staatssammlung München

In the past decades we have all attended many symposia on micro-evolution, biosystematics, the species problem and so on, whereas meetings concerned with the higher categories have been rare events. This may at least partly be due to the conviction of many botanists that most problems here are solved already whereas others may consider the creation of a truly evolutionary system or a generally acceptable classification to be an unattainable aim in our time. As much as I belong to this second group of people I thoroughly enjoyed this Conference—especially as it revealed so many signs of concern.

In the late fifties it was the numerical taxonomists who worried us with the idea that an equally good or even better taxonomy could be made without the preoccupation of phylogeny—an idea which in its basic principle greatly influenced many of us. Nevertheless there was no manifest break with tradition as regards higher taxa and their classification. This was due, I believe, to the fact that at the same time the CRONQUIST and TAKHTAJAN systems were published, the over-whelming success of which may have depended partly on this early sign of détente—so many years before the so-called political one. There was a kind of truce, just slightly disturbed by the steady grumbling voices of MEEUSE, MELVILLE and a few others. But the new serious worry has come from the chemists who as exact scientists could not accept this terrible mixture of "Science and Art" which taxonomists are used to tolerate.

We should admit that we ought to be worried and as a matter of fact we all are—at least in our subconsciousness. HEYWOOD did not simply play his well-known rôle of enfant terrible, castigating us by calling our aims a much overrated pastime, charging us with spiral reasoning, deploring our canalization of conceptual thinking. The same uneasiness has been expressed at this symposium, perhaps less forcefully, by KUBITZKI and MEEUSE as well as by CLIFFORD and SPORNE. Trying to avoid in this Summary Lecture at least some of the inbuilt conflicts resulting from unclear definitions, confusion of aims and so on, I shall simply use the Conference key-words, i.e. Evolution, Classification, Higher Categories, and add a few words about some Higher Taxa, Characters and Methods.

With regard to Evolution I may begin with one of the so-called main problems, the old question of Polyphyly versus Monophyly. My impression was that there is hardly anyone amongst the participants who still believes in an extreme polyphyly in the sense as e.g. Greguss has proposed or as Lam has advocated at least at the zenith of his theory. It may be somewhat symbolic that even the term polyphyly meanwhile has been slightly changed to "polyrheithry". But on the other hand, and more surprising, nobody seems to defend also a strictly monophyletic theory, believing in a common ancestor as having been one or two single plants. There is an ever increasing amount of consent about smaller or larger ancestral groups, an idea expressed by Engler already in 1926, formulated as a theory by Suessenguth (1938), called a "moderate pleiophyly" by Suessenguth & Merxmüller (1954) and strongly stressed by Cronquist who likes to speak in the last years about his "rather loose concept of monophyly". Such concepts fit rather well with examples such as those shown by Huber with his groups of monocot lines; they would be compatible with many facts presented here by Ehrendorfer; and I begin to wonder whether Meeuse's polyrheithric groups really must have had entirely unrelated ancestors in his opinion. We should not forget Simpson's remark that once in the past the orders of today may have been genera only. In this context the old question of poly- versus monophyly could prove in the future to have been an overrated, if not a pseudo-problem.

A much more serious problem can be expressed by the statement that science has made an enormous progress in the understanding of evolutionary processes but that we understand rather little as yet about the evolution of real taxa. There are rather few, if any, cases where we truly succeeded in establishing the real genealogy of a pair of recent species—and this is naturally even more true for the higher taxa. We never know whether the last lines leading towards an extinct base are really converging to one point; Burtt's paper on the Gesneriads offered one more splendid example. In such cases one mostly tries to reconstruct the presumed ancestor by adding together the so-called primitive traits.

Both these words "basic" and "primitive", so often used during this conference, are possibly somewhat dangerous—I already stressed the fact in an earlier paper how much we are sometimes preoccupied by the choice of our terms. There are hardly any "basic groups" living in our time (with the possible exception of some groups leading to future not yet existing ones which we certainly would not recognize as such). On the other hand it is highly improbable that plants possessing "primitive" characters only live to-day or even have lived in the past. We probably should reconsider, in a more generalized way, the term of Heterobathmy which may be an overall characteristic of all living things. In such a context it is not at all surprising when SPORNE's *Theales/Violales* are lower in his correlation scheme than the Magnoliads or when most of these Magnoliads in GOTTWALD's wood anatomical diagrams rank higher than some Dilleniids or Rosids. Even the incredible predominance of unisexuality at the "base" of SPORNE's scheme could be understood more easily with the acceptance of such premises.

As the word "primitive" usually carries with it the connotation of anthropomorphic terms such as "worse", or at least "imperfect", "less progressive", etc. it should be replaced by "conservative" (at least for people who do not interpret this term in the same way as "primitive"). But whatever term we use, we always should keep in our mind that each plant, past or present, is or has been "perfect", that it always lives or has lived in perfect agreement with its environment, that it is or has been fully adapted. STEBBINS has emphasized that it is not the characters but complexes of characters which may be adaptive—and I would add: complexes of conservative and new characters together. That is the reason why I consider it useless to try to reconstruct an extinct plant or group by summing up conservative traits only. Concerning the term "progress" I may add that one of the very few general progressive trends in the evolution of Angiosperms which I can see, is specialization, interdependence of parts and eventually collectivism (as was shown here once more in the *Urticales* by BERG), developments similar to those of ants, bees and man—and it is doubtful if everybody would call this a "progress".

A highlight of this Conference have certainly been the new ideas on evolutionary strategies, based on the interaction of plants with pollinators and predators. The enormous importance of the pollination syndromes is generally agreed on; but the possible sequence of the different flower adaptations to different pollen vectors and the following steps of specialization have rarely been considered in such great detail as was done here by EHRENDORFER and GOTTSBERGER. Likewise CRONQUIST's consideration of the chemotaxonomically fashionable metabolic compounds as a sequence of repellants is a fascinating new idea. These are three concepts which might reasonably and equally be applied to the great traits of the Angiosperm development—but the authors do not always work in the same direction.

CRONQUIST explicitly uses his theory for an explanation of what he thinks to be at least partly a systematic distributions of the metabolites concerned: "The same group of repellants may have been explored and exploited by different taxonomic groups at the same time". Gotts-berger, besides his nice demonstration of early adaptation and speciali-zation in the Magnoliids, produced an interesting interpretation of the secondary polyandry as a secondary adaptation to cantharophily as well as of oligomerization as an adaptation to sucking pollinators. But he admits that this theory does not work for all cases, and we know that at least in the later phases of insect evolution all these factors worked side by side, often even in very small groups.

It is only Ehrendorfer who believes that his theory (the evolu-tionary basis of which seems to be perfectly sound and convincing) is applicable to an overall explanation of the presumed sequence of the four subclasses concerned (if not for *Caryophyllidae* and *Liliopsida*, too). I somewhat doubt this absolute and exclusive validity of his theory. Nature is so intrinsicly interwoven, each small group in such a close contact with its environment, that such events must have oc-curred not once but many times. There may be the danger of producing once more grades, not clades. The same may be feared in the case of Clifford's scheme of the monocots, even if it is, of course, a phenetic, not an evolutionary one. Further considerations would lead us, here too, to the well-known problems of homology, convergence, reversal and so on, which have been treated by Philipson, Sporne and others also—but I now wish to make some remarks on classification.

Classification means "order". In one of his latest papers, Hey-wood sees here the general rôle of systematics: "It has, besides its identificatory rôle, to collate, synthesize and order the data in a syste-matic fashion so that conclusions of all sort may be drawn".

But in the context of order there remains always Gilmour's question: order—to which purpose? If I want a general reference system, a filing and retrieval device, I can live comfortably with the alphabetical one of Willis. For teaching I do not know of any better than our Cron-quist/Takhtajan system whose high didactical standard has contri-buted as much to its great success. as did its few "balloons" which are so easily to handle in the very reduced teaching standards of today. The only phenetic system I know is the certainly still imperfect one of Young & Watson which nevertheless would merit much more considera-tion than it has had so far. Meeuse is the only one who wants to make a truly phylogenetical system (a term the Organizing Committee deli-berately excluded from the program) which reflects the phylogenetic origin and therefore needs the definition of "primitive, basic" features free from typological views. For this postulate we have no answer, I fear.

The rest of us enjoy "evolutionary systems" which means nothing more than the old romantic natural system, with certain arrangements and sequences, perhaps enriched with hyphens and arrows or with boxes and boxes-in-boxes. With regard to the two systematic schemes presented at this Conference (DAHLGREN, THORNE) it was interesting to see that both prefer arrangements where the evolutionary sequences go from the center of the scheme towards its periphery. Compared with TAKHTAJAN's scheme they therefore look somewhat phenetic but both are of course as supposedly evolutionary as TAKHTAJAN's one is. In both schemes I miss a large central hole which ought to represent the extinct ancestral groups; and in both I see some dangers with regard to the very differently shaped groups (one may compare the figures of DAHLGREN's *Saxifragales* and *Campanulales* or of THORNE's *Rosiflorae* and *Lamiiflorae*) which may lead to conceptions quite other than those intended by the authors. Nevertheless, there seemed to be a rather unanimous agreement on the utility, especially of DAHLGREN's scheme.

But there is another point I want to stress here, namely the inflation of categories between order and class. For their 50–70 orders of dicots CRONQUIST and TAKHTAJAN need 6 or 7 subclasses and some 13 superorders; THORNE enumerates 20, DAHLGREN even 27 superorders. This reminds me a little of the old Englerian times in which so many infraspecific categories like proles, subvariety, subforma and so on have been in vogue. We should deliberate whether this superordinal inflation is not another fruitless trial of expressing quite doubtful phylogenetical contexts by box-in-box schemes.

At all events this inflated classification is a perfect means for confusing the non-taxonomists by the fashionable application of non-existing nomenclatural rules to these higher categories. I really wonder how we dare to offer our students such abominable sources of confusion like *Magnoliophyta, Magnoliophytina, Magnoliopsida, Magnolianae* and so on. But really intolerable (as has been mentioned several times in the discussions) is the strict adherence to the typification rule in the case of orders and superorders with a quite different content. At this meeting we have been simply amused by the necessity of defining the order *"Cornales"* every time by the addition of "sensu CRONQUIST" or "sensu HUBER" or "sensu DAHLGREN"—but I may draw attention to the fact (citing HEYWOOD once more) that we are working mainly for non-taxonomists which simply cannot understand that each of these *Cornales*-authors means quite a different group of families and that it is only the *Cornaceae* that is included by all authors in this order.

Here we should try to make some steps towards a solution—and maybe such a solution for both problems could be the increasing use of informal ranks as has been proposed several times during the symposium. HUBER does not like a formal rank to be given to his monocot groups; BURTT wants no formal taxa between his family *Gesneriaceae* and its genera and, if necessary, informal ones only. Could we not try to make informal "order groups", when necessary of course, comparable to our species aggregates and similarly freed from rules and authorships? Such procedures could eliminate also some unnecessary monsters like the monotypic *Rafflesiiflorae* or *Balanophoranae*.

A final word may be said concerning the shape of our higher taxa, including the families, one of the "basic units" of HEYWOOD. CRONQUIST as well as THORNE are in favour of larger entities, mainly for pedagogical reasons as far as I can see. Here I disagree entirely. I am in favour of families (and orders) as clean as possible, well circumscribed, with predictive value, and this is possible only when heterogeneous material has been removed from them. As an example I may recall the *Wellstediaceae* which does not share many characters with the *Boraginaceae* which it nevertheless has been attributed to by MELCHIOR; therefore it did not reappear in any newer scheme. I sometimes wonder what surprises we might expect if numerical taxonomists were to start not with the families of the Syllabus but with its so-called sub-families or even tribes. SPORNE has made some remarks on the same problem.

This does not mean that I want to recommend an inflation of families and orders—why should we not have "free floating" genera and families or even something like the *"Incognitales"* proposed by BAILEY some thirty years ago? By the way, such procedures would at least recall to our minds more often such incredible cases like *Emblingia* which, in our years and in the same paper, has been attributed to the *Capparidaceae, Polygalaceae, Sapindales* and *Goodeniaceae*, respectively. This is a rather bad illustration of what some people like to call Science and Art.

Let me now come to the higher categories, including some special higher taxa. We have agreed, I believe, that the Order is one of the few better and more natural categories in use. As nobody dared to give a definition of this category—as a matter of fact, the definition of categories other than the so much treated species seems to be an important goal of future symposia—one may be inclined to imitate HESLOP-HARRISON defining the order as a group of families which a renowned author has studied thoroughly and declared to be coherent. This is not said for fun only but for adding that the next highlights of this Conference for me have been the presentations of the *Hamamelidales* by ENDRESS and of the *Urticales* by BERG (as well as of the Gesneriads by BURTT). This in my opinion is systematics at its best—and the authors concerned should forgive me for not attempting to sum up in an imperfect manner what they have presented in such a lucid form.

The other (presumably) undisputed category was the Class, which here means the Dicot/Monocot distinction—and which nevertheless offered one of the great problems of this Conference. One may have got the impression that the time is near for re-installing the Monocots in the first position, before the true Dicots—as the older ones amongst us learned it in their early days. WALKER in his latest palynological paper has already inserted the Monocots after his *Magnoliidae* sensu strictissimo and before the *Ranunculidae* and the other subclasses. HUBER has admitted, in the discussion of his present paper, that he also would prefer to put the borderline between the Monocots and the Magnoliids on one side, and the true Dicots on the other side. BEHNKE's scheme of the distribution of his different types and subtypes of sieve-element plastids once more was strongly in favour of such a solution and even the biosynthetic pathways of the cyanogenic compounds (HEGNAUER) lead in the same direction, the tyrosin pathway characterizing once more the Monocots and Magnoliids together. The "undisputed" classes need, I fear, a question mark.

A short remark to the Monocots: we have been offered two different new systems, HUBER's one with the cluster of 12 partly converging partly diverging strains, and CLIFFORD's machine-made one with 10 equally well understandable groups. Both groupings look quite acceptable although both differ greatly from our familiar ones. Any judgement or decision seems to be premature if not impossible today; and this is exactly the situation we are faced with in many other cases.

The situation of the Subclasses has become somewhat deplorable. It certainly will be useful to retain terms like "dilleniid" and so on as a kind of key-words, as hints—but a formal use seems to be scarcely justified. The Magnoliids in the strict sense are just a group of left overs which as we saw should perhaps be treated together with the (informal) Monocots. The Ranunculids are scarcely more than an isolated order or order group, hardly worthy of an higher esteem than the purified *Caryophyllales* which themselves are one of the nicest and most clear-cut orders we have.

With regard to the *Rosidae, Dilleniidae* and even the *Asteridae* the available information makes a clear-cut separation increasingly impossible. I may recall DAHLGREN's remark about the difficulty of shaping subclasses in the case of *Cornales, Ericales* and *Dipsacales* (considered by him and others as rather nearly related albeit treated as members of three different subclasses by others), recall WAGENITZ's and THORNE's presentations treating different evolutionary lines

26*

leading from the *Rosidae/Dilleniidae* complex to different *Asteridae* groups; recall also PHILIPSON's convincing special groups combining *Rosidae* and *Dilleniidae* elements or GOTTWALD's similarly aberrant anatomical schemes.

KUBITZKI cited STEBBINS's remark that, in contrast to animals, in plants subsequently to the initial differentiation of higher taxa, extensive adaptive radiation must have occurred, having led to ample parallelism and convergence. Anthropologists have been using for a long time the term "Tier-Mensch-Übergangsfeld" (zone of transition between animal and man): I begin to wonder whether the less advanced *Rosidae* and *Dilleniidae* and even the early *Asteridae* may also form such a zone of transition—and later on only are diverging into rosid, dilleniid and asterid specializations. Possibly this would fit PHILIPSON's scheme, at least to some extent.

There remains only the so-called *Hamamelididae*. When restricted to ENDRESS's group, it is a perfectly good order or order group of early *Rosidae*, i.e. a further member of the just mentioned zone of transition. When, however, this "subclass" is admitted in the sense of EHREN-DORFER—then it would form a typical other (i.e. prior) zone of transition.

Concerning the systematics of special "superorders" and orders I shall make a few petulant remarks, only as they did not play a prominent rôle in this Conference. Certainly, the stars seem to be in favour of a definite transfer of the *Urticales* to the *Malvales/Euphorbiales* order group; PHILIPSON has shown us very persuasive groupings like his *Theales/Primulales* order group; BEHNKE has produced a convincing scheme of the *Caryophyllales* relationship. But may I be allowed to point out gently that none of the problems nor of the solutions was entirely new. All have been discussed already fifty or even hundred years ago, e.g. by HALLIER to name only one—and have been rejected in more recent systems and therefore forgotten.

But let me turn again to enjoyable matters—and as always, I find it difficult to express my appreciation adequately. I refer to the papers concerning Methods and Characters. One may mention the lucid presentation of numerical methods and their difficulties by CLIFFORD, the likewise clear and pertinent presentation of the advancement index and its implications by SPORNE. The same applies to the fascinating lectures of our chemist friends, HEGNAUER's remarkable introduction to the cyanogenic compounds as possible systematic markers and GOTTLIEB's promising approach to the mapping of metabolites and their graphic presentation. It should be stressed that the latter contribution is one of the few attempts really to collate and to order, in a systematic fashion, the obtainable evidence; its importance also rests upon the fact that it is not based on taxonomic correlations but on chemical and biosynthetical evidence and hence represents what is vitally needed: an independent criterion.

It was likewise impressive to listen to those lectures which presented us with the results of yearlong investigations of single characters, foremost to PHILIPSON's extremely remarkable studies on ovules, to GOTTWALD's overwhelming amount of wood anatomical data, to BEHNKE's convincing résumé on his ultrastructural results. DAHLGREN has shown us a surprisingly high number of new character distribution maps on the base of his useful scheme. For me, these all have been great events and I am sure that the publication of these papers will allow us the welcome opportunity of a more intrinsic study and use of all the data offered.

I began with some words of concern. After a week of papers and discussions I cannot say that this has ceased. But are we in confusion and despair? I certainly do not believe so. The cry for a data bank was loud as always—but it was rightly stressed in the discussion that here also the question "to which purpose" has to be answered first. What I believe we should do is first to keep in mind some of HEYWOOD's words, to give clear distinctions and definitions of what we want to do as well as to indicate clearly and distinctly what we have done. I still consider it extremely important that each of us continues collecting data and searching for further characters, making revisions and monographs—all this being somewhat tedious for great spirits but the only way we can proceed. We certainly shall not refuse further speculation, we gladly shall accept new ideas, but possibly we should not try to insert each new fact into the system as fast as possible. If we do not want to have many different systems we ought to treat carefully our systems of today. As EHRENDORFER said in one of his latest papers, "taxonomy is—and hopefully will remain—a compromise between knowledge, tradition and convenience. This convenience ... necessarily entails a certain stability ... Our system need not reflect immediately every new idea about possible evolutionary affinities." It is not in spite but because of all these worries that I am sure you all share my conviction that this allegedly old-fashioned systematy today is even fuller of fascination than it ever was.

Address of the author: Prof. Dr. H. MERXMÜLLER, Botan. Staatssammlung, Menzinger Str. 67, D-8000 München 19, Federal Republic of Germany.

Index to Plant Taxa

Subject Index